ALSO BY DANIEL E. LIEBERMAN

The Story of the Human Body: Evolution, Health, and Disease

The Evolution of the Human Head

Exercised

Exercised

Why Something We Never Evolved to Do Is Healthy and Rewarding

DANIEL E. LIEBERMAN

PANTHEON BOOKS NEW YORK

Library of Congress Cataloging-in-Publication Data
Name: Lieberman, Daniel, [date] author.
Title: Exercised : why something we never evolved to do is healthy and rewarding / Daniel E. Lieberman.
Description: New York : Pantheon Books, 2020. Includes index.
Identifiers: LCCN 2020009066 (print). LCCN 2020009067 (ebook).
ISBN 9781524746988 (hardcover). ISBN 9781524746995 (ebook)
Subjects: LCSH: Exercise—History. Physical fitness—History.
Physical education and training—History.
Classification: LCC QP301 .L628 2020 (print) |
LCC QP301 (ebook) | DDC 612.044—dc23
LC record available at lccn.loc.gov/2020009066
LC ebook record available at lccn.loc.gov/2020009067

www.pantheonbooks.com

Jacket image based on gerasimov_foto_174/Shutterstock
Jacket design by Linda Huang

Printed in the United States of America
First Edition
2 4 6 8 9 7 5 3 1

To Eleanor

Three Useful Definitions

Physical activity (noun): any bodily movement produced by skeletal muscles that expends energy

Exercise (noun): voluntary physical activity that is planned, structured, repetitive, and undertaken to sustain or improve health and fitness

Exercised (adjective): to be vexed, anxious, worried, harassed

Contents

Prologue

In June 2017, as I was beginning this book, I flew to Kenya, bought a treadmill, and transported it in a Land Cruiser to a remote place called Pemja, a community more than seven thousand feet above sea level in the western part of the country. Pemja lies at the edge of a verdant region of rolling hills and valleys dotted with giant granite outcrops. Scattered everywhere are small fields and simple homesteads, typically one-room houses made from mud and dung and roofed with thatch or tin. Pemja is beautiful but poor, even by Kenyan standards, and far off the beaten path. To drive there from Eldoret, the nearest city, which is just fifty miles away, takes nearly a day on roads that become increasingly treacherous the closer one gets to Pemja. On a good day, the journey requires navigating precipitous, twisting dirt lanes traversed by gullies littered with boulders and other obstacles. When it rains, the route becomes a steep, sticky river of volcanic mud.

Despite the horrendous roads, I have made this trip with my students and Kenyan colleagues nearly every year for the last decade to study how human bodies here are changing as the world rapidly modernizes. The people of Pemja are subsistence farmers who live much as their ancestors did for generations with barely any access to paved roads, electricity, and running water. Most Pemjans lack sufficient means to buy shoes, mattresses, medicine, chairs, and other things I take for granted, and I find it deeply moving to observe how hard they work without assistance from machines to survive and improve their lives, especially the children's. By comparing them with people from the same Kalenjin ethnic group who live in the nearby city of Eldoret, we can study how our bodies change when we sit for much of the day in office jobs and no longer sustain ourselves with daily physical labor, go barefoot, and sit or squat on the ground.

Hence the treadmill. Our plan was to use it to study how efficiently women in this region walk while carrying heavy loads of water, food, and firewood on their heads. But the treadmill was an illuminating mistake. After we invited the women to stand on the machine and the belt started to move, they walked self-consciously, hesitantly, awkwardly. You, too, probably walked strangely the first time you got on one of these bizarre, noisy contraptions that force you to work to get nowhere. Although the women's treadmill-walking skills improved slightly with practice, we realized that to measure how they normally walked with and without loads, we had to abandon the treadmill and ask them to walk on solid ground.

As I grumbled about how much money, time, and effort we had wasted getting a treadmill to Pemja, it struck me how these machines encapsulate the main theme of this book: *we never evolved to exercise.*

What do I mean by that? Well, exercise today is most commonly defined as voluntary physical activity undertaken for the sake of health and fitness. But as such it is a recent phenomenon. Our not-too-distant ancestors who were hunter-gatherers and farmers had to be physically active for hours each day to get enough food, and while they sometimes played or danced for fun or social reasons, no one ever ran or walked several miles just for health. Even the salubrious meaning of the word "exercise" is recent. Adapted from the Latin verb *exerceo* (to work, train, or practice), the English word "exercise" was first used in the Middle Ages to connote arduous labor like plowing a field.[1] While the word has long been used to denote practicing or training to improve skills or health, to be "exercised" also means to be harassed, vexed, or worried about something.

Like the modern concept of exercising for the sake of health, treadmills are recent inventions whose origins had nothing to do with health and fitness. Treadmill-like devices were first used by the Romans to turn winches and lift heavy objects, and then modified in 1818 by the Victorian inventor William Cubitt to punish prisoners and prevent idleness. For more than a century, English convicts (among them Oscar Wilde) were condemned to trudge for hours a day on enormous steplike treadmills.[2]

Opinions differ on whether treadmills are still used for punishment, but they illustrate the odd nature of exercise in the modern, industrialized world. Without seeming like a madman or an idiot, how would I explain to a hunter-gatherer, a farmer in Pemja, or even my great-great-great-grandparents that I spend most of my days sitting in chairs and then compensate for my idleness by paying money to go to a gym to make myself sweaty, tired, and uncomfortable on a machine that forces me to struggle to stay in the same place?

Beyond the absurdity of treadmills, our distant ancestors would also be perplexed by the way exercise has become commercialized, industrialized, and, above all, medicalized. Although we sometimes exercise for fun, millions of people today pay to exercise to manage their weight, prevent disease, and stave off decrepitude and death. Exercise is big business. Walking, jogging, and many other forms of exercise are inherently free, but giant multinational companies entice us to spend lots of money to work out in special clothes, with special equipment, and in special places like fitness clubs. We also pay money to watch other people exercise, and a handful of us even pay for the privilege of suffering through marathons, ultramarathons, triathlons, and other extreme, grueling, or potentially dangerous sporting events. For a few thousand dollars, you, too, can run 150 miles across the Sahara Desert.[3] But more than anything else, exercise has become a source of anxiety and confusion because while everyone knows that exercise is good for their health, the majority of us struggle to exercise enough, safely, or enjoyably. We are exercised about exercise.

Okay, so exercise is paradoxical: salubrious but abnormal, intrinsically free but highly commodified, a source of pleasure and health but a cause of discomfort, guilt, and opprobrium. Why did this realization motivate me to write this book? And why might you wish to read it?

Myths of Exercise

For most of my life, I, too, was exercised about exercise. Like many, I grew up feeling unsure and insecure about my efforts to be physically active. It's a cliché, but as a pint-sized, nerdy kid, I really was picked last for teams in school. Although I dreamed of being more athletic, feelings of inadequacy and embarrassment about my mediocre abilities reinforced my inclination to avoid sports. In first grade, I once hid in a closet during gym. To me, the word "exercise" still summons up anxious memories of being humiliated by physical education teachers who shouted at me as I struggled, ashamed of my body, to keep up with my faster, stronger, and more talented classmates. I can still hear Mr. B—— bellowing, "Lieberman, climb that rope!" I wasn't a total couch potato in school, and throughout my twenties and thirties I occasionally jogged and hiked, but I did not exercise as much as I should, and I was largely ignorant and anxious about what kinds of exercise to do, how often, how vigorously, and how to improve.

Despite my mediocre athleticism, I fell in love with anthropology and evolutionary biology in college and chose to study how and why the human body is the way it is. At the beginning of my career, I focused on skulls, but for various accidental reasons I also became interested in the evolution of human running. That research in turn led me to investigate the evolution of other human physical activities like walking, throwing, toolmaking, digging, and carrying. Over the last fifteen years I have had the opportunity to traverse the globe to observe how hardworking hunter-gatherers, subsistence farmers, and others use their bodies. Because I strive to be adventurous, whenever possible I've tried to participate in these activities. Among other experiences, I've run and carried water on my head in Kenya, tracked musk oxen and kudu with indigenous hunters in Greenland and Tanzania, joined an ancient Native American footrace under the stars in Mexico, played barefoot cricket in rural India, and raced on foot against horses in the mountains of Arizona. Back in my lab at Harvard University, my students and I conduct experiments to study the anatomy, biomechanics, and physiology that underlie these activities.

My experiences and research slowly led me to conclude that because industrialized societies such as the United States fail to recognize that exercise is a paradoxically modern but healthy behavior, *many of our beliefs and attitudes about exercise are myths* (by "myth" I mean a claim that is widely believed but inaccurate and exaggerated). To be clear, I do not contend that exercise isn't beneficial or that everything you have read about exercise is incorrect. That would be silly. I will, however, make the case that by ignoring or misinterpreting evolutionary and anthropological perspectives on physical activity, the contemporary, industrial approach to exercise is marred by misconceptions, overstatements, faulty logic, occasional mistruths, and inexcusable finger-pointing.

Chief among these myths is the notion that we are supposed to *want* to exercise. There is a class of people whom I define as "exercists" who like to brag about exercise and who repeatedly remind us that exercise is medicine, a magic pill that slows aging and delays death. You know the type. According to exercists, we were *born* to exercise because for millions of years our hunter-gatherer ancestors survived through walking, running, climbing, and other physical activities. Even exercists who discredit the theory of evolution think we are fated to exercise. When God expelled Adam and Eve from the Garden of Eden, he condemned them to a life of agrarian drudgery: "By the sweat of your brow you will eat your food until you return to the ground." We are thus nagged to exercise because it is not just good for us but also a fundamental aspect of the human condition. People who don't exercise enough are considered lazy, and physical suffering of the "no pain, no gain" variety is considered virtuous.

Other myths about exercise come in the form of exaggeration. If exercise, as we are told, is really a "magic pill" that will cure or prevent most diseases, why are more people living longer than ever despite being more physically inactive than ever? Are humans fundamentally slow and weak? Is it true that we trade off strength for endurance? Are chairs out to kill us? Is exercise useless for losing weight? Is it normal to be less active as we age? Is drinking a glass of red wine as beneficial as spending an hour in the gym?[4]

Inaccurate, sloppy, contradictory thinking about exercise gives us whiplash and sows confusion and skepticism. On the one hand we are advised to walk ten thousand steps a day, avoid sitting, and never take the elevator, but on the other hand we hear that exercising won't help us shed extra pounds. We are exhorted to spend more time being active and admonished to stop slouching, but then advised to sleep more and use chairs that support the lower back. Expert consensus is that we need 150 minutes of exercise a week, but we also read that just a few minutes of high-intensity exercise a day is enough to make us fit. Some fitness professionals recommend free weights, others prescribe weight machines, yet others reproach us for not doing enough cardio. While some authorities urge us to jog, others warn that running will ruin our knees and promote arthritis. One week we read how too much exercise may damage the heart and that we need comfortable sneakers, but the next we read it is almost impossible to exercise too much and that minimal shoes are best.

Beyond spreading confusion and doubt, the most pernicious consequence of many myths about exercise—especially the one about how it's normal to exercise—is that we fail to help people to exercise and then unfairly shame and blame them for not doing it. Everyone knows they should exercise, but few things are more irritating than being told to exercise, how much, and in what way. Exhorting us to "Just Do It" is about as helpful as telling a drug addict to "Just Say No." If exercise is supposedly natural, why is it that no one, despite years of effort, has found an effective way to help more people to overcome deep-seated, natural instincts to avoid optional exertion? According to a 2018 survey of millions of Americans, about half of adults and nearly three-quarters of teenagers report they don't reach the base level of 150 minutes of physical activity per week, and less than one-third report they exercise in their leisure time.[5] By any objective measure we are doing a lousy job promoting exercise in the twenty-first century in part because we have a muddleheaded approach to physical activity and inactivity.

Enough complaining. How can we do better? And what do I hope you will get out of this book?

Why a Natural History?

The premise of this book is that evolutionary and anthropological perspectives can help us better understand the paradox of exercise—that is, why and how something we never evolved to do is so healthy. I think these perspectives can also help those of us who are anxious, confused, or ambivalent about exercise to exercise in the first place. Consequently, this book is as much for exercise enthusiasts as it is for those who are exercised about exercise and struggle to do it.

Let me start by explaining how I *won't* approach this topic. If you've read any websites, articles, or books on exercise, you'll quickly realize that most of what we know comes from observing people in modern, industrialized countries like the United States, England, Sweden, and Japan. Many of these studies are epidemiological; they look for associations between, say, health and physical activity in large samples of individuals. For example, hundreds of studies have looked for correlations between heart disease, exercise habits, and factors like age, sex, and income. These analyses reveal correlations, not causation. There has also been no lack of experiments that randomly assign people (most often college students) or mice to contrasting treatment groups for short periods of time to measure the effects of particular variables on particular outcomes. Hundreds of such studies have looked, for instance, at the effects of varying doses of exercise on blood pressure or cholesterol levels.

There is nothing inherently wrong with these sorts of studies—as you'll see, I'll make use of them throughout the book—but they view exercise too narrowly. For starters, almost all studies of humans focus on contemporary Westerners or elite athletes. There is nothing wrong with studying these populations, but Westerners such as Americans and Europeans constitute only about 12 percent of humanity and are often unrepresentative of our evolutionary past. Studying elite athletes provides an even more skewed perspective on normal human biology. How many people have ever been able to run a mile in less than four minutes or bench-press more than five hundred pounds? In addition, how similar is your biology to that of a mouse? Just as important, these studies fail to consider how exercise is abnormal without effectively addressing key "why" questions. Large epidemiological surveys and controlled laboratory experiments might elucidate how exercise affects the body, underscore the benefits of exercise, and quantify how many Swedes or Canadians are unmotivated and confused about exercise, but you wouldn't learn much about *why* exercise affects the body as it does, *why* so many of us are ambivalent about exercise, and *why* physical inactivity causes us to age more rapidly and increases our chances of getting sick.

To address these deficiencies, we need to supplement the standard focus on Westerners and athletes with evolutionary and anthropological perspectives. To do that, we will venture beyond college campuses and hospitals in the United States and other industrialized countries to observe a wider range of humanity laboring, resting, and exercising in contexts in which most humans still live. We will look at hunter-gatherers and subsistence farmers in different environments on different continents. We'll also delve into the archaeological and fossil records to better understand the history and evolution of human physical activity even as we compare ourselves with other animals, especially our closest ape relatives. And, finally, we will integrate these diverse lines of evidence about people's biology and behavior into their proper ecological and cultural contexts. To compare how American college students, African hunter-gatherers, and Nepalese porters walk, run, sit, and carry things, and how these activities affect their health, requires knowing something about their different physiologies and cultures. In short, to really understand exercise, let's study the natural history of human physical activity and inactivity.

Accordingly, in the chapters that follow, we will use evolutionary and anthropological perspectives to explore and rethink dozens of myths about physical inactivity, activity, and exercise. Are we born to exercise? Is sitting the new smoking? Is it bad to slouch? Do you need eight hours of sleep? Are humans comparatively slow and weak? Is walking ineffective for losing weight? Does running ruin your knees? Is it normal to exercise less as we age? What is the best way to persuade people to exercise? Is there an optimal kind and amount of exercise? How much does exercise affect our vulnerability to cancer or infectious diseases? The mantra of this book is that *nothing about the biology of exercise makes sense except in the light of evolution, and nothing about exercise as a behavior makes sense except in the light of anthropology.*[6]

For those of you who already love to exercise, I will try to give you new insights into how and why different kinds of inactivity and physical activity affect your body, why exercise really does promote health without being a magic pill, and why there is no optimal dose or type of exercise. For those of you who struggle to exercise, I will explain how and why you are normal, help you figure out how to get moving, and help you evaluate the benefits and drawbacks of different kinds of exercise. But this is not a self-help book. I am not going to hawk "seven easy steps to get fit" or cajole you to take the stairs, run a marathon, or swim the English Channel. Instead, my goal is to explore skeptically and without jargon the fascinating science of how our bodies work when we move and take it easy, how and why exercise affects health, and how we can help each other get moving.

As a natural history, this book has four parts. After an introductory chapter, the first three parts roughly follow the evolutionary story of human physical activity and inactivity, with each chapter spotlighting a different myth. Because we cannot understand physical activity without understanding its absence, part 1 begins with *physical inactivity*. What are our bodies doing when we take it easy, including when we sit and sleep? Part 2 explores physical activities that require *speed, strength,* and *power* such as sprinting, lifting, and fighting. Part 3 surveys physical activities that involve *endurance* such as walking, running, and dancing, as well as their effect on aging. Last but not least, in part 4 we will consider how anthropological and evolutionary approaches can help us exercise better in the modern world. How can we more effectively manage to exercise, and in what ways? To what extent, how, and why do different types and doses of exercise help prevent or treat the major diseases likely to make us sick and kill us?

But we have a long way to go before drawing any conclusions. Let's begin with what you're probably doing right now as you read these words—not moving—to explore more deeply the biggest myth of them all: that it's normal to exercise.

Exercised

ONE

Are We Born to Rest or Run?

MYTH #1 We Evolved to Exercise

> It's true hard work never killed anybody, but I figured why take the chance?
>
> —Ronald Reagan, interview with *The Guardian*, 1987

I neither am nor want to be an exceptional athlete, and I have no desire whatsoever to swim around Manhattan, bike across America, climb Mount Everest, bench-press several hundred pounds, or pole-vault over anything. Among the many tests of extreme strength or endurance I will never attempt is a full triathlon. Not me. But I am curious about demanding athletic challenges. So in October 2012, I eagerly accepted an invitation to travel to Hawaii to observe the legendary Ironman World Championship and attend the sports medicine conference that precedes the race.

Paradoxically, this infamously grueling test of endurance takes place in the paradisiacal setting of Kona, Hawaii, a charming town largely dedicated to helping vacationers relax. In the days leading up to the race, everyone in Kona appears to be engaged solely in the pursuit of pleasure. People swim, snorkel, and surf on picturesque beaches, sip fruity cocktails while watching the sun set, and stroll through town eating ice cream and buying souvenirs and sports equipment. Some also party late into the night in the town's many bars and clubs. If you are looking for a tropical resort in which to relax and be hedonistic, you couldn't do better than Kona.

Then on Saturday at precisely 7:00 a.m. the race begins. As the morning sun paints the sky rose as it emerges from behind the blue silhouette of the volcano that looms above town, about twenty-five hundred ultra-fit people dive off a pier into the Pacific for the race's first leg, a 2.4-mile swim across the bay and back. In case you were wondering, 2.4 miles is the equivalent of swimming seventy-seven lengths of an Olympic-sized pool. Many of the triathletes look apprehensive as they wait for the starting gun, but their spirits are buoyed by a band of Hawaiian drummers, thousands of cheering spectators, and loud, adrenaline-inducing music blaring from car-sized speakers. Once they start, there are so many swimmers churning the water it looks like a shark feeding frenzy.

About an hour later, the lead triathletes make it back to land. As they emerge dripping from the ocean, they rush into a tent, change into high-tech biking gear (including aerodynamic helmets), jump onto ultralightweight bikes that cost upwards of ten thousand dollars, and zoom out of sight for the next leg of the race, a 112-mile course across a lava desert. Because it takes the best riders about four and a half hours to cover this distance, I stroll back to my hotel and savor a tropical breakfast, made all the more pleasant by exercise schadenfreude. Yes, I do think my eggs Benedict and coffee taste better as I think about those two thousand fellow humans out there on the island under the blaring sun trying to bike more than 100 miles as fast as they can and still save enough energy to complete the final leg of their ordeal, a full-length marathon.

Rested and refreshed, I return to the race center to watch the elite triathletes leap off their bikes, lace on running shoes, and then head off on foot to begin their 26.2-mile run along the coast. While the competitors trudge through their marathon in brutally hot and humid conditions (it was 90°F), I enjoy a leisurely lunch and a brief nap. Shortly after 2:00 p.m., I amble back to watch the finish, one of the most exuberant scenes I have ever witnessed. As the first runners arrive back at the town's main street, they are funneled into a chute lined by screaming friends and fans—all whipped into a feverish state by loud, pulsating music. At the finish line, a booming voice greets each finisher (male and female) with the time-honored phrase "YOU ARE AN IRONMAN!" and the crowd goes wild. The elite athletes, who finish about eight hours after they started, cross the line stony-faced, looking more like cyborgs than humans. Later, as the amateurs arrive to complete their ordeal, we glimpse what their achievement means to them. Many weep for joy; others kneel to kiss the ground; some pound their chests and bellow thunderously; a few look dangerously ill and are rushed to the medical tent.

The most dramatic finishes occur near midnight as the seventeen-hour deadline approaches. These intrepid souls desperately will their bodies to overcome crushing pain and fatigue, their minds forcing each leg to take just one more step. As they limp into town, some look near death's door. But the sight of the finish line and the emotional energy from the raucous friends, family members, and fans who line the race's final stretch pull them home. First they hobble; then they shuffle; finally they manage to break into a run to reach the finish line, where they collapse in a state of ecstasy. It is there at midnight where one truly understands why Ironman's motto is "Anything Is Possible."

Ernesto

Watching amateur Ironmen finish near midnight was inspiring. But I flew home with renewed conviction that no amount of money would entice me to do a full triathlon. Further, I couldn't help but feel that what I observed was not just abnormal but also concerning. What would motivate someone to train for hours upon hours a day for years just for the chance to put his or her body through that kind of hell and prove that "anything is possible"? Full triathlons require extreme obsession and money. If you consider airfare, hotel bills, and gear, many Ironmen spend tens of thousands of dollars a year on their sport. Although Ironman attracts diverse participants, including cancer survivors, nuns, and retirees, a large percentage are wealthy Type A personalities who apply the same fanatical devotion to exercise they previously dedicated to their careers. Much as I admire these triathletes, are they damaging their bodies? For every Ironman who qualifies, how many would-be Ironmen were sidelined by crippling injuries? What kind of toll does all the training necessary to do a full triathlon have on the athletes' friends, families, and marriages?

With these and other thoughts percolating in my brain, a few weeks later I packed my bags and headed to the Sierra Tarahumara (sometimes called the Copper Canyons) of Mexico, far from the trappings of the developed world. There I met athletes so different from the triathletes of Kona and observed a competition so different from Ironman that I can only describe the experience as whiplash. And of all the people I encountered, the one who blew my mind the most was an elderly man, Ernesto (that's not his real name), whom I met on a remote mesa, seven thousand feet above sea level.

I had traveled to the Sierra to do research on Tarahumara Native Americans, famous for their long-distance running. Dozens of anthropologists over the last century have written about the Tarahumara, but in 2009 they gained an extra boost of worldwide fame from the best seller *Born to Run*. The book portrays them as a "hidden tribe" of barefoot, ultra-healthy, "superathletes" who routinely run unimaginable distances.[1] Intrigued, and to collect data on how they ran without modern, cushioned running shoes, I traveled up and down four-thousand-foot ravines on perilous switchback roads with a guide, an interpreter, and scientific instruments to measure people's feet and running biomechanics. By the time I met Ernesto, I had interviewed and measured dozens of other Tarahumara men and women and was beginning to have doubts about almost everything I had read about their running. Despite their reputation as extraordinary runners, I hadn't seen a single Tarahumara running anywhere, let alone barefoot. But I did observe them to be hard workers and indefatigable walkers. Most of the people I interviewed said they either didn't run or participated in just one race per year. Not all Tarahumara appeared to be skilled runners, and many of them had paunches or were overweight.

Not Ernesto. A slight man in his seventies who looks twenty or thirty years younger, Ernesto was initially reticent as I measured his height, weight, leg length, and feet, and then used a high-speed video camera to record his running biomechanics on a small track I had set up. Thankfully for me, he gradually became increasingly garrulous and started to tell stories (through an interpreter) about the old days when he hunted deer on foot by running them down and sometimes danced for days in ceremonies. Ernesto told me he was a champion runner in his youth and that he still competed in several races a year. But when I asked him how he trained, he didn't understand the question. When I described how Americans like me keep fit and prepare for races by running several times a week, he seemed incredulous. As I asked more questions, he made it pretty clear he thought the concept of needless running was preposterous. "Why," he asked me with evident disbelief, "would anyone run when they didn't have to?"

Since I had just witnessed the intensity of Ironman triathletes, whose arduous training habits are legendary, I found that Ernesto's question made me both laugh and think. He put the exercise habits of many Westerners, myself included, into stark perspective. If you were a subsistence farmer like Ernesto who grows all his own food without the help of machines, why would you ever spend precious time and calories exercising just for the sake of keeping fit or to prove that anything is possible? Ernesto reinforced my conviction that what I observed at Ironman was bizarre, and he even caused me to question the sanity of my own efforts to train for a marathon. Ernesto also intensified my curiosity about Tarahumara running, which seemed more mythical than actual. Even though Ernesto never trained, and I hadn't seen any Tarahumara running on their own, I had heard and read numerous accounts about how Tarahumara men and women have their own Ironman-like competitions. In the women's race, known as *ariwete,* teams of teenage girls and young women run about twenty-five miles while chasing a cloth hoop. In the men's race, the *rarájipari,* teams of men run up to eighty miles while kicking an orange-sized wooden ball. If the Tarahumara think needless exercise is foolish, why do some of them sometimes run insanely long distances like Ironmen? Just as important, how do they accomplish these feats without training?

Rarájipari Under the Stars

Not long after I met Ernesto, I got some answers to these questions when I had the privilege of witnessing a traditional Tarahumara *rarájipari* footrace. The competition took place on a mountaintop near a tiny Tarahumara settlement, about a two days' walk from the nearest town. The race involved two teams of men, eight on each side. Ernesto's team was captained by Arnulfo Quimare, a champion Tarahumara runner who figured prominently in the book *Born to Run.* The opposing team was captained by Arnulfo's cousin Silvino Cubesare, also a champion runner. By arrangement, the teams had set up two stone cairns about two and a half miles apart, and they agreed that the first team to complete fifteen circuits or to lap the other (in other words, get five miles ahead) would win.

The morning began with a feast. In addition to the runners, about two hundred Tarahumara had assembled from near and far to enjoy the event, socialize, and take a break from working in the fields. At breakfast, the runners somberly tanked up on chicken stew, while the rest of us devoured fresh tortillas, chilies, and an enormous quantity of soup that had been prepared in a former oil drum. The soup contained most of a cow along with corn, squash, and potatoes. In addition to feasting, people placed bets on the two teams, wagering pesos, clothes, goats, corn, and other sundry commodities. Then, at about 11:00 a.m., after several hours of relaxed chaos, the runners started off with no fanfare. As shown in figure 1, the runners wore exactly the same clothes they always wear: a bright tunic, a loincloth, and sandals (*huaraches*) cut from a tire and lashed onto their feet with leather thongs. Each team had its own hand-carved wooden ball, which the runners flick with their toes as far as possible, then run to find it, and kick again without ever using their hands. Although the two teams never stopped, some of the observers (myself included) occasionally jumped in for a lap or two to keep the runners company and offer encouragement by shouting, "*Iwériga! Iwériga!*" (which means both "breath" and "soul"). When the runners were thirsty, their friends offered them *pinole,* a Gatorade-like drink made of powdered corn dissolved in water.

FIGURE 1 Scenes from two different races. The Ironman World Championship in Kona, Hawaii (*top*), and a *rarájipari* in the Sierra Tarahumara, Mexico (*bottom*). The Tarahumara runner (Arnulfo Quimare) is chasing a ball that he has just flicked with his foot. (Photos by Daniel E. Lieberman)

For the first six or so hours of the race, it was impossible to tell who was going to win. Arnulfo's and Silvino's teams went back and forth along the course at a steady, gradual jogging pace of about ten minutes per mile. As the warm December day turned into a chilly, star-filled night, the runners kept going without pause, lighting their way with pine torches. I joined Arnulfo's team then and will never forget the magical feeling of running behind them under that splendid starry sky, a torch in hand, watching Arnulfo and his friends focused intently on that all-important ball, kicking, searching, but always running, running, running. Eventually, however, some of the runners started to drop out from cramps, and finally, around midnight, Arnulfo's team lapped Silvino's and the race was over after about seventy miles. Unlike Kona, there was no applause, no announcer, no triumphant music. Instead, everyone just sat down by an enormous bonfire and drank from gourds of homemade corn beer.

On the surface, that *rarájipari* was the antithesis of Ironman. It was a totally uncommercial, simple community event, part of an ancient tradition that probably dates back thousands of years.[2] There was no timing, no entry fee, and no one wore any special gear. But in other respects, much about the *rarájipari* was familiar. Although there were no trophies or prizes for the winners, the race was a serious competition, and the winning team amassed a small fortune thanks to all the betting. Instead of Gatorade, there was *pinole*. Like the Ironman triathletes, the Tarahumara runners suffered intensely, battling nausea, cramps, and crushing fatigue. And perhaps the most important similarity is that almost all the attendees were bystanders, not runners. While some of the spectators occasionally jumped into the race for a few laps, only a few Tarahumara compete in these races. Most are content to watch rather than run.

The Myth of the Athletic Savage

The races I saw at Kona and the Sierra Tarahumara were inspiring but also perplexing. Who is more normal from an evolutionary perspective: those of us who push our bodies to do nonessential physical activities, sometimes in extreme, or those of us who prefer to avoid unnecessary exertion? And how do some Tarahumara manage to run several back-to-back marathons without training, while Ironmen practice and prepare obsessively for years to accomplish similar feats of endurance?

Answers to these questions typically run the gamut of beliefs about nature versus nurture. On one side of this venerable debate is the view that athletic proclivities and talents are innate. Just as genes make some of us taller or more dark-skinned, there must be genes that influence biological capacities and psychological inclinations to be athletes. If nature is more important than nurture, then to become an extreme athlete, you first need to have the right parents with the right genes. Decades of research have indeed confirmed that genes do play key roles in many aspects of sport and exercise, including our motivation to exercise in the first place.[3] That said, intensive efforts have failed to identify specific genes that explain much about athletic talent including how and why Kenyan and Ethiopian runners currently dominate distance running.[4] In addition, studies of professional athletes who push the limits of endurance reveal that the barriers they must overcome include physiological challenges like generating muscle force effectively, fueling themselves efficiently, and controlling their body temperature, but these competitors are even more challenged by psychological hurdles. To keep going, great athletes must learn to cope with pain, be strategic, and above all believe they can do it.[5] We must therefore look just as much at the other side of the nature versus nurture debate and consider how environment, especially culture, contributes to everyday people's athletic abilities and impulses.

The most widespread and intuitively appealing line of thinking about the effects of nurture on physical activity arises from an idea known as the theory of the natural human. According to this view, championed by the eighteenth-century philosopher Jean-Jacques Rousseau, humans who live in what Rousseau termed a "savage" state of nature reflect our true, inherent selves uncorrupted by civilization. This theory has been warped into many forms, including the myth of the noble savage, the belief that nonwesternized people whose minds have not been polluted by the social and moral evils of civilized society are naturally good and decent. Although widely discredited, this myth has persisted and found new life when applied to exercise in what I label the *myth of the athletic savage*. The essential premise of this myth is that people like the Tarahumara whose bodies are untainted by modern decadent lifestyles are natural superathletes, not only capable of amazing physical feats, but also free from laziness. By contending that the men I observed running seventy miles without training do so naturally, this myth implies that people like you and me who neither can nor will accomplish such feats are, from an evolutionary perspective, abnormal because civilization has turned us into etiolated wimps.

As you have probably divined, I object to the myth of the athletic savage. For one, it stereotypes and dehumanizes people such as the Tarahumara. Since that first trip when I spoke with Ernesto, I have talked with hundreds of Tarahumara all over the Sierra Tarahumara and can assure you no one there wakes up in the morning and thinks, "Gee, what a beautiful day. I think I'll run fifty miles just for fun." They don't even go for needless five-mile runs. When I ask Tarahumara on what occasions they run, the most common answer is "when I chase goats." Instead, I have come to appreciate that the Tarahumara are extremely hardworking, physically fit farmers who never do anything by half and whose culture deeply values running. The reason some Tarahumara run fifty or more miles on rare occasions is not much different from the reason Ironmen do triathlons: they think it is worth it. However, whereas Ironmen subject themselves to full triathlons to test their limits (Anything Is Possible!), Tarahumara run *rarájiparis* because it is a deeply spiritual ceremony that they consider a powerful form of prayer.[6] Many Tarahumara I have interviewed say that the ball-game race makes them feel closer to the Creator. To them, chasing that unpredictable ball for mile upon mile is a sacred metaphor for the journey of life, and it induces a spiritual trancelike state. It is also an important communal event that brings money and prestige. Lastly, I think the *rarájipari* once had a vital practical function. As I watched Arnulfo and his team repeatedly try to find and then kick that dust-colored wooden ball, it struck me that the ball game is a terrific way to learn how to track while running—an essential skill for the way the Tarahumara used to hunt deer on foot.

The myth of the athletic savage mistakenly suggests that humans uncorrupted by civilization can easily run ultramarathons, scale enormous mountains, and perform other seemingly superhuman feats without training. Yes, the Tarahumara and other nonindustrial people rarely if ever "train" as we do by following a course of exercises to develop their fitness and prepare for a specific event. (When I travel to places like the Sierra, I am often the only person who goes for an apparently pointless jog in the morning to the amusement of the locals.) But nearly every day of their lives, hunter-gatherers and subsistence farmers engage in hours of hard physical *work*. Because they lack cars, machines, and other laborsaving devices, their daily existence requires walking many miles in rugged terrain, not to mention doing other kinds of physical labor by hand like plowing, digging, and carrying. When my colleague Dr. Aaron Baggish attached accelerometers (tiny devices, like Fitbits, that measure steps per day) to more than twenty Tarahumara men, he discovered they walked on average ten miles a day. In other words, the training that enables them to run back-to-back marathons is the physical work that is part and parcel of their everyday life.

The myth of the athletic savage also erroneously implies that running an ultramarathon or performing other feats of extraordinary athleticism is somehow effortless for the Tarahumara and other indigenous peoples compared with Westerners. It encourages a racist stereotype akin to the disturbing fiction that Africans raised in the jungle or in slavery did not experience pain in the same way as Europeans do.[7] Moreover, it embraces the fallacy that if only you and I had grown up leading a wholesome lifestyle uncorrupted by sugar and chairs and requiring lots of natural movement, then we could be super-healthy superathletes for whom running a marathon would be child's play. Not only is the myth of the athletic savage an example of truthiness—something that *feels* true because we *want* it to be true—it trivializes the physical and psychological challenges faced by all athletes everywhere, the Tarahumara included. I have observed several *rarájiparis* and *ariwetes* and seen how Tarahumara runners struggle just as much as the Ironmen of Kona to overcome cramps, nausea, bloody toes, and other forms of physical pain. They also suffer mentally and, like other athletes, draw strength from bystanders who urge them on.

It's time to discard once and for all ancient, insidious stereotypes about the physical superiority and virtuousness of people who don't grow up surrounded by laborsaving machines and other modern comforts. But debunking this myth doesn't address the fundamental question: What kind of physical activity and how much of it is normal for a "normal" human being?

Are "Normal" Humans Couch Potatoes?

Imagine you have been asked to conduct a scientific study on how much, when, and why "normal" people exercise. Because we tend to think of ourselves and our societies as normal, you'd probably collect data on the exercise habits of people like you and me. This approach is the norm in many fields of inquiry. For example, because most psychologists live and work in the United States and Europe, about 96 percent of the subjects in psychological studies are also from the United States and Europe.[8] Such a narrow perspective is appropriate if we are interested only in contemporary Westerners, but people in Western industrialized countries aren't necessarily representative of the other 88 percent of the world's population. Moreover, today's world is profoundly different from that of the past, calling into question who among us is "normal" by historical or evolutionary standards. Imagine trying to explain cell phones and Facebook to your great-great-great-grandparents. If we really want to know what ordinary humans do and think about exercise, we need to sample everyday people from a variety of cultures instead of focusing solely on contemporary Americans and Europeans who are, comparatively speaking, WEIRD (Western, educated, industrialized, rich, and democratic).[9]

To go a step further, until a few hundred generations ago, all human beings were hunter-gatherers, and until about eighty thousand years ago everyone's ancestors lived in Africa. So if we genuinely want to know about the exercise habits of evolutionarily "normal" humans, it behooves us to learn about hunter-gatherers, especially those who live in arid, tropical Africa.

Studying hunter-gatherers, however, is easier said than done because their way of life has almost entirely vanished. Only a handful of hunter-gatherer tribes persist in some of the most remote corners of the globe. Further, none are isolated from civilization and none subsist solely on the wild foods they hunt and gather. All of these tribes trade with neighboring farmers, they smoke tobacco, and their way of life is changing so rapidly that in a few decades they will cease to be hunter-gatherers.[10] Anthropologists and other scientists are therefore scrambling to learn as much as possible from these few tribes before their way of life irrevocably disappears.

Of all of them, the most intensely studied is the Hadza, who live in a dry, hot woodland region of Tanzania in Africa, the continent where humans evolved. In fact, doing research on the Hadza has become something of a cottage industry for anthropologists. In the last decade, researchers have studied almost everything you can imagine about the Hadza. You can read books and articles about how the Hadza eat, hunt, sleep, digest, collect honey, make friends, squat, walk, run, evaluate each other's attractiveness, and more.[11] You can even read about their poop.[12] In turn, the Hadza have become so used to visiting scientists that hosting the researchers who observe them has become a way to supplement their income. Sadly, visiting scientists who want to emphasize how much they are studying bona fide hunter-gatherers sometimes turn a blind eye to the degree to which the Hadza's way of life is changing as a result of contact with the outside world. These papers rarely mention how many Hadza children now go to government schools, and how the Hadza's territory is almost entirely shared with neighboring tribes of farmers and pastoralists, with whom they trade and whose cows tramp all over the region. As I write this, the Hadza don't yet have cell phones, but they are not isolated as they once were.

Despite these limitations, there is still much to learn from the Hadza, and I am fortunate to have visited them on a couple of occasions. But to get to the Hadza is not easy. They live in a ring of inhospitable hills surrounding a seasonal, salty lake in northwestern Tanzania—a hot, arid, sunbaked region that is almost impossible to farm.[13] The area has some of the worst roads on the planet. Of the roughly twelve hundred Hadza, only about four hundred still predominantly hunt and gather, and to find these few, more traditional Hadza, you need sturdy jeeps, an experienced guide, and a lot of skill to travel over treacherous terrain. After a rainstorm, driving twenty miles can take most of the day.

Many things surprised me when I first walked into a Hadza camp mid-morning on a torrid, sunny day in 2013, but I remember being especially struck by how everyone was apparently doing *nothing*. Hadza camps consist of a few temporary grass huts that blend in with the surrounding bushes. I didn't realize I had walked into a camp until I found myself amid about fifteen Hadza men, women, and children who were sitting on the ground as shown in figure 2. The women and children were relaxing on one side, and the men on another. One fellow was straightening some arrows, and a few children were toddling about, but no one was engaged in any hard work. To be sure, the Hadza weren't lounging on sofas, watching TV, munching potato chips, and sipping soda, but they were doing what so many health experts warn us to avoid: sitting.

FIGURE 2 What I saw when I first arrived in this Hadza camp. Almost everybody is sitting. (Photo by Daniel E. Lieberman)

My observations since that day along with published studies of their activity levels confirm my initial impression: when Hadza men and women are in camp, they are almost always doing light chores while sitting on the ground, gossiping, looking after children, and otherwise just hanging around. Of course, Hadza men and women head out almost every day to the bush to hunt or gather food. The women typically leave camp in the morning and walk several miles to somewhere they can dig for tubers. Digging is a relaxing and social task that usually involves sitting in a group under the bushes in the shade and using sticks to excavate edible tubers and roots. As Hadza women dig, they eat some of what they extract while chatting and minding their infants and toddlers. On the way there and back, women often stop to collect berries, nuts, or other foods. On the few occasions when I have accompanied Hadza men on hunts, we walked between seven and ten miles. When they are tracking animals, the pace is varied but never so fast that I wasn't able to keep up, and often the hunters stop to rest and look around. Whenever they encounter a honeybee hive, they stop, make a fire, smoke out the bees, and gorge themselves on fresh honey.

Among the many studies of the Hadza, one asked forty-six Hadza adults to wear lightweight heart rate monitors for several days.[14] According to these sensors, the average adult Hadza spends a grand total of three hours and forty minutes a day doing light activities and two hours and fourteen minutes a day doing moderate or vigorous activities. Although these few hours of hustling and bustling per day make them about twelve times more active than the average American or European, by no stretch of the imagination could one characterize their workloads as backbreaking. On average, the women walk five miles a day and dig for several hours, whereas the men walk between seven and ten miles a day.[15] And when they aren't being very active, they typically rest or do light work.

The Hadza, moreover, are typical of other hunter-gatherer groups whose physical activity levels have been studied. The anthropologist Richard B. Lee astonished the world in 1979 by documenting that San hunter-gatherers in the Kalahari spend only two to three hours a day foraging for food.[16] Lee might have underestimated how much work the San do, but more recent studies of other foraging populations report similarly modest physical activity levels as the Hadza.[17] One especially well-studied group is the Tsimane, who fish, hunt, and grow a few crops in the Amazon rain forest. Overall, Tsimane adults are physically active for four to seven hours per day, with men engaging in vigorous tasks like hunting for only about seventy-two minutes a day and women engaging in almost no vigorous activity at all but instead doing mostly light to moderate tasks such as child care and food processing.[18]

All in all, assuming that what hunter-gatherers do is evolutionarily "normal," then comprehensive studies of contemporary foraging populations from Africa, Asia, and the Americas indicate that a typical human workday used to be about seven hours, with much of that time spent on light activities and at most an hour of vigorous activity.[19] To be sure, there is variation from group to group and from season to season, and there is no such thing as a vacation or retirement, but most hunter-gatherers engage in modest levels of physical effort, much of it accomplished while sitting. How different, then, are such "normal" humans from postindustrial people like me (and perhaps you), not to mention farmers like the Tarahumara, factory workers, and others whose lives have been transformed by civilization?

Activity over the Ages

In 1945, in the aftermath of World War II, the United Nations created the Food and Agriculture Organization (FAO) to eliminate hunger, food insecurity, and malnutrition. But when FAO scientists and bureaucrats first tried to figure out how much food the world needed, they realized they didn't know in part because they were ignorant of how much energy people spent being active. Of course, a bigger person must eat more calories per day than a smaller person, but how much more food do you need to eat if you are a factory worker, a miner, a farmer, or a computer programmer? And how do those needs vary if you are male, female, pregnant, young, or elderly?

FAO scientists decided to measure people's energy expenditures using the simplest metric possible, the physical activity level, or

PAL.[20] Your PAL is calculated as the ratio of how much energy you spend in a twenty-four-hour period divided by the amount of energy you would use to sustain your body if you never left your bed. This ratio has the advantage of being unbiased by differences in body size. Theoretically, a big person who is very physically active will have the same PAL as a small person who does the same activities.

Ever since the PAL metric was conceived, scientists have measured the PALs of thousands of people from every walk of life and every corner of the globe. If you are a sedentary office worker who gets no exercise apart from generally shuffling about, your PAL is probably between 1.4 and 1.6. If you are moderately active and exercise an hour a day or have a physically demanding job like being a construction worker, your PAL is likely between 1.7 and 2.0. If your PAL is above 2.0, you are vigorously active for several hours a day.

Although there is much variation, PALs of hunter-gatherers average 1.9 for men and 1.8 for women, slightly below PAL scores for subsistence farmers, which average 2.1 for men and 1.9 for women.[21] To put these values into context, hunter-gatherer PALs are about the same as those of factory workers and farmers in the developed world (1.8), and about 15 percent higher than PALs of people with desk jobs in developed countries (1.6). In other words, typical hunter-gatherers are about as physically active as Americans or Europeans who include about an hour of exercise in their daily routine. In case you are wondering, most mammals in the wild have PALs of 3.3 or more, nearly twice as high as hunter-gatherers.[22] Thus, comparatively speaking, humans who must hunt and gather all the food they eat and make everything they own by hand are substantially less active than average free-ranging mammals.

Here's another, startling way of thinking about these numbers: if you are a typical person who barely exercises, it would take you just an hour or two of walking per day to be as physically active as a hunter-gatherer. Even so, few Americans or Europeans currently manage to achieve those modest levels of activity. The average PAL of industrialized adults in the developed world is 1.67, and many sedentary individuals have even lower PALs.[23] These declines, moreover, are relatively recent and largely reflect changes in how we work, especially the growth of desk jobs that glue us to our chairs. In 1960, about half of all jobs in the United States involved at least moderate levels of physical activity, but today less than 20 percent of jobs demand more than light levels of activity, an average reduction of at least a hundred calories per day.[24] That modest amount of unspent energy adds up to twenty-six thousand fewer calories spent over the course of a year, enough to run about ten marathons. And outside our jobs, we walk less, drive more, and use countless energy-saving devices from shopping carts to elevators that whittle away, calorie by calorie, at how much physical activity we do.

The problem, of course, is that physical activity helps slow aging and promotes fitness and health. So those of us who no longer engage in physical labor to survive must now weirdly choose to engage in unnecessary physical activity for the sake of health and fitness. In other words, exercise.

How Exercise Became Weird

Modern biomedical research relies extensively on millions of mice that spend the entirety of their brief lives in animal facilities where they live in tiny, clear plastic cages eating nothing but mouse chow and never seeing the light of the sun. Because these unlucky animals are naturally social, they are usually housed in groups of about five, and because they are naturally active, it is standard practice to place a little rodent wheel in each cage so they can run in endless circles, not unlike humans on a treadmill. And, boy, do they run. Typical laboratory rodents voluntarily and repeatedly run on their wheels for one- to two-minute bouts, sometimes totaling three miles a night. Curious whether wild rodents would do the same, the Dutch neuroscientist Johanna Meijer placed one of those rodent wheels in an overgrown corner of her garden in 2009 with food as bait, set up a night-vision camera to record what happened, and went to bed. When she viewed the tape the next morning, she found to her delight that dozens of her garden's small wild denizens had run while she slept. After nibbling on the bait, the local mice, rats, shrews, frogs, and even snails (yes, you read that correctly) hopped onto the wheel and enjoyed a few minutes of running in place before disappearing back into the night.[25]

Were these animals exercising, playing, or just running from instinct? No one knows, and the answer partly depends on how we define exercise and play. Samuel Johnson didn't consider either word worthy of an entry in his celebrated dictionary, but subsequent dictionaries generally define "exercise" as a "planned, structured physical activity to improve health, fitness, or physical skills," and "play" as "an activity undertaken for no serious practical purpose." As far as we know, all mammals play when they are young, helping them acquire social and physical skills. Humans are one of the few species that also sometimes play as adults, and uniquely in the context of sports, a distinctive human behavior common to all cultures. Not all sports, however, are exercise (consider darts and auto racing). My opinion is that while many animals are driven by deep instincts to move, sometimes causing pleasure, exercise as we define it—discretionary, planned physical activity for the sake of physical improvement—is a uniquely human behavior. In fact, I think it is fair to make two generalizations about human exercise. First, while youngsters have always played and sports are a human universal, exercise outside the context of sports was extremely rare until relatively recently. Second, as recent technological and social developments have diminished industrialized people's need to be physically active, a growing chorus of experts has never ceased to raise the alarm that we aren't exercising enough.

The first generalization, that adult exercise is modern, is kind of obvious. As we have already seen, early farmers had to toil as hard as if not harder than hunter-gatherers, and for the last few thousand years farmers primarily exercised, often through sports, to prepare for fighting. Ancient texts like *The Iliad,* paintings from pharaonic Egypt, and Mesopotamian carvings testify that sports like wrestling, sprinting, and javelin throwing helped would-be warriors keep fit and hone combat skills. But not all exercise in the ancient world was combat related. If you were wealthy enough to attend one of the great Athenian schools of philosophy, you would have been advised to exercise as part of your overall education. Philosophers like Plato, Socrates, and Zeno of Citium preached that to live the best possible life, one should exercise not only one's mind but also one's body. This idea is not just Western. Confucius and other prominent Chinese philosophers also taught that exercise was equally essential for physical and mental health and encouraged regular gymnastics and martial arts. In India, yoga was developed and popularized thousands of years ago to train both body and mind.[26]

Like so many pursuits, exercise in the Western world took a backseat to other worldly and spiritual concerns after the fall of Rome and didn't have a renaissance until the Renaissance. But primarily for the privileged upper classes. While peasants still toiled in the fields, fifteenth-to-seventeenth-century educators and philosophers such as John Locke, Mercurialis, Cristóbal Méndez, John Comenius, and Vittorino da Feltre advocated exercises like gymnastics, fencing, and horseback riding for the elites to promote vigor, teach character and values, and enrich minds. Then as the middle and upper classes expanded rapidly during the Enlightenment and the Industrial Revolution, Jean-Jacques Rousseau, Thomas Jefferson, and other liberal luminaries enthusiastically extolled the natural value of physical activity and fitness to growing numbers of the newly affluent. Physical culture spread rapidly throughout Europe, the United States, and elsewhere in the nineteenth century, most especially in schools and universities. Exercise and education became inextricably linked.

And yet for the last few centuries, experts have worried incessantly that we aren't exercising enough. Nationalism is one major source of this anxiety. Just as ancient Spartans were required and Romans were urged to be fit enough to fight as soldiers, flag-waving leaders and educators increasingly exhorted ordinary citizens to participate in sports and other forms of exercise as preparation for military service. An especially influential proponent of this movement was Friedrich Jahn, the "Father of Gymnastics." Following Napoleon's humiliating string of victories over German armies in the early nineteenth century, Jahn argued that educators had a responsibility to restore the physical and moral strength of his nation's youth with calisthenics, gymnastics, hiking, running, and more.[27] Later, similar worries in America were spurred by the embarrassing lack of fitness among many men who enlisted or were drafted for World Wars I and II and by the pathetic state of fitness among schoolchildren at the start of the Cold War.[28] National movements to drum up fitness for the sake of the state still occur in China and elsewhere.

The other source of anxiety has been the health consequences of exercising too little. Many people think today's physical inactivity epidemic is a novel crisis, but this state of alarm has been mounting ever since machines started to replace human physical activity. Over the last 150 years, worried physicians, politicians, and educators have regularly raised concerns that the youth of their day are woefully less active, less fit, and thus less healthy than the previous generation. My university, Harvard, is no exception. At the turn of the last century, Dudley Allen Sargent, who founded the modern physical education movement in America (and for years directed the gym where I sometimes work out), worried that "there never was a time in the history of the world when the great mass of mankind could meet the simple exigencies of life with so little expenditure of time and energy as today," and that "without solid physical education programs, people would become fat, deformed, and clumsy."[29] A hundred and twenty years later, a comprehensive survey of college students from Harvard and elsewhere found less than half exercised regularly, thus contributing to "poor mental health and increased stress."[30]

And so we promote exercise. Just as we put wheels in cages for mice in labs, over the centuries we have invented a stunning variety of ways and means for our fellow humans to undertake optional physical activity for the sake of health and fitness. Unsurprisingly, exercise has become increasingly advertised as virtuous and has been commodified, commercialized, and industrialized. To use the weight machines, treadmills, ellipticals, and other contraptions in the gym around the corner from my house costs seventy dollars per month. When I head out for a morning run, I wear specialized running shoes, chafe-preventing shorts, a snazzy moisture-wicking shirt, a washable cap, and an expensive watch that connects to satellites overhead to track my speed and distance. Oscar Wilde once quipped, "I approve of any activity that requires the wearing of special clothing," but I suspect even he would be shocked at the popularity of "athleisure"—workout clothes for everyday activities like sitting that help us look athletic without ever having to break a sweat. Worldwide, people spend trillions of dollars a year on fitness and sportswear.

We have also medicalized exercise. By this I mean we pathologize a lack of physical activity, and we prescribe particular doses and types of exercise to help prevent and treat disease. The U.S. government recommends I engage in at least 150 minutes of moderate or 75 minutes of vigorous exercise a week and weight train at least twice a week.[31] Epidemiologists have calculated that this level of activity will reduce my risk of dying prematurely by 50 percent and lower my chances of getting heart disease, Alzheimer's, and certain cancers by roughly 30 to 50 percent.[32] Insurance companies offer me incentives to exercise, and entire professions have sprung up to motivate me to exercise in the first place, help me work out, and fix me up once I get injured.

There is nothing wrong with medicalizing, commercializing, and industrializing exercise. In fact, these trends are necessary. But they rarely make exercise more fun. To me, the apotheosis of what's good and bad about contemporary exercise is the treadmill. Treadmills are incredibly useful, but they are also loud, expensive, and occasionally treacherous, and I find them boring. I sometimes use treadmills to exercise but struggle as I trudge monotonously under fluorescent lights in fetid air with no change of scene, staring at those little flashing lights informing me how far I've gone, at what speed, and how many calories I've supposedly burned. The only way I endure the tedium and discomfort of a treadmill workout is by listening to music or a podcast. What would my distant hunter-gatherer ancestors have thought of paying lots of money to suffer through needless physical activity on an annoying machine that gets us nowhere and accomplishes nothing?

I have little doubt they would have considered it abnormal to exercise this way. But to understand what kinds of physical activities we did evolve to do and how they affect our health requires, counterintuitively, first grappling with what our bodies are doing when we are physically *in*active.

PART I

Inactivity

TWO

Inactivity: The Importance of Being Lazy

MYTH #2 It Is Unnatural to Be Indolent

> Six days you shall labor and do all your work, but the seventh day is a Sabbath to the Lord your God. On it you shall not do any work, neither you, nor your son or daughter, nor your male or female servant, nor your ox, your donkey or any of your animals, nor any foreigner residing in your towns, so that your male and female servants may rest, as you do.
>
> —Deuteronomy 5:13–14

Why is the Jewish God so insistent that once a week we rest and do absolutely no work whatsoever? Among the many explanations, one possibility is that taking an occasional day off was a wise idea back in the Iron Age, when the Sabbath laws were written. The first Jews were subsistence farmers whose survival depended on regular hard labor. In an age without machines and with little commerce, they had to produce with their own blood, sweat, and tears just about everything they consumed and used. In addition to plowing, sowing, weeding, and harvesting their crops, they tended their livestock, made clothing, fabricated tools, built houses, schlepped water, and more. Were the physical demands of their existence so extreme they occasionally needed to take time off to allow their bodies to rest and avoid injury and illness? Or maybe the Sabbath helped them to be fruitful and multiply?

Regardless of why observing the Sabbath became a sacred obligation, a weekly day of rest certainly wouldn't make sense for hunter-gatherers who don't store provisions and must head into the bush every day to find food for themselves and their families. As we have already seen, hunter-gatherers typically spend less than half the day foraging and otherwise do light work or rest. For perpetually hungry hunter-gatherers who never exercise but have to be physically active to acquire every calorie they eat, a Sabbath would not only be unnecessary but also cause a weekly day of hunger.

And here is something to ponder: How feasible would a Sabbath day of rest be for our closest cousins, the great apes?

Most of us see gorillas and chimpanzees in zoos or documentaries, but it is possible to observe these highly endangered animals in the wild if you are willing to spend a lot of time, effort, and money to travel to the remote rain forests that straddle the African equator where they live. The least accessible wild apes are the mountain gorillas that live high up on the slopes of dormant volcanoes in the highlands of Rwanda and Uganda. To get to these animals, you must first hike up through hand-tilled fields into a steamy rain forest dominated by giant leafy trees and stinging nettles that gradually transitions into bamboo, and then a cooler, dense forest of African redwoods and vines. The trek is hard work because the terrain is steep and there are no trails through the overgrown tangle of vines, shoots, ferns, and thorns that cover the steep and slippery forest floor. Any thoughts of the strenuous effort required to reach gorillas, however, are quickly dispelled by the calm that comes from watching these predominantly sedentary animals.

Because gorillas live in what is best described as a giant salad bowl, they spend their days mostly sitting on their butts and eating the food that literally surrounds them. While infants sometimes play and climb trees, adults lounge placidly amid the shrubbery, munching, scratching, grooming, and napping. In fact, a typical gorilla troop travels only a mile per day.[1] On the rare occasion, however, when big males fight or threaten others in the troop, things can get intense, and you appreciate their strength and power. One of the scariest moments of my life occurred when a four-hundred-pound silverback gorilla charged two females that provoked his ire. To pursue them, he ran on his hind legs at full speed within a few inches of me, thumping his chest. To this day, I don't know how I managed to hold still and not soil myself. Yet, for better or worse, such bouts of vigorous activity are extremely rare, and most of the time adult gorillas are phlegmatic and inactive.

On several occasions I have also been to the forests alongside Lake Tanganyika to observe chimpanzees in the wild, and frankly they aren't much more active than gorillas. When chimpanzees travel on the forest floor, it can be a challenge to keep up with them, but they spend most of their day either feeding or digesting. Chimpanzees typically devote about half their waking hours to filling their stomachs with highly fibrous food, and for much of the rest of the day they rest, digest, groom each other, and take long naps.[2] On an average day, they climb only about a hundred meters and walk just two to three miles.[3] To be sure, chimpanzees are highly social animals, and they occasionally fight, copulate, and do other exciting things, but for most of the time our closest ape relatives are sluggards that live a sort of perpetual Sabbath.

Though hunter-gatherers like the Hadza don't work incredibly hard and they spend many daily hours being physically inactive, apes make them seem like workaholics. And because we evolved from apelike ancestors who largely resembled chimpanzees and gorillas, that suggests it is evolutionarily normal humans who are unusual in terms of how much they work and rest.[4] This revelation raises a host of questions about how and why nonindustrial humans (hunter-gatherers and farmers) are not that active but nonetheless are typically more active than wild apes. To answer this question, however, we first need to head to the lab to appreciate what our bodies are doing and how much energy we expend when we are inactive.

The Cost of Doing Nothing

Imagine you foolishly agreed to participate in an experiment in my lab. As you enter, the first thing to catch your attention would be the enormous treadmill with no visible controls in the center of the room. You'd also notice various cameras and instruments, but the apparatus you should regard with the most apprehension is a bluish silicone face mask connected to a long, flexible tube that is suspended from the ceiling. The tube runs to a large metal box with lots of dials, switches, and displays. In a typical experiment, we fit the mask snugly around your nose and mouth, and a pump sucks all the air you exhale into the mask through the tube to the box, which then measures how much oxygen and carbon dioxide you breathed out. The mask is annoying and uncomfortable, especially when you are running, but the measurements it affords yield a treasure trove of information. Just as a stove burns gas or wood, your body uses oxygen to burn fat and sugar while giving off carbon dioxide. By quantifying how much oxygen you consume and carbon dioxide you exhale, we can calculate precisely how much energy your body is using at any given moment.[5]

Although most experiments in my lab measure your energy expenditure while walking or running, the first thing we ask you to do after donning the mask is to stand or sit quietly for at least ten minutes while we assess your oxygen consumption and carbon dioxide production at rest. This is a crucial step because to measure the energetic cost of walking or running, we need to subtract how much energy you spend when being physically inactive. The unit of measurement we use is a calorie. (Confusingly, the "calorie" used on food labels is actually one kilocalorie, the amount of heat required to raise the temperature of a kilogram of water 1°C, and I will follow the same convention here.)[6]

If you are an average adult American male who weighs 180 pounds (82 kilograms), your rate of energy expenditure while resting quietly in a chair is approximately seventy calories per hour. This is your resting metabolic rate (RMR), so named because your resting metabolism comprises all the chemical reactions going on in your body while you are not being physically active. Based on your RMR, we can calculate that if you do nothing but sit in a chair for the next twenty-four hours, your body will expend about seventeen hundred calories.

Seventeen hundred calories is a lot. Even when you are sitting, you are not entirely at rest. Some of that energy is being spent digesting the last meal you ate, regulating your body temperature, and preventing your body from slumping to the floor. To correct for these expenses, we could measure your energy expenditure in bed just after you woke up from an eight-hour sleep in a dark 70°F room following a twelve-hour fast. That measurement, your basal metabolic rate (BMR), would be roughly 10 percent lower than your RMR (in our example, 1,530 calories). Your BMR is the energy you use to maintain the most basic processes of your body necessary to stay alive in a nearly coma-like state.

How does the amount of energy you spend at rest compare with your total energy budget? To compute this ratio, we next need to measure your total daily energy expenditure (DEE), the overall number of calories you spend over the course of twenty-four hours doing everything you do including moving, reading, sneezing, talking, and digesting. Until recently, we'd estimate your DEE by measuring your oxygen consumption during different tasks such as sitting, eating, walking, and running. By knowing how much energy and time you spent doing these and other tasks, we could add them up to get an approximate estimate of your DEE. As you can imagine, curious scientists interested in human energetics have doggedly followed people around with oxygen masks to assess the cost of doing almost every activity imaginable including digging, sewing, making a bed, and working on a car assembly line.[7] Several studies even tried to measure how much energy it costs to think.[8] These methods, however, are cumbersome, inaccurate, and challenging to perform, especially in remote parts of the world.

You'll be relieved to know that to measure your DEE, we no longer follow you around all day with a gas mask. Instead, we measure your pee. To be more precise, we'd ask you to drink some very expensive, harmless water that contains a known quantity of rare (heavy) hydrogen and oxygen atoms and then collect samples of your urine over the next few days. This sounds like a creepy magic trick, but by measuring the rate at which these heavy atoms become less abundant in urine, we can calculate the rates at which both the hydrogen and the oxygen atoms leave the body from sweating, urinating, and breathing. Because hydrogen exits the body only in water but oxygen leaves in both water and carbon dioxide, the difference in the concentration of these two atoms in urine allows us to compute exactly how much carbon dioxide someone produced from breathing, hence how much energy he or she used.[9] Thousands of people's metabolisms have been measured this way, providing a remarkable database on human energy expenditure. If you weigh 180 pounds, your DEE is probably about twenty-seven hundred calories a day. Because we already learned your RMR is about seventeen hundred calories a day, that means nearly two-thirds (63 percent) of the energy you expend each day is spent just on your resting metabolism. Who would have thought that being a couch potato was so expensive?

Is doing nothing as costly for hunter-gatherers like the Hadza as it is for industrialized people in countries like the United States? Fortunately, Herman Pontzer and colleagues have intrepidly measured many Hadza men and women using the methods described above. According to their analyses (which involve some estimates), Hadza basal metabolisms are no different from yours or mine after correcting for the fact that they are relatively smaller and thinner. To give you the numbers, the average Hadza man weighs about 115 pounds and spends an estimated 1,300 calories on his BMR; the average Hadza woman is about 100 pounds and spends 1,060 calories on her BMR.[10] Because fat is a relatively inert tissue that does not contribute much to metabolism, we need to account for the fact that Hadza men and women have about 40 percent less body fat than average Westerners.[11] When we make this correction, it becomes evident that adult humans spend about 30 calories every day for each kilogram of fat-free body mass just to maintain their bodies regardless of whether they spend the day staring into computer screens in New York, manufacturing shoes in a Chinese factory, growing corn in rural Mexico, or hunting and gathering in Tanzania. Of the more than 20 trillion calories consumed today by human beings on Earth, the majority are devoted to paying for the most basic needs of their bodies at rest.

In sum, even if you are a highly active person, you probably spend more energy maintaining your body than doing stuff. Understandably, this fact seems counterintuitive. As I sit here writing these words, there is little visible evidence that every one of my body's systems is working industriously to keep me alive apart from the fifteen to twenty gentle breaths I take every minute. Yet my heart is contracting sixty times a minute to pump blood to every corner of my body, my intestines are digesting my last meal, my liver and kidneys are regulating and filtering my blood, my fingernails are growing, my brain is processing these words, and countless other cells in every tissue of my body are busily replenishing themselves, repairing damage, fending off infection, and monitoring what's going on.

Are all these functions really that costly? Do we need to spend so much energy doing nothing?

One way to address this question is to perform a "stress test" and see how the body handles the challenge of not having enough energy. You stress your body this way whenever you diet and consume fewer calories than you expend for days, weeks, or months. Effective diets, however, tend to be gradual, helping you lose weight slowly by burning just a little extra fat every day. A more demanding, hence revealing, stress test for your metabolism requires a more extreme reduction in energy intake: starvation. For good reason, it is unethical and illegal to starve people in the lab for the sake of science. Yet carefully designed, controlled experiments to study the effects of starvation on human metabolism were conducted on humans in Minnesota during the waning months of World War II.

Will You Starve That They Be Better Fed?

Between fifty and eighty million people died in World War II. Twenty million of these casualties were soldiers, but at least as many civilians perished slowly from starvation as the conflict destroyed crops and disrupted supply lines.[12] During the siege of Leningrad, a thousand people starved to death every day. As the war dragged on and the magnitude of this staggering humanitarian problem became evident, Dr. Ancel Keys, a researcher at the University of Minnesota, started to worry about how to help these victims. Keys was intensely aware that scientists knew almost nothing about the effects of long-term food deprivation on the human body. Helping multitudes of starving people would require a better understanding of what was happening to their bodies. Keys and others also worried that when the war ended, millions of hungry people would be more susceptible to fascism or communism. Thus, for both humanitarian and geopolitical strategic reasons, the U.S. government gave Keys money to assemble a team of scientists to comprehensively study the effects of starvation and rehabilitation on volunteers. Armed with an eleven-page brochure, Keys appealed to conscientious objectors who had refused military service but wanted to help others by being human guinea pigs. The brochure's cover featured a photo of three starving French children with empty bowls and the question in bold letters "Will you starve that they be better fed?"

Despite having read accounts of the Minnesota Starvation Experiment, including the two-volume monograph published in 1950 with all the results and photos, I still find it hard to imagine being one of those thirty-six volunteers at the start of the experiment in November 1944.[13] For the first twelve weeks, the experiment wasn't so bad, because this initial control phase standardized the men to a generous diet of 3,200 calories a day. Every week, the volunteers also had to walk twenty-two miles and do fifteen hours of typical physical work like laundry and chopping wood. During this period, Keys and his team measured just about everything they could, including height, weight, amount of body fat, resting pulse, red blood cell count, physical stamina, strength, hearing, psychological state, even sperm count. Then, on February 12, 1945, the volunteers' diet was suddenly halved to 1,570 calories a day. Just as important, the volunteers had to maintain the same physical activity levels, including twenty-two miles of walking per week. Keys required these activities because starving people rarely have the luxury of doing nothing, but instead must work to obtain food and survive.

Although 1,570 calories a day is close to a man's typical resting metabolic rate, theoretically enough to maintain normal bodily function, the starvation diet combined with the exercise requirements quickly became an excruciating physical and mental ordeal. In addition to losing weight rapidly and feeling perpetually ravenous, the starving men became lethargic, depressed, and frequently angry. Many suffered from horrific nightmares, and one fellow cut three fingers off his hand while chopping wood (whether this was intentional, an accident, or the result of delirium is unknown). Gradually, their bodies wasted, strength and stamina dwindled, legs swelled, and heart rates dropped. Sitting on increasingly skinny buttocks became agonizingly painful. As these and other changes transformed the volunteers' bodies and minds, Keys and his team ceaselessly, carefully, and comprehensively measured the ravages of starvation. Finally, after twenty-four weeks when the slowly starving men had lost precisely 25 percent of their initial weight, Keys increased their daily rations gradually over twelve weeks so they could start regaining weight. On October 20, 1945, the men were released, nearly two months after the war officially ended.

We learned much about starvation and rehabilitation from this extreme stress test, but let's focus on what we learned about resting metabolism. As expected, the starving men's bodies survived primarily through weight loss and inactivity. As their metabolic demands continued to exceed their caloric intake, the human guinea pigs tapped into their fat reserves. Approximately 15 percent of a typical thin man's body is fat (thin women average 25 percent body fat). That fat has several functions, but the most essential is to be an enormous reservoir of calories that can be burned when needed. In the case of the men Keys starved, their body fat reserves plummeted by 70 percent over those twenty-four harrowing weeks, from an average of twenty-two to seven pounds.[14] Just as important, while their bodies wasted away, the volunteers became severely lethargic and pared down their physical activity to the bare minimum. When not doing their required walks and chores, they often just lay in bed doing as little as possible to conserve energy. Their concentration plummeted, and their sex drive evaporated.

But there is more. The conscientious objectors who starved themselves for science also survived thanks to another vital set of adaptations that were less easy to observe: *their bodies transformed to use less energy even while resting*. After twenty-four weeks of starving, the volunteers' resting and basal metabolic rates plummeted by 40 percent, far more than expected from the weight they lost. According to the measurements Keys and his colleagues took, the average volunteer's basal metabolic rate decreased from 1,590 calories a day to 964 calories a day, the basal metabolic rate you expect for an eight-year-old child who weighs fifty-five pounds!

The key lesson to digest from the starving men's dramatically lower resting metabolic rates is that *human resting metabolisms are flexible*. Most critically, resting metabolism is what the body has *opted* to spend on maintenance, not what it *needs* to spend. One of the main ways the starving volunteers spent less energy was to skimp on maintenance. Basically, their metabolisms slowed down and cut back on costly physiological processes that keep the body in a state of balance. Their heart rates decreased by one-third, and their body temperatures dropped from a normal 98.6°F to 95.8°F, causing them to feel constantly cold, even in well-heated rooms. Their bodies also spent less energy replacing cells in their skin and other organs that are normally replenished on a regular basis. Their skin became flaky, their sperm counts fell, and they manufactured fewer blood cells. Keys's comprehensive measurements even showed they produced less earwax.

The other way the starving volunteers saved energy was by downsizing: they reduced the size of costly organs that account for a large percentage of resting metabolism. By measuring how much blood and oxygen go in and out of organs, physiologists can approximate how much energy different body parts consume. Such measurements indicate that nearly two-thirds of a person's resting metabolism is spent on just three very expensive tissues: brain, liver, and muscle. Your brain and liver each consume about 20 percent of your resting metabolism, and if you are a typically strong human, your muscles expend 16 to 22 percent of your resting metabolism.[15] The remaining 40 percent accounts for everything else including your heart, kidneys, guts, skin, and immune system. If you are sitting while reading this, for every five breaths you take, one pays for your brain, another for your liver, a third for your muscles, and the last two pay for the rest of your body.

Keys's data showed that the bodies of the men he starved saved calories in a manner similar to the way most people economize when they confront a severely reduced income: they prioritized "essential" organs like the brain but abandoned "expendable" costs like reproduction and drastically cut back on "reducible" functions like staying warm, active, and strong. By shrinking their muscles by 40 percent, they saved about 150 calories a day, leaving the starving men feeble and easily fatigued. Their hearts also got smaller by an estimated 17 percent, and their livers and kidneys shrank similarly.[16]

It took five years to analyze and publish the results from the Minnesota Starvation Experiment—too late to help any victims of World War II. However, one of the principal insights we learned from those courageous volunteers is that *resting is not just a state of physical inactivity.* When we seem to do nothing, our bodies are still actively expending a lot of energy on many dynamic and costly processes. Just as important, because we cannot spend a calorie more than once, resting is a crucial way our bodies engage in *trade-offs.* As you read these words, you are spending roughly sixty calories an hour (the energy in a typical orange) tending to your brain, liver, muscles, kidneys, intestines, and more. If you decide to toss aside this book and climb a mountain, you will necessarily divert some calories from those basic functions to ascend and descend that mountain. Then, when you get home, you'll probably eat and rest to replenish those spent extra calories.

If resting and physical activity are just different ways of using energy, which is normally limited, then how many calories should we spend taking care of our bodies versus walking, running, and doing other physical activities? That problem partly depends on our goals, which could include trying to fight an infection, lose weight, get pregnant, or train for a marathon. In the grand scheme of things, however, the way our bodies allocate resources has been molded by a much larger process: evolution by natural selection. We use energy as we do in large part because of the way Darwinian evolution acted on millions of generations of our ancestors.

To illustrate how Charles Darwin's powerful and insightful theory explains why and when we trade off precious calories on physical activity versus other functions, let's turn to the keen observations of another great English writer, Jane Austen.

The Truth About Trade-Offs

Of Jane Austen's six novels, *Mansfield Park* is my least favorite. The heroine, Fanny Price, is priggish, and the middle of the book drags on tediously. But its plot insightfully revisits a classic problem of intense interest to evolutionary biologists: trade-offs. To set the scene, Fanny's mother is one of three sisters who fared differently in life. One of Fanny's aunts, Lady Bertram, married the wealthy Sir Thomas Bertram of Mansfield Park and had four children, all raised in luxury. Her other aunt, Mrs. Norris, married a clergyman and was childless but helped raise her rich nieces and nephews. Fanny's mother, Mrs. Price, rebelled by marrying a penniless, inebriate sailor, lived in squalor and poverty, and struggled to raise ten children with no help from her husband.

Austen died when Charles Darwin was only eight years old, but Mrs. Price and her sisters' different reproductive strategies exemplify a fundamental but sometimes underappreciated prediction of Darwin's theory of natural selection. As a quick refresher, this theory—among the most thoroughly scrutinized and tested theories ever proposed—is that over generations heritable features that cause organisms to have more surviving offspring will become more prevalent while features that impair reproductive success will become rarer.[17] For example, if longer legs make you faster, and speed helps you escape predators (or be a better predator), then selection will favor long legs. But because speed is obviously always beneficial, why don't more species have long legs? The explanation is trade-offs. Because variations almost always involve limited alternatives, natural selection has to act on competing costs and benefits. If you are long-legged and big, then you can't be short-legged and small, which has other advantages depending on your situation. Selection will inevitably favor whichever alternative or compromise most improves your reproductive success in your environment.

That principle brings us back to Fanny's family because organisms constantly trade off how they spend limited calories. A key trade-off that Fanny's mother and aunts illustrate is between the quantity and quality of offspring. One strategy is to have as many children as possible without investing highly in any one of them; alternatively, one can invest highly in just a few children to make sure they grow up and have offspring of their own. Crucially, and as Jane Austen illustrates, the optimal strategy from the viewpoint of natural selection depends on one's circumstances. If you are like Lady Bertram and can protect, invest in, and nurture your children, you can afford to opt for quality over quantity. If, however, you are like Fanny's impoverished mother, whose children are less likely to survive and thrive, the best strategy is quantity over quality. Finally, if you are childless, like Aunt Norris, your only option is to help raise your sisters' children because each niece and nephew has one-quarter of your genes. Jane Austen's story of the three sisters' reproductive strategies exemplifies the importance of trade-offs when energy is limited.

No one, I hope, makes or advocates decisions about how many children to have using Darwinian theory. But without knowing it, our bodies are constantly making plenty of other consequential trade-offs—many involving energy—that have been selected over millions of generations. Critically, among these trade-offs, one of the most fundamental is whether to spend precious calories being physically inactive or active.

To appreciate trade-offs between inactivity and activity, it bears repeating that a calorie can be spent only once. In fact, as figure 3 illustrates, you can spend a given calorie in just five ways: growing your body, maintaining your body (resting metabolism), storing energy (as fat), being active, or reproducing. The compromises your body makes among these functions depend on your age and energetic circumstances. For example, if you are young and still growing, you probably won't have enough energy to reproduce, which is why animals usually start having offspring only after they stop growing. If you climb a mountain today, you will have less energy to spend on maintenance, storing fat, and (perhaps) reproducing. If you go on a diet, you will have less energy to be active or reproduce. And so on. But, remember, not all trade-offs are equal in the eyes of natural selection. Like an unsentimental novelist such as Jane Austen, natural selection doesn't care if we are happy, nice, or wealthy; it just favors heritable traits, including trade-offs, that enable us to have more children.

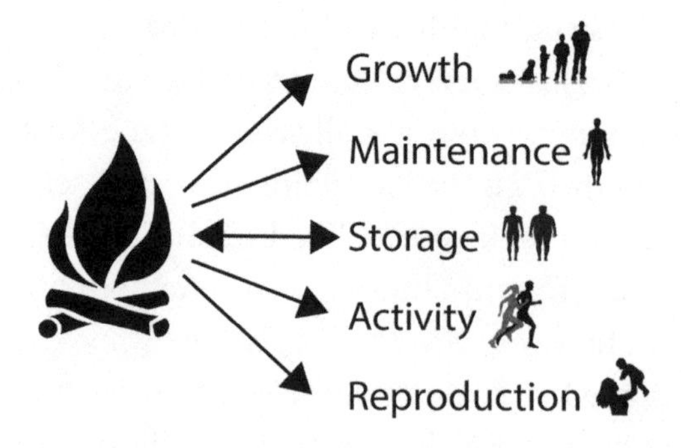

FIGURE 3 Energy allocation theory: the different, alternative ways the body can use energy from food.

Which brings us back to physical inactivity. From the perspective of natural selection, when calories are limited, it always makes sense to divert energy from nonessential physical activity toward reproduction or other functions that maximize reproductive success *even if these trade-offs lead to ill health and shorter life spans.*

Stated simply, we evolved to be as inactive as possible. Or to be more precise, our bodies were selected to spend *enough but not too much* energy on nonreproductive functions including physical activity. Note that I included the qualifier "as possible" because, obviously, one cannot survive or thrive without moving. As a young hunter-gatherer, you would need to play to develop athletic skills, gain strength, and build stamina. As an adult, you would have no choice but to seek out food, do chores, find mates, and avoid being killed. You would probably also need and want to participate in important social rituals like dancing. But whenever energy is scarce, as it usually was, any gratuitous physical activity would reduce how much energy you could devote to survival and reproduction. No sensible adult hunter-gatherer wastes five hundred calories running five miles just for kicks.

A trade-off perspective explains the transformations we saw in the Minnesota Starvation Experiment. The bodies of those starving volunteers engaged in compromises necessary for survival. Although they were forced to exercise a little, they avoided unnecessary physical activity, they reduced how much energy they spent to maintain their bodies, and they completely gave up any interest in reproduction. Fortunately, those trade-offs were temporary responses to a brief, unusual crisis. Starvation also turns out to be rare in preagricultural societies because hunter-gatherers live in tiny populations within enormous territories, they are not dependent on crops that can fail, and when times are tough, they move to find food. Decades of research show that hunter-gatherers generally manage to avoid starvation and maintain about the same weight throughout the year.[18] That doesn't mean hunter-gatherers don't face tough times. They do. In fact, they frequently complain of being hungry. But one of the ways hunter-gatherers survive is by not foolishly squandering scarce calories on unnecessary activity.

So if, as you read these words, you are seated in a chair or lounging in bed and feeling guilty about your indolence, take solace in knowing that your current state of physical inactivity is an ancient, fundamental strategy to allocate scarce energy sensibly. Apart from youthful tendencies to play and other social reasons (topics for later chapters), the instinct to avoid nonessential physical activity has been a pragmatic adaptation for millions of generations. In fact, compared with other mammals, humans might have evolved to be especially averse to exercise.

Born to Be Lazy?

One of the nicest places to take a walk near my home in Cambridge, Massachusetts, is our town reservoir, Fresh Pond. This tranquil spot is surrounded by woods and a two-mile-long path where residents walk, run, and bicycle year-round. Because well-behaved dogs are allowed off leash, my wife and I regularly take our dog, Echo, for walks around the pond. Echo adores the place. The instant we unleash her, she dashes off as fast as possible, reveling in her freedom and enjoying her speed and agility. My wife and I frustrate Echo because we just stroll along behind her at a casual, sensible pace while she scampers to and fro. Eventually, however, Echo starts to fatigue, and by the time we have circled the pond, she inevitably lags behind us, tuckered out and ready for a nap.

Our contrasts with Echo call attention to how dogs are capable of more speed and less endurance than humans, but they also make me feel like a lazy plodder. Why don't I have the urge to bolt out of the car like Echo? Does she dash off recklessly at her first opportunity because she finds it intensely pleasurable to gallop at top speed, because she lacks the foresight to conserve energy, or because she needs to release pent-up energy? Perhaps all three explanations are true, but the different ways dogs and humans circumnavigate the reservoir highlight how humans—with the important exception of children—tend to be cautious about spending calories. I never see adults leap out of their cars in the Fresh Pond parking lot and sprint as fast as possible until they gasp for air. Further, for every human exercising, many more are back at home taking it easy. Unlike children and dogs, adult humans routinely need to be persuaded or coerced to get out of our chairs and exercise.

People who avoid exercise are commonly labeled lazy, but aren't these exercise avoiders behaving normally? As we just saw, it's sensible to be cautious about wasting scarce energy on discretionary physical activity—that is, exercise. But there is reason to speculate that humans tend to avoid exercise more than most animals because we trade off energy differently. Perhaps by investing more energy in essential active pursuits, we have less to spare for inessential activities?

To explore this idea, figure 4 compares how hunter-gatherers (again, the Hadza), Westerners, and chimpanzees spend their calories being physically active.[19] The left panel shows each group's total active energy expenditure averaged for males and females.[20] As you can see, chimpanzees spend far fewer calories per day being active than either of the two human groups do. By this metric, all humans are high-energy, "gas-guzzling" creatures compared with our ape cousins. You might be surprised to see in the left panel that Westerners spend about 80 percent as many calories as the Hadza doing things, even though the Hadza are considerably more active than Westerners. That similarity in overall active energy expenditure, however, is partly a function of body size. On average, the Hadza weigh about 60 percent as much as Westerners and have about one-third as much body fat. Because bigger individuals expend more energy but fat tissue uses almost no calories, the right panel shows active energy expenditures

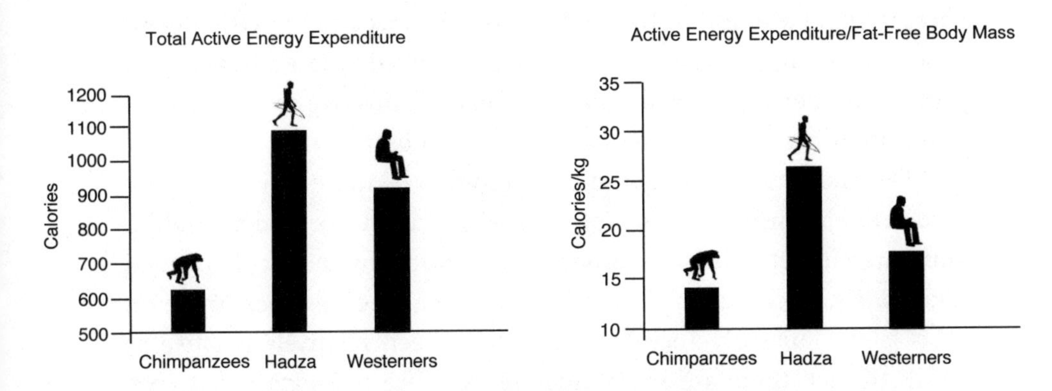

FIGURE 4 Total active energy expenditure (*left*) and active energy expenditure divided by fat-free body mass (*right*) in chimpanzees, Hadza hunter-gatherers, and Westerners. Males and females are averaged. (Hadza data are from Pontzer, H., et al. [2012], Hunter-gatherer energetics and human obesity, *PLOS ONE* 7:e40503; chimpanzee and Westerner data are from Pontzer, H., et al. [2016], Metabolic acceleration and the evolution of human brain size and life history, *Nature* 533:390–92)

divided by fat-free body mass. According to this admittedly crude correction (which does not factor in the slope of the relationship between size and energy expenditure), the Hadza spend twice as many calories per pound of fat-free body mass as chimpanzees, and even sedentary Americans spend about one-third more calories per day per pound than chimpanzees.

So despite the evidence we already saw that hunter-gatherers don't work particularly hard, they are still considerably more physically active than chimpanzees. In fact, over the course of a year, an average Hadza woman spends an impressive 115,000 calories more than a similar-sized chimpanzee female on everything from walking and foraging to preparing food and taking care of her offspring. That's roughly enough energy to run twelve hundred miles, about the distance from New York to Miami.

One possible interpretation of figure 4 is that humans, especially hunter-gatherers, are unusually hardworking animals. Not so. Most wild mammals have physical activity levels of 2.0 to 4.0, and the Hadza fall at the bottom of that range (Hadza women average 1.8, and men average 2.3).[21] Instead, it is chimpanzees and sedentary Westerners who have unusually low PALs of about 1.5 and 1.6. Other great apes like orangutans also have low PALs.[22] Stated differently, apes and sedentary industrialized people are unusually inactive compared with most mammals, and hunter-gatherers are in-between.

Evidence that hunter-gatherers evolved to be more active than our indolent ape ancestors has profound ramifications for understanding human energetics because evolution is contingent on what happened previously. Because we evolved from chimpanzee-like apes (more on this later), our early ancestors must have been relatively inactive as well. In fact, there are many lines of evidence to suggest that apes were specially selected to have unusually low levels of physical activity to help them thrive in the rain forest. As we saw, apes usually don't need to travel far to get food, and their highly fibrous diet requires them to spend much time resting and digesting between bouts of feeding. In addition, their adaptations to climb trees make them outlandishly inefficient at walking. A typical chimpanzee spends more than twice as much energy to walk a mile as most mammals, including humans.[23] When walking is so calorically costly, natural selection inevitably pushes apes to spend as little energy as possible schlepping about the forest so they can devote as much energy as possible to reproduction. Apes are adapted to be couch potatoes.

If we evolved from chimpanzee-like apes that are unusually sedentary, what happened to make humans so much more active, and how does that legacy affect how much we move? The answer, which we will learn about in later chapters, is that climate change spurred our ancestors to evolve an unusual but extremely successful way of life, hunting and gathering, that demands more work. In terms of physical activity, hunter-gatherers are only very active for a few hours a day, but they nonetheless walk five to ten miles a day, carry food and infants, dig for many hours, sometimes run, and perform myriad other tasks to survive. In order to cooperate, communicate, and make tools, our ancestors were also selected to have large expensive brains. Last but not least, we evolved to be highly active to fuel a unique and unusually exorbitant reproductive strategy.

The expensive energetic strategy of hunter-gatherers is so important for understanding the evolution of human activity it behooves us to compare human and chimpanzee energy budgets in more detail. When a typical chimpanzee female becomes a mother at the age of twelve or thirteen years and starts to nurse, she needs to consume about 1,450 calories a day, which she acquires by foraging mostly for raw fruits. On this budget, she manages to have a new baby every five or six years.[24] In contrast, a typical hunter-gatherer woman takes about eighteen years to mature into an adult, and when she becomes a mother, she needs at least 2,400 calories a day to take care of her body plus nurse an infant. To get more energy than the chimpanzee mother, the human mother eats a higher-quality, more diverse diet of fruits, nuts, seeds, meat, and leaves, some of which she gathers herself but some of which is provided by her husband, her mother, and others. Further, by cooking her food, she gets more energy than by eating it raw.[25] Thanks to that extra energy, instead of having a baby every five or six years, the human mother weans her infants earlier and typically has another baby about every three years. Thus, human mothers typically care for and feed more than one dependent child at a time. In addition, humans take 50 percent longer than chimpanzees to reach maturity, magnifying the time and cost of raising each offspring.

The bottom line is that humans evolved to acquire and expend much more energy than chimpanzees. As we'll see in later chapters, by walking long distances, digging, sometimes running, processing our food, and sharing, we spend a lot more energy being active every day than chimpanzees, but that effort yields more calories that enable us not only to be more physically active but also to reproduce at about twice the rate. The extra energy also enables us to have bigger brains, store more fat on our bodies, and do other good things. But there is a cost. The more calories we need, the more we are vulnerable to not having enough. Although the hunter-gatherer strategy is a boon to our reproductive success, it selects against wasting calories on discretionary physical activity.

Of course, this logic applies to all animals. Whether you are a human, ape, dog, or jellyfish, natural selection will select against activities that waste energy at a cost to reproductive success. In this regard, all animals should be as lazy as possible. However, the evidence suggests that humans are more averse to needless physical activity than many other species because we evolved an unusually expensive way of increasing our reproductive success from an unusually low-energy-budget ancestor. When your expenses are high, every penny saved is valuable.

In Praise of Inactivity

Yesterday I drove to the supermarket and, without thinking, waited for a car to pull out from a prime parking spot right next to the store's entrance so I didn't have to walk another thirty feet. As I grabbed a shopping cart to hunt and gather within, I remonstrated myself for being lazy. But then I wondered if I was being one of those annoying exercists who nags people (in this case myself) to sneak in more physical activity by parking at the back of the lot. How did something as normal and instinctive as saving energy become associated with the sin of slothfulness?

I might be lazy, but spiritually I am in the clear. The mortal sin of sloth comes from the Latin word *acedia,* which means "without care." To early Christian thinkers like Thomas Aquinas, sloth had nothing to do with physical laziness, but instead was a sort of mental apathy, a lack of interest in the world. By this definition, sloth is sinful by causing us to neglect the pursuit of God's work. It was only later that sloth came to mean the avoidance of physical work, perhaps because almost no one back then, apart from a few elites, could avoid regular physical labor. Laziness, the disinclination to carry out an activity because of the effort involved, has become today's version of sloth, but it shouldn't carry the same spiritual overtones. Saving a few calories by parking in a prime location is hardly preventing me from fulfilling my duties to anyone. It's just an instinct.

If you've ever doubted humanity's deep-seated tendencies to save energy, spend a few minutes in a mall or airport standing below one of those escalators alongside a stairway. How many people elect to take the stairs, not the escalator? Somewhat naughtily, I once conducted this experiment informally at the annual meeting of the American College of Sports Medicine, a conference filled with professionals dedicated to the idea "Exercise Is Medicine." For this admittedly unscientific study, I stood for ten minutes at the bottom of the stairs and counted how many took the escalator versus the stairs. Of the 151 people I observed during those ten minutes, only 11 took the stairs, about 7 percent. Apparently, people who study and promote exercise are no different from the rest of us. Worldwide, the average is just 5 percent.[26]

Today, many people have jobs that involve little to no manual labor, requiring us to choose to be physically active through exercise. Whether we take the stairs, jog, or go to the gym, we need to override ancient, powerful instincts to avoid unnecessary physical activity, and it should hardly be surprising that most of us—hunter-gatherers included—naturally avoid exercise. Until recently, those instincts helped us maximize how many children we had who themselves survived and reproduced. Energy wasted on a senseless ten-mile run is energy not spent on our offspring. Maybe that's another reason the Jewish God was so insistent that overworked Israelites in the Iron Age observe the Sabbath. In addition to helping the early Jews spiritually and physically, a weekly day of rest probably helped them fulfill another one of God's commands, to be fruitful and multiply.

So let's banish the myth that resting, relaxing, taking it easy, or whatever you want to call inactivity is an unnatural, indolent absence of physical activity. Let's also refrain from stigmatizing anyone for being normal by avoiding nonessential exertion. Unfortunately, we have a long way to go. According to a 2016 survey, three out of four Americans think obesity is caused by a lack of willpower to exercise and control appetite.[27] Despite stereotypes of non-exercisers as lazy couch potatoes, it is deeply and profoundly normal to avoid unnecessarily wasting energy. Rather than blame and shame each other for taking the escalator, we'd do better to recognize that our tendencies to avoid exertion are ancient instincts that make total sense from an evolutionary perspective.

The problem, however, is that until recently only great kings and queens could enjoy taking it easy whenever they wanted. Today in a bizarre reversal of the human condition, voluntary physical activity for the sake of health—a.k.a. exercise—has become a privilege for the privileged. In addition to being surrounded by laborsaving devices, billions of people have jobs and commutes that prevent them from being physically active by requiring them to sit for most of the day. In fact, it is more likely than not you are sitting as you read this. It is also more likely than not you've heard that sitting too much can ruin your health. How can something so ancient, universal, and ordinary as sitting be so unhealthy?

THREE

Sitting: Is It the New Smoking?

MYTH #3 Sitting Is Intrinsically Unhealthy

Unhappy Theseus, doom'd for ever there,
Is fix'd by Fate on his eternal chair.

—Virgil, *The Aeneid,* book 6 (trans. John Dryden)

Spring break, for those of my students and colleagues who can afford the money and time, is traditionally a chance to escape the end of a long, dreary winter by migrating south for a week in March to someplace warm. For these lucky folks, a quintessential spring break involves spending hour after hour on a warm, sunny beach relaxing in that most fundamental posture of physical inactivity: sitting.

I, too, usually sit for hour after hour that week, but working at my desk at home. However, in early March 2016, I found myself at the start of spring break getting off an airplane in Kangerlussuaq, Greenland, a few degrees north of the Arctic Circle. As I deplaned, it was windy and –35°C (–31°F)—conditions that can cause frostbite in minutes. My terror was soon heightened as my Inuit hosts, Jahenna and Julius, issued me the outfit I'd need to survive my upcoming trip to the frigid interior of Greenland. In addition to three layers of underwear and socks along with enormous fur-lined boots, they lent me a massive, smelly sealskin suit including gloves, pants, and a hooded parka. That night, as the wind howled outside the tiny airport building that also serves as the town's hotel and restaurant, I barely slept wondering how I would survive the expedition I had foolishly agreed to make. I was there with my Danish colleague Chris MacDonald to travel by dogsled up a frozen fjord and into the snowy mountains bordering Greenland's central glacier to experience how the Inuit of Greenland used to survive. With Jahenna and Julius's guidance, we were going to hunt musk oxen (a kind of giant arctic sheep), fish, and camp outdoors in bone-chilling conditions as part of a documentary exploring how changing lifestyles are affecting people's health.

Although I expected my arctic safari to be physically demanding, the one challenge my imagination never prepared me for was sitting on that dogsled for days. The traditional Inuit sled, a *qamutik,* is a six-foot-long wooden platform lashed to two curved wooden runners. In the bitter morning cold, we tied the supplies onto the sled with ropes; then Julius harnessed our team of thirteen snarling sled dogs, and as they started to pull with gusto, we jumped on. While Julius sat in front managing the dogs, my job was to sit at the back of the sled. Sounds easy, right? In reality, as the icy terrain and hours slipped by, I found it increasingly harrowing to sit on that sled. For one, it was –30°F and windy, so being inactive was numbingly cold. Even more grueling, however, was sitting for hours without any back support. Although I consider myself a champion sitter, the chairs in which I typically sit have backrests. As the dogs pulled us relentlessly through that intensely cold gray landscape of snow and ice, my back started to ache and then spasm from fatigue. While Julius sat upright at the front of the sled, I slumped behind him, writhing in agony, trying to enjoy the experience while controlling my posture and not tumbling off into a frozen oblivion.

Ironically, even though I have spent much of my life sitting, I apparently don't sit very well, and I repeatedly read that I sit too much. As a white-collar worker, I have little choice but to sit as I do my job, and I also sit when traveling in cars, eating, and watching TV. Despite Ogden Nash's observation that "people who work sitting down get paid more than people who work standing up," a growing chorus of exercists who nag us to exercise condemn sitting as a modern scourge.[1] One prominent physician has declared that chairs are "out to get us, harm us, kill us" and that "sitting is the new smoking."[2] According to him, the average American sits for an unacceptable thirteen hours a day, and "for every hour we sit, two hours of our lives walk away—lost forever." This admonition is obviously hyperbole, but other well-publicized studies estimate that sitting more than three hours a day is responsible for nearly 4 percent of deaths worldwide and that every hour of sitting is as harmful as the benefit from twenty minutes of exercise.[3] By some estimates, replacing an hour or two of daily sitting with light activities like walking can lower death rates by 20 to 40 percent.[4] As a result, standing desks are all the rage, and many people now wear sensors or use their phones to keep track of and limit their sitting time. We have become exercised about sitting.

If we wish to understand how we evolved to be both physically inactive and active, and why that matters, then we need to understand sitting. Most fundamentally, if we evolved to avoid unnecessary physical activity (a.k.a. exercise), how can sitting be so deadly? Does the average American really sit for thirteen hours a day, and how does that compare with our ancestors' habits? My Inuit companions, Julius and Jahenna, sat as much as I did on our trip through Greenland, and hunter-gatherers like the Hadza sit for hours in camp every day hanging out, doing light chores, talking, and otherwise staying off their feet. Do the dangers of sitting arise from the sitting itself, the way we sit, or the time we spend not exercising? As a first step toward addressing these and other potential seated perils, let's use the dual lenses of evolution and anthropology to gain some perspective on why, how, and how much we sit.

If you are not sitting already, find a comfy chair, and read on. . . .

How and Why We Sit

Among the many reasons I love my dog, Echo, is that she helps expose my hypocrisies. When not chasing squirrels, barking at the mailman, or going for walks, Echo is a lazy creature who spends untold hours

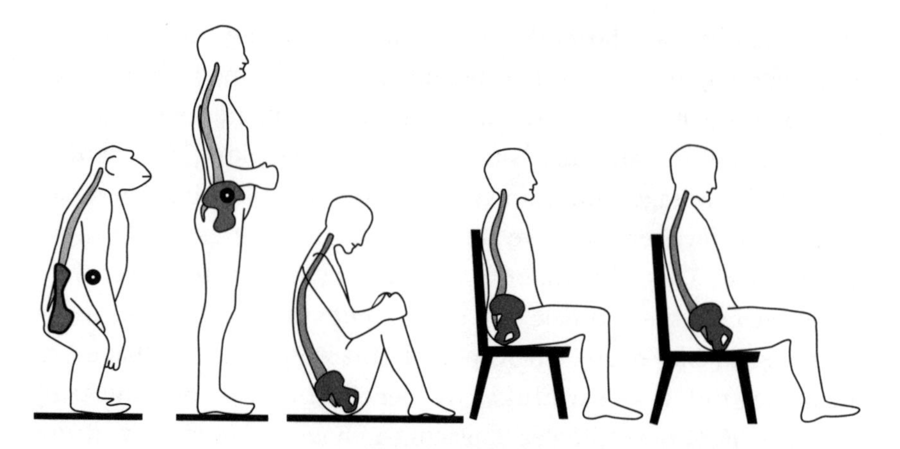

FIGURE 5 The spine and pelvis during standing and sitting. Compared with the chimpanzee (*left*), the human lower spine (the lumbar region) has a curvature (a lordosis) that positions our center of mass (*circle*) above our hips when we stand. When we squat on the ground (the way people often sat for millions of years) or slouch when sitting in a chair with a backrest, we tend to rotate the pelvis backward and flatten the lower spine, reducing this lordosis. (Note that I have shown just a few of the many postures people adopt when sitting.)

reposing. She sometimes sleeps on the hard floor, but she prefers to relax on carpets, couches, upholstered chairs, and any other warm, soft place she can find (including our bed). When I tease her about her slothful ways, she gazes back wordlessly, and I know what she is thinking: "How are you any different from me?" Yes, I too travel about the house sitting in various comfy, pleasant spots, and much as I try to stand more often, I am no less of a compulsive sitter than Echo.

And for good reason: sitting is less tiring and more stable than standing. Studies that compare the energy spent standing with that spent sitting report that standing costs about 8 to 10 percent more calories than sitting quietly in a simple desk chair.[5] For a 175-pound adult, this difference is a modest eight calories per hour—the number of calories in an apple slice. Over time, these calories could add up: a typical white-collar worker who stands rather than sits on the job potentially burns an extra sixteen thousand calories a year.[6] In case it makes you feel better, bipedal humans apparently stand more efficiently than bipedal birds or large four-legged animals like cows and moose (yes, someone measured how much energy moose spend standing).[7] As figure 5 depicts, humans also stand more efficiently than apes because we can straighten our hips and knees and our lower spine has a backward curve (a lordosis) that positions our torsos mostly above rather than in front of our hips.[8] Even so, muscles in our feet, ankles, hips, and trunk must work intermittently when we stand to prevent us from swaying too much or toppling over.[9]

Regardless of how efficiently we stand, the benefits of conserving a few calories per hour by sitting accrue so much over time that the instinct to sit is no less a universal habit among humans than among other animals like Echo. Further, like other creatures, humans until recently sat without chairs. Hunter-gatherers rarely make furniture, and in many parts of the non-Western world people still sit often on the ground.[10] In a delightfully comprehensive study, the anthropologist Gordon Hewes documented more than a hundred postures that humans from 480 different cultures adopt when they sit without a chair.[11] Chairless humans often sit on the ground with their legs stretched out, cross-legged, or to the side; sometimes they kneel on one or two legs; and frequently they squat with their knees so bent that their heels either touch or come close to the backs of their thighs. If you are like me, you rarely squat, but that avoidance is a modern Western peculiarity. Because squatting creates tiny smoothed regions on ankle bones known as squatting facets, we can see that humans for millions of years, including *Homo erectus* and Neanderthals, regularly squatted.[12] Squatting facets also indicate that many Europeans squatted habitually until furniture and stoves became common after the Middle Ages.[13]

Although it is more evolutionarily normal to squat than to sit in a chair, I squat pathetically. My inept squatting skills were never more evident than late one afternoon in front of a Hadza fire. On that occasion, some men had brought back to camp a live tortoise, which they gave to the women because tortoises are women's food that men are forbidden to eat. Curious, I joined the women who casually threw the poor animal on the fire and roasted it alive as they sat on the ground and gossiped, unperturbed by the tortoise's silent dying agonies.

Because I was the only guy present, I decided I'd better behave like a man and squat casually while taking some photos. After all, how long does it take to roast a tortoise?

The answer is much longer than I can squat. Because I rarely squat, my calves were too tight to allow me to keep my feet flat on the ground, and my foot muscles started to ache, as did my calves and quads. After a few minutes, my feet and legs felt as if they were on fire, and then my lower back started to hurt. I needed to move but realized my spasming legs lacked the strength to stand up, and because the fire was just to my right, the only way to extricate myself was to roll left directly onto the elderly Hadza woman sitting close beside me. *"Samahani"* (excuse me), I said as sincerely as possible as I extricated myself while she and all the other women burst into laughter. Their giggles continued for quite some time; I have no idea what they said about me, but they kindly offered me some of the roasted tortoise (which tasted like rubbery chicken).

Aside from causing me humiliation, my lack of stamina while squatting, not to mention sitting on a sled, highlights how addicted I am to chairs, especially those with backrests. Whenever I sit on the ground or use a stool without a backrest, muscles in my back and abdomen must do a little work to hold up my torso, and when I squat, muscles in my legs, especially my calves, are also sometimes active. To be sure, the muscular effort isn't great: squatting and standing use about the same degree of muscle activity.[14] But over long periods of time those muscles require and develop endurance. My colleagues Eric Castillo, Robert Ojiambo, and Paul Okutoyi and I found that rural teenagers in Kenya who rarely sit in chairs with backrests have 21 to 41 percent stronger backs than teenagers from the city who regularly sit in the sorts of chairs you and I usually use.[15] We can't prove that rural Kenyans have stronger backs solely because of their chair habits, but other studies show that backrests demand less sustained muscular effort.[16] It is reasonable to conclude that those of us who regularly sit in chairs with backrests have weak back muscles that lack endurance, making it uncomfortable to sit for long on the ground or on stools. The result is a vicious cycle of chair dependency.

A reliance on chairs with backrests is unquestionably recent. Archaeological and historical evidence suggests that in most cultures chairs with backs, wherever they first appear, were used primarily by high-ranking personages, while peasants, slaves, and other laborers mostly had to sit on stools and benches. In art from ancient Egypt, Mesopotamia, Mesoamerica, China, and elsewhere, the only folks seated on comfy chairs with backrests are gods, royalty, and priests. In Europe, it was not until the late sixteenth century that chairs with supportive backs started to become common among the growing middle and upper classes who could afford furniture.[17] Then, during the Industrial Revolution, a German manufacturer, Michael Thonet, figured out how to mass manufacture bentwood chairs with backrests that were light, strong, beautiful, comfortable, and affordable to the needy masses. In 1859, Thonet perfected his archetypal café chair, which sold like hotcakes and remains popular in coffeehouses. As chairs with backrests became less expensive and more common, some experts condemned them. To quote one alarmed physician in 1879: "Of all the machines which civilization has invented for the torture of mankind . . . there are few which perform their work more pertinaciously, widely, or cruelly than the chair."[18]

Despite these and other warnings, the popularity of chairs has continued to exceed concerns, especially as workplaces have shifted from forests, fields, and factories to offices. An entire profession, ergonomics, was invented to help us cope with modern industrial environments including chairs. Today, billions of people have no choice but to sit much of the day, and then after an exhausting day of sitting, they follow deep-seated instincts to save a few more calories by sitting at home to relax. But for how long do we really sit on any given day?

How Much Do We Sit?

If I google how many hours per day Americans sit, dozens of websites pop up with factoids that range from as little as six hours to as much as thirteen hours per day. What is the correct answer within this more than twofold range? And what about you or me? Typically, I get up around 6:00 a.m., walk the dog, make coffee, go for a run, but otherwise work at my desk with short breaks for lunch and dinner, and collapse in bed by 10:00 p.m. Given that my job mostly requires me to stare at computer screens and I live half a mile from my office, I suspect I sit as much as twelve hours on an ordinary day. That said, I fidget and pace frequently, often do small errands, sometimes eat lunch standing up (to my wife's dismay), and occasionally use a standing desk. So maybe I sit less?

The fastest, least expensive way to get information on how much people sit is simply to ask them, as I just asked myself. Many studies use self-reported estimates, but we tend to be inaccurate, biased judges of our activities, sometimes claiming activity levels as much as four times higher than reality.[19] Today scientists can easily collect more objective and reliable data using wearable sensors that measure heart rate, steps, and other movements. Curious, I decided to measure how much I actually sit by wearing a tiny accelerometer. These thumb-sized devices record how forcefully a body is moving in different directions every second over the course of days or weeks.[20] They record minimal accelerations if one is sedentary, moderate accelerations during walking, and high accelerations during vigorous activities like running. My lab has dozens of these accelerometers, so for a week I clipped one to my waist from the moment I woke up to the time I went to bed. I then downloaded the data to quantify how much time I spent sitting or moving at light, moderate, and vigorous levels.

The results of my week of self-measurement surprised me. First, I was less sedentary than I expected. On average, I spent 53 percent of my wakeful hours sitting or otherwise immobile. My sitting time, however, varied enormously from day to day. On my most active day, I sat for three hours, but on my least active day I sat for nearly twelve hours. My average was eight and a half hours. My percentage of sedentary activity, it turns out, isn't much different from that of many Americans. High-quality studies that measured thousands of people find that average adult Americans are sedentary 55 to 75 percent of the time they are awake.[21] Given that most Americans are in bed about seven hours a night, the average time spent being effectively immobile adds up to between nine and thirteen hours a day. Keep in mind these averages mask considerable variation from person to person and over time. Unsurprisingly, Americans tend to be more active on weekends and become increasingly sedentary with age. Whereas young adults tend to sit about nine to ten hours a day, older individuals average slightly more than twelve hours a day.

Although not all latter-day Americans sit as much as some alarmists suggest, we are more sedentary than earlier generations. There is evidence that the total time Americans spent sitting increased 43 percent between 1965 and 2009, and slightly more for people in England and other postindustrial countries.[22] So I probably spend two to three hours more in chairs during a given day than my grandparents did when they were my age. My grandparents, however, were not much more sedentary than most hunter-gatherers and subsistence farmers. Researchers have used accelerometers, heart rate monitors, and other sensors to measure activity levels in hunter-gatherers in Tanzania,[23] farmer-hunters in the Amazonian rain forest,[24] and several other nonindustrialized populations.[25] In these groups, people tend to be sedentary between five and ten hours a day. The Hadza, for example, spend about nine "non-ambulatory" hours on a typical day, mostly sitting on the ground with their legs in front of them, but also squatting about two hours a day and kneeling an hour a day.[26] So while nonindustrial people engage in considerably more physical activity than average industrialized and postindustrialized people, they also sit a lot.

One critique of these statistics is that they classify activity levels rather coarsely as either sitting or not sitting. Standing isn't exercise, and sitting isn't always totally inactive. What if I am playing a violin or making an arrow while sitting? Or standing while listening to a lecture? A solution to this problem is to classify activity levels based on percentage of maximum heart rate. By convention, your heart rate during sedentary activities is between its resting level and 40 percent of maximum; light activities such as cooking and slow walking boost your heart rate to between 40 and 54 percent of maximum; moderate activities like rapid walking, yoga, and working in the garden speed your heart rate to 55 to 69 percent of maximum; vigorous activities such as running, jumping jacks, and climbing a mountain demand heart rates of 70 percent or higher.[27] Large samples of Americans asked to wear heart rate monitors indicate that a typical adult engages in about five and a half hours of light activity, just twenty minutes of moderate activity, and less than one minute of vigorous activity.[28] In contrast, a typical Hadza adult spends nearly four hours doing light activities, two hours doing moderate-intensity activities, and twenty minutes doing vigorous activities.[29] Altogether, twenty-first-century Americans elevate their heart rates to moderate levels between half and one-tenth as much as nonindustrial people.

Although more people today are couch potatoes compared with their ancestors, we can take solace by comparing ourselves with apes. For the last three decades, Richard Wrangham and his team have assiduously followed a group of chimpanzees in the Kibale Forest of Uganda, recording just about everything these animals do every day and for how long. On a given day, they know when a chimpanzee woke up, went to sleep, and how much time it spent eating, traveling, grooming, fighting, having sex, or doing anything else of interest. According to this extraordinary database, adult male and female chimpanzees spend on average 87 percent of every day in sedentary activities such as resting, grooming, feeding quietly, and nesting. Over a twelve-hour day, chimps are physically inactive for almost ten and a half hours. On their most active days, chimpanzees rest for almost eight hours; on their least active days, they rest for more than eleven hours. In either case, as the sun goes down after a tiring day of mostly sitting around, they build a nest and sleep for another twelve hours until sunrise.[30]

All things considered, even sedentary American couch potatoes are wildly more active than wild chimpanzees. If being idle is a normal, adaptive part of the human and ape conditions, then why and how can many daily hours of sitting really be so unhealthy?

There are three major, related health concerns about long periods of uninterrupted sitting. The first is what we are otherwise *not* doing. Every hour spent resting comfortably in a chair is an hour not spent exercising or actively doing things. The second concern is that long periods of uninterrupted inactivity harmfully elevate levels of sugar and fat in the bloodstream. Third and most alarmingly, hours of sitting may trigger our immune systems to attack our bodies through a process known as inflammation. Don't panic, but as you sit comfortably reading this, your body may be on fire.

On Fire

I dread the initial signs of an oncoming cold because I know what miseries will come next. For several days, my throat will swell, my nose will become a faucet of mucus, and I will cough painfully and otherwise feel cruddy, tired, and headachy. To add insult to injury, everything I hate about colds is caused not by the viruses that invaded me but by my body's inflammatory efforts to combat them. Technically, "inflammation" describes how the immune system first reacts after it detects a harmful pathogen, something noxious, or a damaged tissue. In most cases, inflammation is rapid and vigorous. Whether the offenders are viruses, bacteria, or sunburns, the immune system quickly launches an armada of cells into battle. These cells discharge a barrage of compounds that cause blood vessels to dilate and become more permeable to white blood cells that swoop in to destroy any invaders. This extra blood flow brings critically needed immune cells and fluids, but the swelling compresses nerves and causes the four cardinal symptoms of inflammation (which literally means "to set on fire"): redness, heat, swelling, and pain. Later, if necessary, the immune system activates additional lines of defense by making antibodies that target and then kill specific pathogens.

Until recently, no one in their right mind would ever associate sitting comfortably in a chair with the immunological fire your body ignites in response to a microbe or an injury. How could a few relaxing hours on a couch reading a book or watching TV have anything to do with defending my body against an infection?

The answer has recently become apparent thanks to new technologies that accurately measure minuscule quantities of the more than one thousand tiny proteins that cells pump into our bloodstreams. Several dozens of these proteins, termed cytokines (from the Greek *cyto* for "cell" and *kine* for "movement"), regulate inflammation. As scientists started to study when and how cytokines turn inflammation on and off, they discovered that some of the same cytokines that ignite short-lived, intense, and local inflammatory responses following an infection also stimulate lasting, barely detectable levels of inflammation throughout the body. Instead of blazing acutely in one spot for a few days or weeks, as when we fight a cold, inflammation can smolder imperceptibly in many parts of the body for months or years. In a way, chronic, low-grade inflammation is like having a never-ending cold so mild you never notice its existence. But the inflammation is nonetheless there, and mounting evidence indicates that this slow burn steadily and surreptitiously damages tissues in our arteries, muscles, liver, brain, and other organs.

The discovery of low-grade inflammation and its effects has simultaneously created new opportunities to combat disease and unleashed new worries. In the last decade, chronic inflammation has been strongly implicated as a major cause of dozens of noninfectious diseases associated with aging, including heart disease, type 2 diabetes, and Alzheimer's. The more we look, the more we find the fingerprints of chronic inflammation on yet more diseases including colon cancer, lupus, multiple sclerosis, and just about every medical condition with the suffix "-itis" including arthritis.[31]

Chronic inflammation is a hot topic that merits serious attention, but we need to be wary of reacting overheatedly. The most egregious claim is that we can efficaciously prevent or treat almost any disease—from autism to Parkinson's—by simply avoiding "pro-inflammatory" foods like gluten and sugar or by wolfing down "anti-inflammatory" foods like turmeric and garlic. If these miracle diets seem too good to be true, they are.[32] But such quackery shouldn't distract us from genuine concerns. The bad news is that chronic inflammation plays a role in many serious diseases. The good news is that the biggest causes of chronic inflammation are largely avoidable, preventable, or addressable: smoking, obesity, overconsumption of certain pro-inflammatory foods (a chief one being red meat), and—surprise, surprise—physical inactivity. Which brings us back to the topic of sitting. How would an innocuous few hours relaxing in a chair inflame my body?

Smoldering as I Sit?

The most widely accepted explanation for how a surfeit of sitting ignites chronic inflammation is that it's fattening. Before explaining how fat can inflame us, we first need to partly exonerate this much misunderstood substance.[33] Most fat is not just harmless but salubrious. In healthy, normal human adults, including hunter-gatherers, fat constitutes about 10 to 25 percent of body weight in men and about 15 to 30 percent in women. The majority of that fat (about 90 to 95 percent) is subcutaneous, so named because it is stored in billions of cells distributed in buttocks, breasts, cheeks, feet, and other nameless places just below the skin.[34] These fat-filled cells are efficient storehouses of energy that help us cope with long-term shortages of calories (as we saw from the Minnesota Starvation Experiment). Subcutaneous fat cells have other functions too, especially as glands that produce hormones regulating appetite and reproduction. The other major type of fat is cached in cells in and around our bellies and other organs including the heart, liver, and muscles. There are many terms for this fat including "visceral," "abdominal," "belly," and "ectopic," but I will use the term "organ fat." Organ fat cells are dynamic participants in metabolism and, when activated, can quickly dump fat into the bloodstream. Organ fat in moderate quantities (about 1 percent of total body weight) is thus normal and beneficial as a short-term energy depot for times when we need rapid access to a lot of calories such as when we walk or jog a long distance.

Although most fat is healthy, obesity can turn fat from friend into inflammatory foe. The biggest danger is when fat cells malfunction from overswelling. The body has a finite number of fat cells that expand like balloons. If we store normal amounts of fat, both subcutaneous and organ fat cells stay reasonably sized and harmless. However, when fat cells grow too large, they distend and become dysfunctional like an overinflated garbage bag, attracting white blood cells that trigger inflammation.[35] All bloated fat cells are unhealthy, but swollen organ fat cells are generally more harmful than subcutaneous fat cells because they are more metabolically active and more directly connected to the body's blood supply. So when organ fat cells swell, they ooze into the bloodstream a great many proteins (cytokines) that incite inflammation. Telltale signs of excess organ fat are a paunch or an apple-shaped body. Disconcertingly, it is also possible to be "skinny fat" with significant deposits of organ fat in and around one's muscles, heart, and liver without necessarily having a potbelly figure.

The mechanisms by which too much fat, especially in and around organs, can ignite low-grade, chronic inflammation suggest that too much sitting may be hazardous simply because it causes weight gain. It bears repeating that sitting in a comfy chair barely taxes one's muscles. In contrast, even squatting or kneeling requires some muscular effort, just standing burns about eight more calories per hour, and light activities like folding laundry can expend as much as a hundred calories per hour more than sitting.[36] These calories add up. By merely engaging in low-intensity, "non-exercise" physical activities for five hours a day, I could spend as much energy as if I ran for an hour. So if I sit instead of move, the calories I consumed at lunch are more likely to be converted to fat rather than burned. In one alarming experiment, Danish researchers paid a group of healthy young men to sit like veritable couch potatoes and take no more than fifteen hundred steps a day (about a mile) for a fortnight. As the before-versus-after scans of their bellies in figure 6 show, these men added 7 percent more organ fat in just two weeks.[37] Alarmingly, as these volunteers gained fat, they started to exhibit the classic signs of chronic inflammation including less ability to take up blood sugar after a meal. Note, however, that this experiment implicates sitting only indirectly. No one is claiming that sitting itself caused these Danish guys to gain weight: it was the combination of physical inactivity plus excess calories that caused them to stockpile excess organ fat, which in turn lit the smoldering fire of chronic inflammation. In addition, these volunteers added mostly organ fat, which suggests they were stressed, and there are plenty of physically inactive people who are not overweight but suffer from inflammation. So what else about sitting might promote chronic inflammation?

A second way lengthy periods of sitting may incite widespread, low-grade inflammation is by slowing the rate we take up fats and sugars from the bloodstream. When was the last time you had a meal? If it was within the last four or so hours, you are in a *postprandial* state, which means your body is still digesting that food and transporting

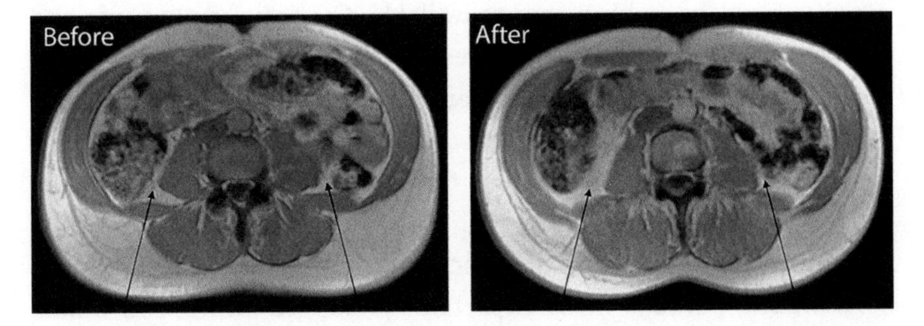

FIGURE 6 MRI scans of a man's abdomen before (*left*) and after (*right*) two weeks of nearly nonstop sitting and otherwise being physically inactive. The amount of organ fat (highlighted by the arrows), which shows up as white in the MRI, increased by 7 percent. For more details, see Olsen, R. H., et al. (2000), Metabolic responses to reduced daily steps in healthy nonexercising men, *Journal of the American Medical Association* 299:1261–63. (Photo courtesy of Bente Klarlund Pedersen)

its constituent fats and sugars into your blood. Whatever fat and sugar you don't use now will eventually get stored as fat, but if you are moving, even moderately, your body's cells burn these fuels more rapidly. Light, intermittent activities such as taking short breaks from sitting and perhaps even the muscular effort it takes to squat or kneel reduce levels of fat and sugar in your blood more than if you sit inertly and passively for long.[38] Such modest extra demands appear to be beneficial because although fat and sugar are essential fuels, they trigger inflammation when their concentrations in blood are too high.[39] Put simply, regular movement, including getting up every once in a while, helps prevent chronic inflammation by keeping down postprandial levels of fat and sugar.

Another worrisome way sitting can provoke inflammation is through psychosocial stress. I hope you are reading these words contentedly on a beach or some other pleasant place and not fretting about sordid things like swollen fat cells and inflammation. Sadly, sitting is not always relaxing. Long hours of commuting, a demanding desk job, being sick or disabled, or otherwise being confined to a chair can be stressful situations that elevate the hormone cortisol. This much-misunderstood hormone doesn't cause stress but instead is produced when we are stressed, and it evolved to help us cope with threatening situations by making energy available. Cortisol shunts sugar and fats into the bloodstream, it makes us crave sugar-rich and fat-rich foods, and it directs us to store organ fat rather than subcutaneous fat. Short bursts of cortisol are natural and normal, but chronic low levels of cortisol are damaging because they promote obesity and chronic inflammation. Consequently, long hours of stressful sitting while commuting or a high-pressure office job can be a double whammy.

Last, and perhaps most important, prolonged sitting can kindle chronic inflammation by allowing muscles to remain persistently inactive. In addition to moving our bodies, muscles function as glands, synthesizing and releasing dozens of messenger proteins (termed myokines) with important roles. Among other jobs, myokines influence metabolism, circulation, and bones, and—you guessed it—they also help control inflammation. In fact, when researchers first started to study myokines, they were astonished to discover that muscles regulate inflammation during bouts of moderate to intense physical activity similarly to the way the immune system mounts an inflammatory response to an infection or a wound.[40] Without going into too many details, we have learned that the body first initiates a proactive inflammatory response to moderate- or high-intensity physical activity to prevent or repair damage caused by the physiological stress of exercise and subsequently activates a second, larger anti-inflammatory response to return us to a non-inflamed state.[41] Because the anti-inflammatory effects of physical activity are almost always larger and longer than the pro-inflammatory effects, and muscles make up about a third of the body, active muscles have potent anti-inflammatory effects. Even modest levels of physical activity dampen levels of chronic inflammation, including in obese people.[42]

The discovery that using your muscles inhibits inflammation provides yet another compelling explanation for why endless hours of sitting are associated with many chronic diseases.[43] By remaining inert for hour upon hour, our bodies never extinguish that faint inflammatory fire that may otherwise be smoldering in the background. In fact, none of the mechanisms that inflame us—swollen fat cells, too much fat and sugar in the bloodstream, stress, and inactive muscles—are caused by sitting per se. Instead, they result from the absence of being sufficiently physically active, which usually means a lot of sitting. Given that sitting for hours a day is an utterly normal behavior both in the past and today, are there better, less inflammatory ways to sit?

Active Sitting

A useful German word with no English equivalent is *Sitzfleisch*. Its literal translation is "butt flesh," but figuratively it refers to the ability to sit patiently for long hours to accomplish something challenging. *Sitzfleisch* connotes perseverance and endurance. To win a chess game, solve a complex math problem, or write a book requires *Sitzfleisch*. The word is generally a compliment, but calls attention to an important principle about dosage: for some things, how often and when you do them are just as important as how much. If I gulp down four cups of coffee all at once, I'll become jittery and get a headache, but if I drink them over the course of a day, I'll be fine. Is the same true for sitting? Additionally, does a daily bout of hard exercise negate the effects of sitting for the rest of the day?

Given the convoluted relationship between physical activity and sitting, how much people exercise voluntarily doesn't necessarily correspond to how much they sit. Surprisingly, marathon runners who train regularly sit just as much as less athletic individuals.[44] In fact, because these avid runners are often exhausted, they might sit more. Because you and I may sit the same number of hours but in different ways and contexts, we need to consider how different patterns of sitting—extended versus interrupted—potentially affect chronic, low-grade inflammation. What about the office worker who is sedentary at work but goes to the gym for an hour every day? What about someone who sits for most of the day but whose sitting is interrupted by abundant tiny breaks?

Most efforts to address the effects of different sitting patterns are epidemiological studies that look backward in time for correlations between people's health, their sitting time, how much they interrupt their sitting with breaks, and how active they are when not sitting. These studies cannot test for causation, but they help assess risks and generate hypotheses. They also bring bad news for those of us who think an hour in the gym erases the negative effects of a day otherwise spent in a chair. One large study collected ten years of data on sitting time, fitness, and other variables from more than 900 men.[45] As expected, the fitter, more active men were considerably less likely to have heart disease, type 2 diabetes, and other chronic diseases than their unfit and sedentary counterparts. But among the men who were fit, those who sat the most had a 65 percent higher risk of inflammatory-related diseases like type 2 diabetes than those who sat the least. An even larger study based on survey data from more than 240,000 Americans found that time engaged in moderate and vigorous activity lowered but did not erase the risk of dying associated with being sedentary.[46] Even those who engaged in more than seven hours per week of moderate or vigorous exercise had a 50 percent higher risk of dying from cardiovascular disease if they otherwise sat a lot. Altogether, these and other alarming studies suggest that even if you are physically active and fit, the more time you spend sitting in a chair, the higher your risk of chronic illnesses linked to inflammation, including some forms of cancer.[47] If these results are correct, then exercise alone doesn't counter all the negative effects of sitting.

I find these data downright scary. However, before I throw away my desk chair, what about the hypothesis that interrupted or "active" sitting is less harmful than uninterrupted sitting? Should I get up and do a few jumping jacks every ten minutes? Thankfully, here there is encouraging news. A multiyear analysis of almost five thousand Americans found that people who broke up their sitting time with frequent short breaks had up to 25 percent less inflammation than those who rarely rose from their chairs despite sitting the same number of hours.[48] A more morbid study put accelerometers for a week on a diverse sample of eight thousand Americans above the age of forty-five and then tallied up who died over the next four years—about 5 percent of the sample.[49] Predictably, those who were more sedentary died at faster rates, but these rates were lower in people who rarely sat for long, uninterrupted bouts. In fact, people who rarely sat for more than twelve minutes at a time had lower death rates, and those who tended to sit for half an hour or longer at a stretch without getting up had especially high death rates. One flaw with this study is that people who are already sick are inherently less able to get up and be active, but the results nonetheless suggest that the risk of death increases both from the total hours we spend sitting and from whether those hours are accrued in short or long intervals.

I am by nature and profession a skeptic, but the more I read these worrisome statistics, the more I have tried to modify my habits. I have been striving to get up more regularly to do little errands and pet my dog. I also have been using my standing desk more often. But epidemiological studies don't test causation. Further, they cannot correct for other factors that may blur the relationship between health and sitting time. For example, sitting time at home watching TV is more strongly associated with health outcomes than sitting time at work.[50] Because wealthy folks watch less TV, get better health care, and eat healthier food, they are less at risk. To really persuade me to abandon the comforts of my chair, I want to know *how* and *why* active sitting may be better than uninterrupted sedentary behavior. I want mechanisms, not just statistical associations.

As we have already discussed, a likely explanation is that short bouts of activity wake up our muscles and thus keep down levels of blood sugar and fat. When we squat, periodically stand, or do light activities like pick up a child or sweep the floor, we contract muscles throughout the body, setting in motion their cellular machinery. Like turning on a car engine without driving anywhere, these light activities stimulate muscle cells to consume energy, turn on and off genes, and perform other functions. It bears repeating that washing the dishes or doing other light chores can burn as much as a hundred additional calories per hour beyond just sitting. To supply the energy needed for these low-intensity movements, muscles extract and then burn sugar and fat from the bloodstream.[51] These activities aren't serious exercise, but experiments that ask people to interrupt long periods of sitting even briefly—for example, just a hundred seconds every half hour—result in lower levels of sugar, fat, and so-called bad cholesterol in their blood.[52] In turn, less circulating blood sugar and fat prevent inflammation as well as obesity. In addition, small and occasional bouts of moving stimulate muscles to quench inflammation and reduce physiological stress.[53] Finally, muscles, especially in the calves, act as pumps to prevent blood and other fluids from building up in the legs, not just in veins, but also in the lymph system, which functions like a series of gutters to transport waste throughout the body. It's good to keep these fluids moving. Sitting for long hours without moving increases the risk of swelling (edema) and developing clots in veins.[54] For this reason, squatting and other more active forms of sitting may be healthier than sitting in chairs by requiring intermittent muscle activity, especially in the calves, thus recirculating blood in the legs.

Another way to sit actively is to fidget, or do what researchers drily term "spontaneous physical activity." As an inveterate squirmer, I marvel at people who can sit inanimately for hours without going crazy. Apparently, the propensity to fidget may be partly heritable, and the effects can be substantial. In a famous 1986 study, Eric Ravussin and colleagues asked 177 people to spend twenty-four hours (one at a time) in an enclosed ten-by-twelve-foot chamber that measured precisely how many calories they spent. To the researchers' surprise, the individuals who fidgeted spent between one hundred and eight hundred calories more per day than those who sat inertly.[55] Other studies have found that simply fidgeting while seated can expend as much as twenty calories an hour as well as promote beneficial levels of blood flow to restless arms and legs.[56] One study even found a 30 percent lower rate of all-cause mortality among people who fidget after adjusting for other forms of physical activity, smoking, diet, and alcohol consumption.[57]

When all is said and done, *Sitzfleisch* may boost productivity, but it doesn't foster health. Yet in the postindustrial world, more and more jobs require us to stare for hours at screens. Should we all rush out and buy standing desks?

How and How Much Should I Sit?

Among the many hyperbolic statements written about sitting, maybe the most extreme is that sitting is the new smoking. While cigarettes are novel, addictive, expensive, smelly, toxic, and the world's number one killer, sitting is as old as the hills and utterly natural. More truthfully, the problem isn't sitting itself, but hours upon hours of inactive sitting combined with little to no exercise. If our ancestors from generations ago behaved like today's hunter-gatherers and farmers, then they likely sat for five to ten hours a day, as much as some but not all contemporary Americans and Europeans.[58] But they also got plenty of physical activity when not sitting, and when these chairless ancestors plunked themselves down, they didn't rest in supportive chairs with seat backs; instead, they squirmed as they squatted, kneeled, or sat on the ground, using about the same degree of muscle activity in their thighs, calves, and backs as when they stood. When the Hadza sit, squat, or kneel, they typically do so for only fifteen minutes at a time.[59] Further, if our ancestors resembled nonindustrial peoples today, then when sitting they often simultaneously did household chores, minded children, and frequently had to get up. Overall and by necessity, their sitting was less inert and less sustained and didn't come at the expense of several daily hours of physical activity.

Because desk jobs are here to stay for the foreseeable future, standing desks have been widely advertised as a panacea for excess sedentariness. Such marketing deceptively confuses not sitting with physical activity. Standing is not exercise, and as yet no well-designed, careful study has shown that standing desks confer substantial health benefits. Keep in mind also that while numerous epidemiological studies have found that people who sit for twelve or more hours tend to have higher mortality rates than those who sit less, prospective studies have yet to show that people who sit more at work (occupational sitting) have elevated mortality rates. One massive fifteen-year-long study of more than ten thousand Danes found no association between time spent sitting at work and heart disease.[60] An even bigger study on sixty-six thousand middle-aged Japanese office workers yielded similar results.[61] Instead, leisure-time sitting best predicts mortality, suggesting that socioeconomic status and exercise habits in mornings, evenings, and weekends have important health effects beyond how much one sits during weekdays at the office.[62]

And while we are at it, other exaggerated statements about sitting may also be myths. How often have you been admonished to stop slouching and sit up straight? This old chestnut dates back to the late-nineteenth-century German orthopedic surgeon Franz Staffel.[63] As the Industrial Revolution caused more people to work long hours in chairs, Staffel worried these sitters were ruining their posture by sliding their buttocks forward and straightening their lower backs. Alarmed, Staffel opined that a person's spine should maintain the same characteristic double-S curve when sitting as when standing normally, and he advocated chairs with lower back supports to force us to sit upright (like the second fellow from the right in figure 5). Decades later, Staffel's opinions were backed up by the Swedish ergonomics pioneer Bengt Åkerblom and his students, who X-rayed people in chairs while measuring their muscle activity.[64] As a result, most Westerners, including a majority of health-care professionals, think we can avoid back pain by sitting with a curved lower back and an unrounded upper back.[65]

Scientific evidence discredits this modern cultural norm. A big clue is that while chairs with backrests do facilitate slouching, chairless people worldwide also commonly adopt comfortable postures that straighten the lower back and round the upper back, as evident in figure 5.[66] Many biomechanical arguments against slouching have also been disproved.[67] But most convincing to skeptics are the dozens of careful meta-analyses and systematic reviews that have combed through and rigorously evaluated every study published on the relationship between sitting posture and back pain. When I sat down to read these papers, I was frankly astonished: nearly all high-quality studies on this topic fail to find consistent evidence linking habitual sitting in flexed or slouched postures with back pain.[68] I was also surprised to read there is no good evidence that people who sit longer are more likely to have back pain,[69] or that we can lessen the incidence of back pain by using special chairs or getting up frequently.[70] Instead, the best predictor of avoiding back pain is having a strong lower back with muscles that are more resistant to fatigue; in turn, people with strong, fatigue-resistant backs are more likely to have better posture.[71] In other words, we've confused cause and effect. As the back pain expert Dr. Kieran O'Sullivan told me, "Good posture is primarily a reflection of environment, habits, and mental state and is not a talisman against back pain."

So if you are feeling guilty or concerned because you are sitting now, perhaps slouching, keep in mind you evolved to sit just as much as you evolved to be active. Instead of vilifying chairs and remonstrating yourself for slouching or not squatting, try to find ways to sit more actively without being inert for too long, squirm shamelessly, and don't let sitting get in the way of also exercising or otherwise being physically active. Such habits prevent or lessen chronic inflammation that provokes ill health, and it bears repeating that the scary statistics we read about sitting are primarily driven by how much we sit when not at work.

The more I learn about the benefits of light activity over prolonged physical inactivity like sitting, however, the more one question puzzles me: If sustained periods of not moving slowly cause harm, why are we simultaneously warned not to sit too much but also advised to spend more time—a good eight hours, nearly one-third of our lives—barely moving in a semi-comatose state of sleep?

FOUR

Sleep: Why Stress Thwarts Rest

MYTH #4 You Need Eight Hours of Sleep Every Night

> But [Pooh] couldn't sleep. The more he tried to sleep, the more he couldn't. He tried Counting Sheep, which is sometimes a good way of getting to sleep, and, as that was no good, he tried counting Heffalumps. And that was worse. Because every Heffalump that he counted was making straight for a pot of Pooh's honey, and eating it all. For some minutes he lay there miserably, but when the five hundred and eighty-seventh Heffalump was licking its jaws, and saying to itself, "Very good honey this, I don't know when I've tasted better," Pooh could bear it no longer.
>
> —A. A. Milne, *Winnie the Pooh*

Like sitting, sleep is a quintessential state of inactivity. But unlike sitting, which too many of us supposedly enjoy too much, sleep is a biological necessity that too many of us supposedly enjoy too little. If humans evolved to rest as much as possible, why do so many of us skimp on sleep?

Self-imposed deprivation certainly describes my approach to sleep in college. Like many twenty-year-olds, I loved staying up until the wee hours. When I finally crawled into bed, I tossed and turned between the sheets. Even after sleep eventually arrived, I rarely got enough because some nasty part of my brain insistently woke me up every dawn. No matter how late I drifted off, ZAP, I'd be wide awake at 6:00 or 7:00 a.m. Being sleep deprived set in motion a vicious cycle. Anxiety about not sleeping enough kept me awake, causing me to be even more stressed about not falling sleep. I tried buying over-the-counter sleeping medications, but they didn't work. Eventually, I was so stressed I sought professional help.

I'll never forget the sympathetic doctor who saw me. I'm sure she had spoken to hundreds of students like me with similar woes. Nevertheless, she listened compassionately as I poured out my anxieties about insomnia, school, and everything else. I spared no detail because I desperately wanted her to prescribe me a powerful pill to knock me out. Instead, she patiently used the Socratic method to make me realize that I was far better at falling asleep than I gave myself credit for. Did I fall asleep in class? Yes. Did I sleep when studying in the library? Yes. Did I sleep better at home during school breaks? Yes. Having made her point, she then explained how levels of hormones, especially cortisol, fluctuate throughout the day to regulate my alertness and that, like it or not, I was condemned to being a morning person for the rest of my life. Although we never discussed exercise, she did make a radical suggestion I had never considered: Why not go to bed earlier?

That was not the advice I wanted to hear. To a college student like me, late night was the best time of the day. Sometimes I would study well past midnight, and my social life—on the rare occasions when I had one—often didn't start until 9:00 or 10:00 p.m. Was getting eight hours of sleep worth sacrificing the best hours of the day?

Sleep-deprived college students illustrate how sleep is not just an essential form of rest but also an inescapable trade-off. Calories come and go, but the arrow of time is inexorable. Because you can never relive any precious minute of your life, time asleep is time wasted in a state Virginia Woolf described as a "deplorable curtailment of the joy of life."[1] Or as Margaret Thatcher more contemptuously remarked, "Sleep is for wimps." To be sure, parents of newborns, people who work night shifts, and others who suffer from chronic stress are often cheated of wanted sleep, but conventional wisdom suggests that Woolf's and Thatcher's opinions are increasingly commonplace in the modern world. According to this line of thinking, ever since we harnessed fire in the Stone Age, the human species has been dreaming up technologies to stave off sleep and have fun after the sun goes down. Thomas Edison proudly nicknamed the engineers in his laboratory the "insomnia squad."

Ostensibly, we have a crisis. The consensus opinion among experts is that sleep has declined in tandem with physical activity and that prior to the Industrial Revolution, people used to get more sleep, up to nine or ten hours a day, but the modern world's "brutish treatment" of sleep has reduced this average to seven hours, with 5 percent of us sleeping less than five hours.[2] The result is an "epidemic" of sleep deprivation that supposedly afflicts one in three people in industrial nations around the globe.[3] You have probably heard that lack of sleep promotes obesity, shortens lives, causes more than 20 percent of auto accidents, and precipitates disasters like the Chernobyl nuclear meltdown, the crash of the *Exxon Valdez,* and lethal mistakes by sleepy doctors.[4] Just as we are exhorted to exercise, we are admonished not to skimp on sleep, and millions of people spend billions of dollars on comfortable mattresses, earplugs to block out distracting noises, thick curtains to darken bedrooms, machines that lull us to sleep, and of course drugs that induce drowsiness.

Are we getting good information and advice? And how do we explain the contradiction between our supposed tendencies to avoid both exercise and sleep? If the instinct to avoid needless physical activity is so strong that we need to be pulled out of our chairs and forced to move, why aren't we just as strongly inclined to enjoy as much restful sleep as possible?

Adequate sleep is profoundly important for health, and in no way do I wish to trivialize the real and serious problems of those who cannot or do not get enough sleep, but I wonder if our treatment of sleep suffers from a lack of evolutionary and anthropological perspective similar to that we saw with sitting. As a human being I want to get enough sleep, but as an evolutionary biologist I'd like to know more about the causes, costs, and benefits of variations in sleep, and as an anthropologist I'd like to know what we are missing by not looking beyond modern Western sleep habits. What is "normal" sleep for a "normal" human? Last but not least, I frequently read that lightbulbs, televisions, smartphones, and other newfangled inventions have robbed us of our requisite eight hours of sleep, but I am also curious about the effects of physical inactivity on sleep. Everyone knows that exercise expedites and sustains a good night's sleep, so to what extent does lack of physical activity hamper sleep?

As a first step toward addressing these questions, let's begin by evaluating what sleep is and why we need it.

A Good Night's Rest for the Body or the Brain?

As I write these words, my dog is asleep on the couch next to me, snoring. At least I suspect she is asleep. Echo is curled into a ball with her eyes closed, her breaths are slow and regular, and she is tuned out. She doesn't even react when I utter the magic words "walk" and "biscuit." Given Echo's slothful, stress-free existence and her habit of sleeping nearly half the day and most of the night, I doubt she is napping to rest her weary bones, let alone to knit up her "ravell'd sleave of care." Even so, I empathize with her drive to get plenty of shut-eye. If I sleep poorly tonight, I will suffer tomorrow. In addition to being sluggish and drowsy, my attention span will decline, I'll forget things, my judgment will deteriorate, my senses will dull, and I'll be crankier than usual. If, heaven forbid, I go for several days and nights without sleep, my cognitive function will decline precipitously. I cannot imagine intentionally depriving oneself of sleep for more than a night, let alone trying to set a world record for consecutive hours spent awake (an achievement no longer tracked by *Guinness World Records* because of its dangers). Astoundingly, such masochists exist, but their attempts cause harrowing cognitive dysfunction, paranoia, and hallucinations.[5]

Any creature with a brain engages in some form of sleep, which is recognizable both behaviorally and physiologically. In terms of behavior, whether you are a fish, frog, whale, or human, sleep is a rapidly reversible state of reduced physical activity and sensory awareness, usually in a resting posture. To arouse sleeping animals requires loud noises, bright lights, or forceful pushes. Physiologically, however, sleep is more complex and varied, especially in terms of brain activity. Measures of the brain's electrical output reveal two general phases of sleep, shown in figure 7. At first, we go through several progressive stages of "quiet" NREM (non-rapid eye movement) sleep. With each stage, we become increasingly unconscious, metabolism slows, body temperature falls. The brain's electrical signals during NREM sleep are mostly characterized by slow waves with high voltages, and our eyes stay still or roll slowly behind our eyelids. Eventually, we enter a different, more "active" REM (rapid eye movement) stage of sleep. During REM sleep, when we mostly dream, the brain's electrical output is characterized by fast waves with low voltages, and our eyeballs rotate swiftly. Other characteristics of REM sleep include less regular heart rate and breathing, temporary paralysis, and spontaneous swelling of the clitoris or penis. During a typical full night's sleep, we go through this entire cycle of NREM and then REM sleep four or five times, with the intensity and duration of the REM phases increasing. If all goes well, as dawn's rosy fingers approach, our dreams become more intense.

Sleep is obviously vital for the brain, but it's also associated with decreased physical activity. All creatures, even bacteria, have nearly twenty-four-hour internal clocks that generate circadian rhythms—about (circa) a day long—that slow them down or speed them up at different times. These ubiquitous cycles have led to the notion that sleep evolved to help animals save energy when it is sensible to be less physically active and divert calories toward repair and growth. If I climb a mountain or run a marathon today, I'll sleep extra long and hard tonight, and if I don't sleep, I'll feel unrested tomorrow. During sleep, our metabolic rate drops about 10 to 15 percent; about 80 percent of growth also occurs during NREM sleep.[6]

Sleep is restful, but I am skeptical it evolved as an adaptation to rest. Metabolism probably drops during sleep because it is beneficial for organisms to save energy when they are inactive. But we don't actually need to sleep to save energy, repair tissues, and otherwise recuperate; we could just sit still. In addition, sleeping involves substantial

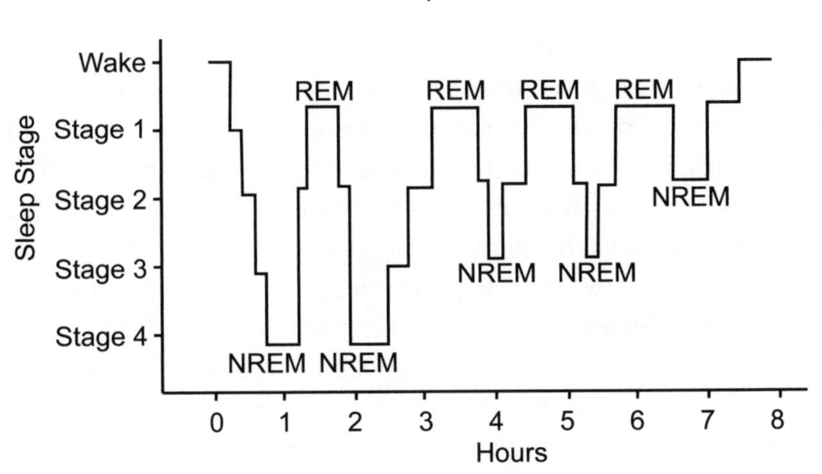

FIGURE 7 Cycles of NREM (non-rapid eye movement) and REM (rapid eye movement) sleep during a typical night's sleep.

costs and risks. Whenever we are asleep, we aren't accomplishing any of the tasks natural selection most cares about like finding mates, getting food, and, above all, avoiding being someone else's food. The first night I tried to sleep by a campfire in the African savanna under the stars, I didn't get much rest because I was frightened by the distant, eerie whoops of hyenas and the throaty calls of lions. Eventually, I learned not to be scared by the sounds of these nocturnal predators, which stay away from campfires and humans, but a few million years ago, before we tamed fire, it must have been terrifying to try to fall asleep to these bloodthirsty animals' cries. Sleep is such a vulnerable state that animals like zebras sleep only three or four hours a day because they are in constant fear of lions, whereas lions that eat the zebra typically enjoy about thirteen.[7] Today, few humans worry about being eaten by carnivores after the sun goes down, but the night is still fraught with perils.

It doesn't take a lot of brainpower to realize that sleep is mostly about the brain. Over the last few decades, researchers have spent many sleepless nights to reveal how the neurological advantages of sleep outweigh its costs. One conspicuous benefit is cognitive: sleep helps us remember important things and helps synthesize and integrate them. It sounds like magic, but while we sleep, our brains file and then analyze information. I sometimes experience this phenomenon when I stay up late trying to comprehend complex information (like how sleep affects the brain). As the night progresses, my brain becomes increasingly muddled, and eventually I give up and go to bed. But then in the morning, almost miraculously, everything seems to make sense. What happened while I was asleep?

To appreciate how sleep helps us think, consider how from an evolutionary perspective the only benefit of memory is to help us cope with the future.[8] If a zebra witnesses a human hunter kill her sister with a gun, that awful memory will help her only if she recalls that incident the next time she sees a gun-toting human and then runs away. Effective cognition, however, requires organisms to sort through all the memories they generate every day, throw out the inconsequential ones, store the important ones, and make sense of them.[9] Elegant experiments using sensors that peer into the brain of people before, during, and after they have slept (or been deprived of sleep) reveal that these functions often occur during sleep.[10] As the day marches on, we store memories in a region of the brain called the hippocampus, which functions as a short-term storage center like a USB drive. Then, during NREM sleep, the brain triages these memories, rejecting the innumerable useless ones (like what color socks the man sitting next to me on the subway wore) and sending the important ones to long-term storage centers near the surface of the brain. The brain apparently also tags and sorts memories, identifying and strengthening ones we may need. And, fantastically, the brain may also analyze certain memories during REM sleep, integrating them and looking for patterns. Critically, however, the brain has limited abilities to multitask and cannot perform these cleaning, organizing, and analytical functions as effectively when we are awake and alert.[11]

An even more vital function of sleep for the brain is janitorial. The zillions of chemical reactions that make life possible inevitably create waste products known as metabolites, some highly reactive and damaging.[12] Because the power-hungry brain uses one-fifth of the body's calories, it generates abundant and highly concentrated metabolites. Some of these garbagy molecules such as beta-amyloid clog up neurons.[13] Others such as adenosine make us sleepy as they accumulate (and are counteracted by caffeine).[14] Getting rid of these waste products, however, is a challenge. Whereas tissues like liver and muscle wash out metabolites directly into blood, the brain is tightly sealed off from the circulatory system by a blood-brain barrier that prevents blood from coming into direct contact with brain cells.[15] To rid itself of waste, the brain evolved a novel plumbing system that relies on sleep. During NREM sleep, specialized cells throughout the brain expand the spaces between neurons by as much as 60 percent, allowing cerebrospinal fluid that bathes the brain to literally flush away this junk.[16] These opened spaces also admit enzymes that repair damaged cells and rejuvenate receptors in the brain for neurotransmitters.[17] The only catch, however, is that the brain's interstitial pathways are like single-lane bridges that let cars pass in only one direction at a time. Apparently, we cannot think while cleansing our brains. We thus must sleep to flush out the cobwebs left behind by the day's experiences.

Sleep is therefore a necessary trade-off that improves brain function at the cost of time. For every hour spent awake storing memories and amassing waste, we need approximately fifteen minutes asleep to process those memories and clean up. That ratio, however, is highly variable: some people like the elderly sleep less, while others, especially children, need more. As every parent knows, a missed nap can turn even the sweetest child into the toddler from hell. Thankfully, sleep-deprived adults are usually less troublesome than children, but in the end none of us escape the inevitable trade-off between time spent asleep versus awake. Late nights and early mornings avoiding sleep can be fun or profitable, but we pay a price, sometimes disastrous, in terms of memory, mood, and long-term health. Apart from the damaging effects of sleep deprivation on health, an estimated six thousand car accidents per year in the United States are caused by drowsy drivers.[18]

So did you get the eight hours of sleep you needed last night?

The Myth of Eight Hours

One novelty of the modern world is our tendency to medicalize certain behaviors by prescribing them in specific doses. Commonly recommended doses include a minimum of 150 minutes of physical activity per week, twenty-five grams of fiber per day, and eight hours of sleep per night. No one knows exactly when and where that eight-hour prescription originated, but during the late nineteenth century striking factory workers marched through city streets shouting, "Eight hours for work, eight hours for rest, eight hours for what we will!" And as Ben Franklin sanctimoniously opined, "Early to bed, early to rise, makes a man healthy, wealthy, and wise." I, too, have spent most of my life under the impression that I ought to sleep eight hours a night, and won't pretend I haven't occasionally been slightly smug about being a morning person. Despite these commonly held opinions, the world is full of people—my students especially—who love staying up late and who apparently survive, often intentionally, on far less than eight hours of sleep. Are they abnormal products of our electrified, time-obsessed modern world? And how do we compare with other animals?

Even a cursory look around should convince you that when it comes to sleep, there is no single pattern among humans or mammals. Donkeys sleep only two hours a day, but armadillos sleep as much as twenty hours. Some animals like giraffes nap frequently, but other species sleep in one uninterrupted bout. A few large animals like elephants can nap standing up, and, most extraordinarily, marine mammals such as dolphins and whales evolved the ability to put just one half of their brain to sleep at a time while they swim.[19]

Because sacrificing one-third or more of one's life to sleep is an exceptional trade-off, it should hardly be surprising that natural selection has fostered a breathtaking variety of sleeping patterns and norms. Efforts to make sense of this variation, however, have revealed only a few weak associations. The strongest correlation is that vulnerable prey animals tend to sleep less than the carnivores that want to eat them.[20] Perhaps "the wolf shall dwell with the lamb and the leopard will lie down with the kid" (Isaiah 11:6), but the farm animals probably won't get much sleep in the company of predators. In addition, bigger-bodied animals that have to spend more time getting food tend to sleep less. Otherwise, there appears to be little rhyme or reason to why some animals sleep more or less than others. Regardless of what factors explain this puzzling diversity, the majority of mammals sleep between eight and twelve hours a day, and most primates sleep between nine and thirteen hours. Chimpanzees, our closest relatives, apparently average eleven to twelve hours of rest per night.[21]

What about humans? Predictably, information about human sleep patterns comes mostly from people in the United States and Europe, where most adults report getting between seven and seven and a half hours of sleep per night, but one in three say they regularly get less than seven hours.[22] Self-reported estimates of sleep, however, are infamously unreliable.[23] New sensor technologies that monitor sleep objectively indicate that the average adult in the United States, Germany, Italy, and Australia tends to sleep about six and a half hours in the summer when it is warm and light and between seven and seven and a half hours in the colder, darker winter months.[24] Altogether and despite much variation, most adult Westerners probably average about seven hours a night, a good hour (13 percent) less than the eight hours we supposedly need.

But is that normal? And where does the holy grail of eight hours come from?[25] A major premise of this book is that most people from modern westernized societies, myself included, are hardly representative of humanity prior to the Industrial Revolution. How much are my sleep patterns contaminated by alarm clocks, lights, smartphones, and other enemies of sleep such as jobs, train schedules, and the nightly news?

Fortunately, researchers have woken up to these problems, and new technologies have made possible a surge of high-quality data on sleep in nonindustrial populations. The most electrifying study by far was by the UCLA sleep researcher Jerome Siegel and his colleagues, who affixed wearable sensors to ten Hadza hunter-gatherers from Tanzania, thirty San forager-farmers from the Kalahari Desert, and fifty-four hunter-farmers from the Amazon rain forest in Bolivia. None of these populations have electric lights, let alone clocks or internet access. Yet to Siegel's astonishment, they slept less than industrialized people did. In warmer months, these foragers slept on average 5.7 to 6.5 hours a day, and during colder months they slept on average 6.6 to 7.1 hours a night. In addition, they rarely napped. Studies that monitored Amish farmers who shun electricity as well as other nonindustrial populations such as rural Haitians and subsistence farmers in Madagascar report similar average sleep durations, about 6.5 to 7.0 hours a day.[26] Thus, contrary to what we are often told, there is no evidence that nonindustrial populations sleep more than industrial and postindustrial populations.[27] What's more, when you look closely, there is little empirical evidence that average sleep duration in the industrial world has decreased in the last fifty years.[28] The more we look, the less we can profess eight hours to be normal.[29]

If you are reading this skeptically (as you should), you might be thinking that just because nonindustrial foragers and farmers typically sleep less than eight hours doesn't mean their habits are optimal for health. Many hunter-gatherers also smoke. Yet in 2002, the sleep world was rocked by a massive study by Daniel Kripke and colleagues that examined the health records and sleep patterns of more than one million Americans.[30] According to these data, Americans who slept eight hours a night had 12 percent higher death rates than those who slept six and a half to seven and a half hours. In addition, heavy sleepers who reported more than eight and a half hours and light sleepers who reported less than four hours had 15 percent higher death rates. Critics pounced on the study's flaws: the sleep data were self-reported; people who sleep a lot may already be sick; correlation is not causation. Yet since then, numerous studies using better data and sophisticated methods to correct for factors like age, illness, and income have confirmed that people who sleep about seven hours tend to live longer than those who sleep more or less.[31] In no study is eight hours optimal, and in most of the studies people who got more than seven hours had shorter life spans than those who got less than seven hours (an unresolved issue, however, is whether it would be beneficial for long sleepers to reduce their sleep time).

The need for eight hours might be a myth, but what about patterns of sleep? You and I may sleep the same number of hours but differently. While some of us are "larks" who go to bed and rise early, others are "owls" who stay up late and sleep well past dawn when possible. These contrasting tendencies turn out to be remarkably heritable and hard to overcome.[32] In addition, as we age, we sleep less and wake up more easily, and while many of us sleep through the night, others sometimes wake up for as much as an hour or two before going back to sleep. Debate over the normality of these varying patterns was triggered by the anthropologist Carol Worthman and the historian Roger Ekirch.[33] These scholars argued that it was normal prior to the Industrial Revolution for people to wake up for an hour or so in the middle of the night before going back to sleep. In between "first sleep" and "second sleep," people talk, work, have sex, or pray. By implication, electric lights and other industrial inventions might have altered our sleep patterns. However, sensor-based studies of nonindustrial populations reveal a more complex picture. Whereas most foragers in Tanzania, Botswana, and Bolivia sleep through the night, subsistence farmers in Madagascar often divide their sleep into first and second segments.[34]

In truth, most biological phenomena are highly variable, and sleep is no exception. Thanks to differences in circadian rhythms and the way our bodies regulate wakefulness and drowsiness, sleeping schedules are as variable in humans as they are in other species.[35] The lack of any single pattern of sleep, moreover, applies to populations surrounded by lights in New York and Tokyo, or without electricity in the African savanna or the Amazonian rain forest. When the anthropologist David Samson measured sleep activity in a camp of twenty-two Hadza hunter-gatherers for twenty days, he found so much variation in terms of who was asleep at different times that he estimated at least one person in the camp was awake for all but eighteen minutes per night.[36] From an evolutionary perspective, such variation is probably adaptive because we are most vulnerable when asleep in the dangerous night. Having at least one alert sentinel, often an older individual, would have reduced the dangers of sleep in a world full of leopards, lions, and other humans who wish us harm.[37]

So if you sometimes wake up in the middle of the night or sleep seven rather than eight hours a night, relax. In fact, humans appear to be adapted to sleep less than our ape relatives, including chimpanzees. This reduction possibly evolved about two million years ago when our ancestors apparently lost many of the features that help us climb trees, which offer a safe place to sleep in the wilds of Africa. As slow, unsteady bipeds who had to sleep on the dangerous ground, we must have been easy pickings for leopards, lions, and saber-toothed tigers before we harnessed fire. Under such conditions, our vulnerable forebears might have gone extinct had they not slept lightly, minimally, and in staggered bouts so that someone in the group was always awake to raise the alarm.

Another benefit of not sleeping so much—then and now—was having more hours a day to be social. Just as our ancestors probably used the close of day to gossip, sing, dance, and otherwise interact around the fire, today we enjoy gathering in the evening over dinner, in a bar, or some other well-lit place. Eventually, however, the urge to sleep overcomes all other desires, and many of us retreat to a dark and tranquil room, crawl into a soft and warm bed, lay our heads on a plush pillow, and fall into the arms of Morpheus. And in this respect, sleep really has become weird.

Sleep Culture

One night in December 2012, while traveling off the beaten track in the mountains of northern Mexico, my colleagues and I didn't reach the adobe hut that was to be our shelter until nighttime. The stars were out, it was cold, and I was physically exhausted and desperately wanted to hit the sack. By the time I had peed and brushed my teeth and was ready to sleep, four of my fellow travelers had already piled onto the only bed in the tiny hut, a queen-sized platform with a paper-thin mattress. Because they were snoring contentedly and there was no more room, I slept on the hard, dirt floor, wrapped in a few blankets. Actually, I was relieved not to have to sleep in that overpopulated, noisy bed of people, none of whom had showered for days, and I slept like a dog, even though I would have preferred a comfortable bed with clean sheets and a pillow. From a cultural perspective, how normal is my prudish preference to sleep alone or with just my wife?

Anthropologists have long studied how people sleep, producing a rich body of evidence on sleep practices and attitudes from around the world. If there is any one generalization to make, it is that people's approach to sleep varies impressively from culture to culture, and nowhere is sleep viewed as just a remedy to sleepiness. Many cultures consider sleep a social occasion. The Maori of New Zealand, for example, used to sleep communally in longhouses and still sleep this way at funerals to accompany a corpse on its journey from this world to the next.[38] The Asabano of New Guinea never let a stranger sleep alone because of the dangers of nighttime witchcraft, and the Warlpiri of Central Australia sleep under the stars in rows whose order is determined by strict social rules.[39] In numerous cultures, people consider it normal to talk or have sex next to sleeping neighbors, and only in modern, westernized families do mothers not always sleep with their infants.[40] Sleeping with others is also a great way to stay warm.

If these communal habits seem exotic, consider that before the Industrial Revolution made beds less expensive, Americans and Europeans routinely shared beds not just with family members and guests at home but with strange bedfellows when they traveled.[41] At the beginning of Melville's *Moby-Dick,* Ishmael, the narrator, first meets his fellow sailor Queequeg in bed in a New Bedford inn. At first, Ishmael is horrified by the prospect of sleeping next to a murderous tattooed cannibal, but he decides it's "better [to] sleep with a sober cannibal than a drunken Christian." In the morning, he awakes with Queequeg's arm thrown over him in the "most loving and affectionate manner."

Sleep in the industrial world has become not only more private but also more comfortable. Before the invention of the spring-coil mattress in the 1880s, only the wealthy in Europe and America could afford a comfortable mattress stuffed with feathers or hair. Most mattresses were thin, lumpy, straw-filled pads. Sheets and soft pillows, now ubiquitous, were also luxuries enjoyed by only the privileged. As the photos in figure 8 illustrate, for millions of years in most parts of the world, almost everyone slept without pillows on hard, unyielding surfaces with just grass, straw, skins, bark, leaves, or anything else for insulation. Such arrangements may seem uncomfortable, but I can assure you it takes little effort to get used to sleeping on the floor. In addition, traditional forms of disposable bedding are more hygienic than straw-filled mattresses, which are ideal homes for lice, fleas, and bedbugs, the "unholy trinity of early modern entomology."[42] Medieval habits of sleeping in groups on straw mattresses helped spread contagions like the plague.

As sleep has become more luxurious and isolated, it has also become quieter and darker. You will probably spend about one-third of your life on beds more comfortable than kings of yore ever enjoyed and in bedrooms designed to keep out light, noise, and other disturbances, maybe also heated or cooled to an "ideal" temperature. Sleeping in this kind of sensory insulation is uncommon outside the modern industrialized world. As a rule, foragers sleep in conditions that border on bedlam. People sleep in groups usually near a fire in relatively busy environments with no barriers to block out noise or light. As they fall asleep, others in camp may be talking, nursing, cavorting, or doing chores, and often one hears animals in the distance. In my opinion, the worst offenders of the night in Africa are not humans or hyenas but tree hyraxes, cat-sized, tree-dwelling ungulates (distant relatives of elephants) whose hair-raising nocturnal calls resemble the screams of someone being throttled. Hyraxes notwithstanding, modern preferences for sleeping in dark and tranquil environments are culturally prescribed. If you require quiet and dark to fall asleep, you are evolutionarily unusual.

To modern sensibilities, the chaos of Stone Age sleeping conditions seems antithetical to a good night's rest, but the anthropologist Carol Worthman has proposed that the reverse may be true.[43] As we go through the initial stages of NREM sleep, we become gradually less aware of our environment. This progressive tuning out may be adaptive because our brain is monitoring the world around us as we fall asleep, possibly to assess whether it is dangerous to sleep. Slowly receding perceptions of nearby friends and family talking, a crackling fire, infants crying, and the fact that those hyenas are far away signal to the brain that it is safe to enter a deeper, unconscious stage of sleep. Ironically, by insulating ourselves so effectively from these comforting stimuli, we may be making ourselves more prone to becoming stressed about sleep.

I can think of no peculiarity of modern sleeping culture that more counterproductively promotes privacy at the expense of stress than

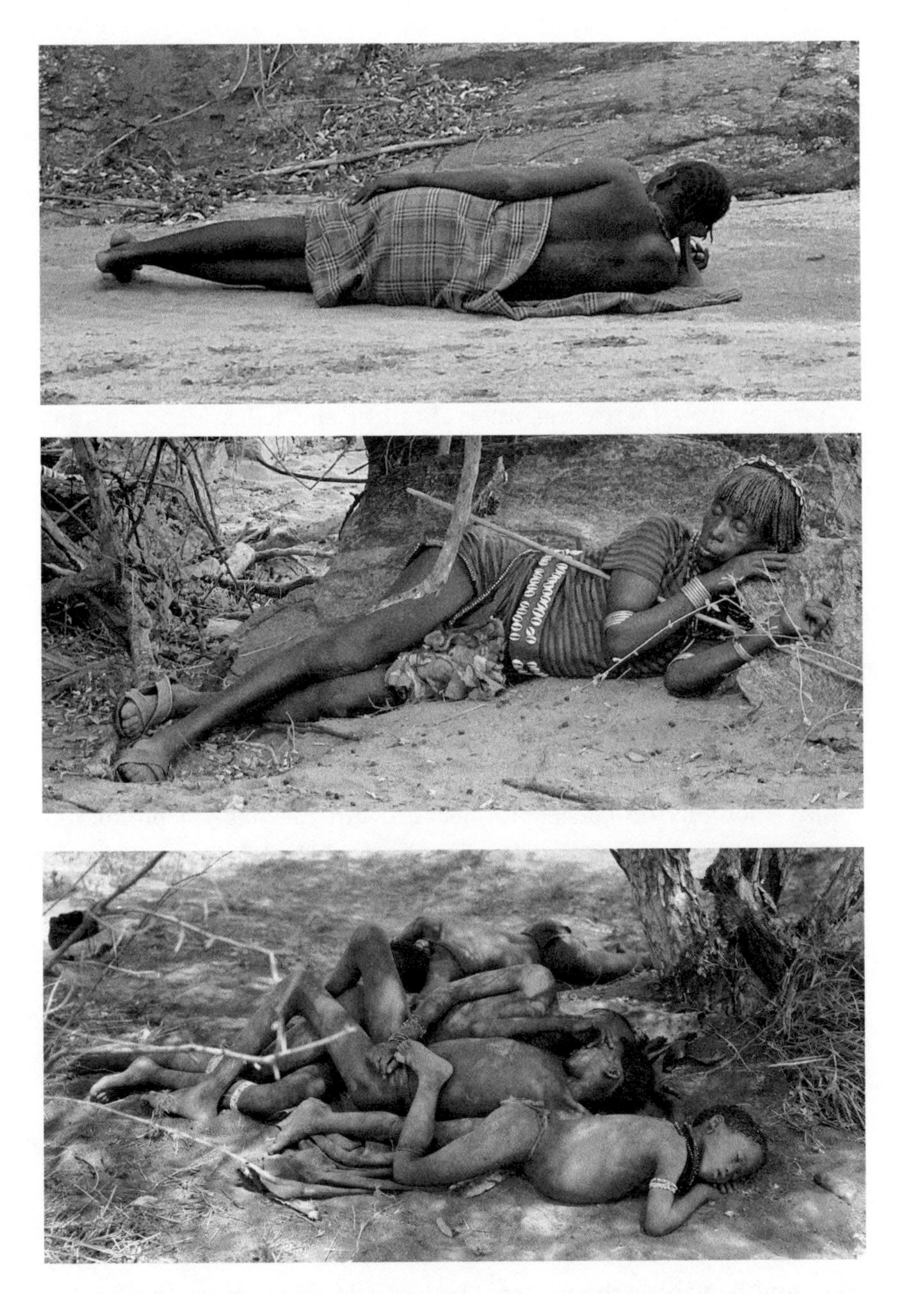

FIGURE 8 Examples of nonindustrial sleeping. *Top*, Hamer man asleep in Ethiopia (photo by Daniel E. Lieberman); *middle*, Hamer woman asleep in Ethiopia (photo by Daniel E. Lieberman); *bottom*, group of San children asleep in the Kalahari (photo gift of Laurence K. Marshall and Lorna J. Marshall © President and Fellows of Harvard College, Peabody Museum of Archaeology and Ethnology, PM2001.29.14879).

banishing children from beds and bedrooms. In every culture until recently, infants slept with their mothers. Many cultures consider not sleeping with your child a form of child abuse.[44] Yet when my wife and I first became parents, many books and strangers advised us against co-sleeping with our daughter. Naively, we followed the advice of Dr. Richard Ferber, whose infamous method of "Ferberizing" involved abandoning our daughter in her crib in a separate room while she cried her lungs out to be with us.[45] According to his prescription, we were supposed to visit our howling, panicked child at progressively increasing intervals until she fell asleep and thus learned to "self-soothe" herself. What a disaster. After a harrowing week of torturing our daughter and ourselves, we decided to do what humans have always done and sleep with her in our bed. Co-sleeping not only helps mothers and infants sleep better; it helps mothers and infants coordinate their sleeping and feeding and provides a wealth of positive, nurturing interactions.[46] Although bed sharing with a parent who smokes, drinks, or takes drugs involves risks to the infant, most especially sudden infant death syndrome, misinformation has scared many parents from co-sleeping.[47]

My thoughts on co-sleeping aside, I am inevitably a product of the culture in which I was born and raised, and I don't ever want to treat sleeping as a social activity. I prefer to sleep in a quiet, dark room on a comfortable mattress with just my wife by my side and our dog at the foot of the bed. Maybe many of our forebears would have preferred such conditions. Appreciating the extent to which my sleep preferences are culturally constructed is comforting when I try to sleep in less than ideal conditions. The worst, by far, is an airplane. As I attempt to drift off strapped into a cramped seat with the engines roaring, people chatting, toilets flushing, and babies crying, I remind myself that we evolved to sleep communally in chaotic, noisy contexts, albeit not in a metal tube traveling five hundred miles per hour thirty thousand feet above the planet. The key is to avoid stress, the desperate enemy of sleep. Why are so many people today stressed out by a behavior as intrinsically restful and effortless as falling sleep? And how does physical activity help?

Stressed About Sleep

If there is any ideal laboratory for studying the effects of too little sleep, it must be a university. Today's generation of students seems no less sleep deprived than I was at their age. Last year, an extremely conscientious undergraduate came to my office hours because she had done poorly on an exam. As we discussed her test, I was impressed by how well she knew the material, and it soon became evident that lack of sleep might have impaired her performance. She confessed she routinely gets only four hours of sleep, and the night before the exam was no exception. When I asked her why she sleeps so little, I felt I was listening to a recording of myself at her age. She typically goes to bed at two or three in the morning, struggles to fall asleep, and then wakes up early before she feels rested. During the day, she has trouble not dozing off in lectures, and then at night after long hours in the library fueled by caffeine she has trouble falling asleep.

College students are a special breed of humans, in part because so many of them are enjoying their first taste of being grown up without yet shouldering the responsibilities of being adults. Most of my sleep-deprived students will have no choice but to settle down and get more sleep once they leave the ivory tower, but some will stay sleep deprived. According to some studies, about 10 percent of American adults have diagnosable insomnia (that is, repeatedly taking more than half an hour to fall asleep or persistently being unable to sleep through the night), and almost one-third think they don't sleep enough.[48] Elsewhere, the prevalence of insomnia is similar.[49] Predictably, many of these sufferers, about 5 percent of Americans, resort to sleeping pills.[50] Why do so many people excel at resting too much during the day but then fail to rest enough at night?

To answer this question, we need to consider the two major biological processes that interact in the brain to regulate wakefulness and sleep.[51] When these processes function normally, we wake up in the morning feeling refreshed, stay pleasantly alert for most of the day, and then fall gently asleep at night. When they are disrupted, we nap inappropriately during classes or meetings, struggle miserably to fall asleep at night, wake up too early, or spend tortured hours lying awake with a dismal headache.

The first system is our nearly twenty-four-hour circadian cycle regulated by a specialized group of cells within a region of the brain known as the hypothalamus.[52] (The sleep-inducing name for this cluster of cells is the suprachiasmatic nucleus.) These cells wake us up in the morning by signaling to the glands atop our kidneys to produce cortisol, the major hormone that stimulates the body to spend energy. Then as darkness falls, the hypothalamus directs the pineal gland, another structure in the brain, to produce melatonin, the "Dracula hormone," which helps induce sleep. Clocklike, the circadian system is re-synced every day by light levels and other experiences. As anyone who has suffered from jet lag knows, circadian rhythms can be reset slowly (about an hour a day) by light and other environmental cues.

It would be a problem if our bodies relied solely on circadian clocks to regulate sleep. Imagine not being able to sleep late after several days of sleep deprivation, or being unable to stay up late even if you are well rested. For this reason, our sleep-wake states are modulated by a second system that is tightly linked to activity levels. This homeostatic system functions like an hourglass that counts how long we've been awake, slowly building up pressure for us to sleep. The longer we stay awake, the more sleep pressure we accrue from the accumulation of molecules such as adenosine left behind when the brain expends energy. Then by sleeping, we reset the hourglass, primarily through NREM sleep. Overall, the homeostatic system helps balance the time we spend awake versus asleep, and if we are up too long, it will eventually override our circadian systems and help us recover lost sleeping time.

Under normal circumstances the circadian and homeostatic systems work in concert to maintain a routine sleep-wake cycle. But life isn't always routine. What if your house catches fire, a pack of hungry hyenas escapes the zoo and invades your neighborhood, or your mother-in-law announces she is moving in with you? These and other life-threatening crises activate your "fight and flight" system to induce a state of hyperarousal. In a trice, your body unleashes a cascade of hormones including epinephrine and cortisol that speed up your heart, dump sugar into your blood, halt your digestive system, and raise your level of alertness. Obviously, these hormones also counter the processes that permit sleep, a critical adaptation to maintain CONSTANT VIGILANCE.[53] Tonight you'll barely sleep a wink as you cope with the emergency. Then if all goes well and the fire is doused, the hyenas are captured, or your mother-in-law departs, your equilibrium will return and you'll sleep like a log the next night thanks to the sleep debt you accumulated.

The effects of the fight-and-flight response (technically, the sympathetic nervous system) on sleep explain how and why exercise has such important, well-known effects on sleep. If you run a mile at top speed or lift heavy weights just before going to bed, you'll probably have a hard time falling asleep because vigorous physical activity turns on this system, stimulating arousal. In contrast, a good dose of physical activity earlier in the day like a game of soccer, an hour or two of gardening, or a long walk helps sleep come more easily. These activities increase sleep pressure, and they stimulate the body to counter the initial fight-and-flight response with a deeper "rest and digest" response (technically the parasympathetic nervous system). Among other benefits, recovery from exercise gradually lowers basal cortisol and epinephrine levels, depresses body temperature, and even helps re-sync the circadian clock.[54] Although physical activity doesn't prevent or cure all sleep problems, a multitude of studies demonstrate that a single bout of exercise (but not immediately before bed) usually helps people sleep, and regular exercise is even better.[55] One survey of more than twenty-six hundred Americans of all ages that controlled for factors like weight, age, health status, smoking, and depression found that those who regularly engaged in at least 150 minutes of moderate to vigorous activity a week not only reported a 65 percent improvement in sleep quality but also were less likely to feel overly sleepy during the day.[56] In turn, getting enough sleep helps people be active and improves athletic performance by allowing the body to have sufficient time to rest and repair.[57] Adolescents who sleep less than six hours have twice the injury rate of those who sleep eight or more.[58] And finally, adults who are persistently physically inactive are more vulnerable to suffering from insomnia.[59]

Insomnia, which is a long-term condition and not a night or two of poor sleep in response to an emergency, is especially cruel because it often triggers a vicious cycle. If underlying chronic stress from too much time commuting, social conflicts, or endlessly tough homework assignments elevates stress hormones like cortisol above normal levels, we become more alert at night when we'd otherwise become drowsy, or we wake up after one or two NREM and REM cycles.[60] Then as we become chronically sleep deprived, we produce more cortisol, especially at night, which can then inhibit sleep, keeping the problem going and promoting insomnia.[61]

Sadly, stresses that elevate cortisol levels and cause sleep deprivation can also slowly erode our health in other ways by depressing the immune system and directing the body to store more organ fat. Lack of sleep also wreaks havoc with the hormones that regulate appetite, increasing levels of a hormone called ghrelin that makes us hungry and simultaneously depressing levels of another hormone called leptin that inhibits the desire to eat.[62] I certainly snack more when I am short on sleep, as do millions of sleep-deprived college students whose midnight cravings are conveniently met by late-night stores near college campuses that sell cookies and other energy-rich snacks. And to add insult to injury, chronic sleep deprivation promotes chronic inflammation and disrupts the normal nighttime release of growth hormone.[63] Altogether, sleep deprivation helps promote obesity and its associated conditions like type 2 diabetes and heart disease; it is also associated with cancer.[64] Then in a cruel twist of fate, overweight people are at higher risk of having trouble breathing while asleep (apnea), which further disrupts their sleep.

Because they are routinely physically active and don't have lawyers, hunter-gatherers probably get less insomnia than most industrialized people, but they assuredly experience stress and occasional sleepless nights. If so, they almost certainly don't fall into the modern trap of treating the symptoms rather than the underlying causes of their sleep deprivation. One of the most common forms of modern symptomatic sleep treatment is self-medication with alcohol, which can initially induce drowsiness but disrupts neurotransmitters that maintain sleep.[65] Even more insidiously, we have become prey to what has been termed the sleep-industrial complex. People stressed about sleep are enticed to spend a fortune on hi-tech mattresses, sound machines, noise-canceling headphones, light-blocking curtains, gizmos to halt their bedfellow's snoring, eye masks, and something called high-performance bedding. These mostly harmless gadgets would no doubt amuse our ancestors who slept on the ground by a fire, but we should be downright alarmed by the abuse of sleeping pills. Sleeping pills, which are highly habit forming, are a multibillion-dollar industry. Not counting over-the-counter medications, prescriptions for these pills in the United States have more than tripled since 1998.[66]

Despite their popularity, sleeping pills are dangerous. According to one observational study of more than thirty thousand people, American adults who took sleeping pills on a regular basis increased their risk of dying over the subsequent two and a half years by almost fivefold.[67] Many other studies also report strong associations between sleeping pills and depression, cancer, respiratory problems, confusion, sleepwalking, and other dangers.[68] And if these cautions were not damning enough, several studies report that most of the benefits of sleeping pills are placebo effects. Insomniacs and healthy controls prescribed popular sleep medications (for example, Sonata and Lunesta) slept on average the same number of hours (about six hours and twenty minutes) as those prescribed a placebo, and they fell asleep only fourteen minutes faster, despite sometimes also reporting memory lapses the next day.[69] To quote Jerome Siegel, "In twenty years, people will look back on the sleeping-pill era as we now look back on the acceptance of cigarette smoking."[70]

Exercised About Sleep

Let's conclude by returning to the question I posed at the beginning of this chapter: If humans evolved to rest as much as possible, why do so many of us skimp on sleep? I don't wish to turn a blind eye to the evidence that too many people are sleep deprived, thus harming their own health and endangering others (especially when behind a wheel), but by failing to consider evolutionary and anthropological perspectives on sleep, some alarmists mischaracterize everyday people's sleeping behavior as abnormal, not unlike the way we have demonized sitting. Fearmongering about sleep can be profitable, and our society tends to be judgmental about behaviors involving physical activity and inactivity. We label sitting as bad and sleeping as good. In truth, both ways of resting are utterly normal but highly variable behaviors with complex costs and benefits that are strongly influenced by our environment and contemporary cultural norms.

If you are unsure about your own sleep health, sleep researchers suggest you ask yourself five simple questions:[71]

Are you satisfied with your sleep?
Do you stay awake all day without dozing?
Are you asleep between 2:00 and 4:00 a.m.?
Do you spend less than thirty minutes awake at night?
Do you get between six and eight hours of sleep?

If your answers to these questions are "usually or always," then you should sleep contentedly knowing that you generally get enough sleep. If not, I hope you get some relief through well-studied, sensible, effective approaches like cognitive behavioral therapy, good habits like sticking to a regular sleep schedule, and—of course—exercise. It bears repeating that sleep and physical activity are inextricably linked: the more physically active we are, the better we sleep because physical activity builds up sleep pressure and reduces chronic stress, hence insomnia. In that sense, physical activity and sleep are not trade-offs but collaborators.[72] Maybe it is not so paradoxical that the same well-intentioned people who nag us to exercise sometimes also badger us to spend more time in bed.

PART II

Speed, Strength, and Power

FIVE

Speed: Neither Tortoise nor Hare

MYTH #5 Normal Humans Trade Off Speed for Endurance

> And there again, shrill and inevitable, was the ululation sweeping across the island. At that sound he shied like a horse among the creepers and ran once more till he was panting.
>
> —William Golding, *Lord of the Flies*

I think the fastest I ever ran was at a place called Olorgesailie, Kenya, about twenty-five miles southwest of Nairobi. Apart from a small population of Maasai herders, few people live in this hot, dusty, desolate area. However, between 1.2 million and 400,000 years ago, a large lake, long since gone, sustained life for early humans and other animals such as hippos, elephants, monkeys, and zebras. When I was twenty-four and just starting graduate school, I had the wonderful opportunity to spend several weeks at Olorgesailie helping to excavate timeworn fossils and stone tools that provide enigmatic clues of what happened there long ago. In addition to being scientifically intriguing, Olorgesailie is a stunningly beautiful place to live, work, and learn. We camped in tents on a little promontory overlooking a vast, arid scrubland that extends to an extinct volcano in the distance. Every morning, before we headed out to excavate Stone Age bones and stones, I would rise early to enjoy the sunrise. Sometimes, I would see a few hyenas loping home from a night of mischief, carrying bones or animal legs in their massive, powerful jaws. Although the hyenas' den was not far from our camp, we two species left each other alone.

Then, late one morning, I was concentrating so intently on my work I failed to notice the powerful stench that should have been a warning sign. I still recall vividly the instant I noticed the beady eyes, black snout, and putrid smell of a reeking hyena staring at me just a few feet away. Terrified, I dropped my clipboard and bolted for my life. I am sure I set a personal record in that mad dash, which I think lasted about thirty seconds. When I finally turned around, gasping for breath and my legs on fire, I was relieved to see the hyena running slowly away in the opposite direction. Perhaps the hyena was as eager to get away from me as I was to escape it. To this day, whenever I smell anything hyena-like, I have an intense memory of that momentary panic.

As the years pass, I am increasingly grateful the hyena decided not to chase me, because I would have had no chance of outrunning it. Hyenas can reportedly run about forty miles an hour.[1] I have always been more of a tortoise than a hare, and I reckon that in my prime I was never able to exceed fifteen miles per hour for even a minute. In addition, unlike the hyena, I have no claws, paws, and fangs with which to defend myself. Had it wanted to, that hyena could have easily mauled or killed me.

Although most humans sensibly steer clear of large, wild carnivores, every July hundreds of grown men willingly risk being mauled or overrun by large, untamed bulls in Pamplona, Spain. On eight consecutive days during the festival of San Fermín, a dozen bulls are released at precisely 8:00 a.m. into the town's twisty, narrow medieval streets. Over the next few minutes, these dangerous beasts chase a crowd of colorfully dressed daredevils half a mile (825 meters) to the town's bullfighting ring. By no means are the men and bulls evenly matched. The bulls are ten times heavier, they have deadly horns, and as they stampede through the streets, they easily overtake the human runners, especially those who are drunk or hungover.[2] It is not uncommon for bulls to trample or gore unlucky *hombres* who slip or otherwise fail to get out of their way. Dozens of people get injured every year, and every few years someone dies from being impaled.[3] Whatever your opinion of the running of the bulls, it highlights the same, obvious difference between me and my hyena friend: we humans are slow, weak, vulnerable creatures more dependent on brains than brawn.

But are all humans really that slow? What if a really speedy, well-trained sprinter had been racing that hyena or those bulls instead of a bunch of amateur thrill seekers?

How Slow Is Usain Bolt?

As the athletes enter the stadium in London, the crowd erupts. It's the 2012 Olympic final for the hundred-meter dash, the world's premier footrace. Seemingly oblivious to the eighty thousand exuberant spectators, the eight finalists stretch and warm up, their thoughts focused inward as they silently rehearse their race strategies. At last, when everyone is ready, the officials direct the sprinters to the starting blocks, where they are introduced one by one to the crowd and the television cameras. Although these guys are supposedly the fastest eight men on the planet, everyone's attention is directly primarily at the tall Jamaican phenomenon Usain Bolt, the world record holder. At six feet five inches, Bolt towers over the other runners, and he drinks in the raucous applause, smiling broadly. Then, as if by magic, a tense silence descends as the athletes gingerly place themselves in the starting blocks with their strongest leg in the forward pedal, their knees on the ground, and their fingertips millimeters behind the starting line. Much of the world is now focused on just these eight men, especially Bolt. Then, at the umpire's command, they bring up their knees. Seconds later, bang, the gun goes off!

Within milliseconds, the runners straighten their hips and knees, pushing their bodies upward and forward from the blocks, their trunks at a forty-five-degree angle to the ground, one arm driving forward, the other backward. Bolt is among the last to leave the blocks. As the sprinters accelerate, they slowly raise their torsos, and by ten to fifteen steps into the race all eight men are fully upright and abreast of one another. Then, as the runners continue to accelerate, Bolt pulls ahead. At this point, the sprinters are trying to maintain perfect form by landing just behind the ball of each foot, touching the ground slightly in front of their hips, and then raising that knee as fast as possible before accelerating the foot and shank down to the surface again as if hammering the ground. Although their legs are their primary engines, the runners pump their arms parallel to their torsos, which they keep as relaxed as possible with their shoulders down, avoiding unnecessary rotations. By fifty meters, when they have reached their maximum speed of twenty-five to twenty-six miles per hour, Bolt is clearly in the lead. Only now do the sprinters take their first breath while continuing to focus on one thing and one thing only: staying on track without slowing.

By seventy meters, it is Bolt's race to lose as he is now several steps in front. Although it is impossible to see, everyone including Bolt is now slowing down slightly, and the racers know that victory will go to whoever slows the least. The athletes must dig deep here and focus on getting their knees up and staying relaxed. True to his reputation, Bolt barrels forward, and just as he puts his foot on the finish line, he leans slightly forward with his torso crossing the line 9.63 seconds after the gun went off. A new Olympic record! As the crowd erupts, Bolt celebrates exuberantly, draping himself in the Jamaican flag and posing as a lightning bolt.

Bolt retired in 2017 after a spectacular career including many world records and Olympic gold medals, but how did he manage to dominate his sport like no other sprinter? Because speed is the product of stride length and stride rate (a stride being a full cycle from the time a foot hits the ground to the next time the same foot hits the ground), one can go faster by taking longer strides, by taking more rapid strides, or some combination of the two.[4] Because Bolt has longer legs but moved them nearly as quickly as shorter-legged sprinters, he ran faster than his competitors.[5] In a typical hundred-meter race, Bolt took only forty steps, whereas the rest of the field took about forty-five. However, for Bolt to drive his legs that fast required incredible strength. Just as it takes more force to swing a longer baseball bat, it takes more force to accelerate a longer leg. All in all, Bolt's long legs combined with his ability to generate high forces meant he spent more

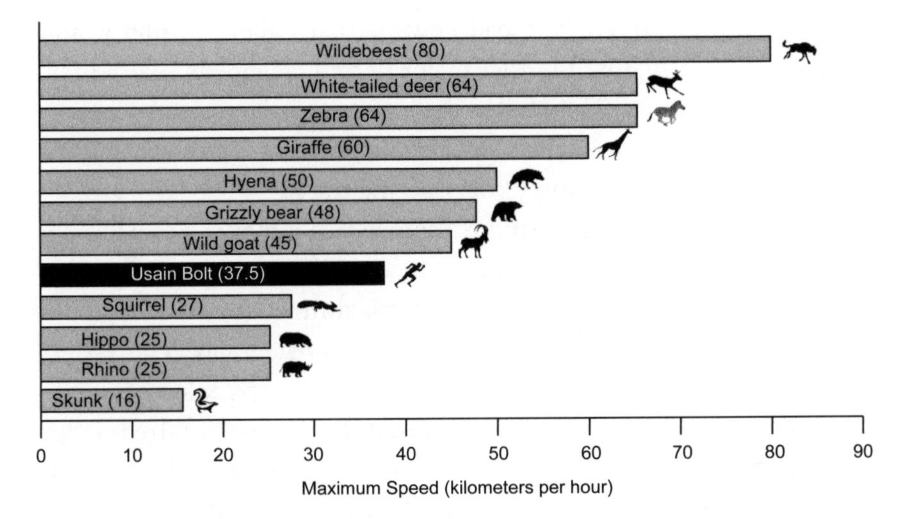

FIGURE 9 Maximum running speeds of Usain Bolt versus purported maximum speeds of various mammals. Keep in mind that the maximum speeds of many animals are difficult to measure and validate, so some of these maximum speeds need to be taken with a grain of salt. Remember also that most reasonably fit humans cannot run much faster than twenty-four kilometers per hour, about the speed of a hippo. (Animal data mostly from Garland, T., Jr. [1983], The relation between maximal running speed and body mass in terrestrial mammals, *Journal of Zoology* 199:157–70)

time flying through the air. In his fastest run ever (Berlin, 2009), he averaged twenty-three miles per hour and briefly achieved a top speed of twenty-eight miles per hour.

Sometimes I run on a track and try to work on my speed. As I huff and puff, swifter runners easily blow by me at velocities that seem unimaginable. And these are just amateur runners, making me wonder what it must feel like to be passed by the likes of Bolt. Yet, as fast as Bolt and other elite sprinters can go, these great sprinters are unimpressive compared with ordinary four-legged animals. Keeping in mind the challenges of measuring the running speeds of wild animals accurately and the impossibility of knowing if the fastest speeds we measure are really their top speeds, figure 9 puts Usain Bolt's world record of 23.3 miles per hour (37.5 kilometers per hour) into perspective using purported maximum speeds for a range of quadrupedal mammals.[6] The good news is that Bolt could outsprint skunks, rhinos, hippos, and most tiny rodents including the common gray squirrels in my garden. The bad news is that's about it. Bolt would have no chance against the vast majority of quadrupeds like zebras, giraffes, wildebeests, white-tailed deer, or even wild goats. As for carnivores, even slower predators like grizzly bears and hyenas could eat Bolt's lunch—not to mention Bolt—on the track.

One problem with sprinting is that we run out of gas quickly. No animal can maintain peak speed for long, and humans are no exception. Elite sprinters like Bolt sustain maximum velocity for about twenty seconds but then must slow down considerably. The fastest thousand-meter time (currently 2:11) was run at 27.3 kilometers per hour, less than three-quarters the speed of the fastest hundred-meter dash, and the world's fastest five-kilometer time (12:37:35) was run even slower, at 23.7 kilometers per hour. Cheetahs in the wild also run for a maximum of about thirty seconds before slowing down.[7] However, because many mammals can run so much faster than humans, they can chase us or flee at submaximal speeds they can sustain for longer.

Sprinting humans also turn pathetically. In the real world, animals never line up on one painted line and then dash as fast and as straight as possible on a flat surface to another painted line. Instead they zig and zag, which slows them down. If you watch a cheetah chase a gazelle, the gazelle (whose top speed of about 80 kilometers per hour is perilously slower than the cheetah's maximum of 110 kilometers per hour) will desperately try to turn rapidly and unpredictably to slow its deadly pursuer.[8] The strategy often works, but if a cheetah ever chases you, zigzagging might be a bad idea. As veterans of Pamplona will attest, the most dangerous parts of the course are the turns where the two-legged humans become even slower and less stable than the four-legged bulls.[9]

The world's fastest humans, then, are no match for much of the animal world. Consider also that we have been comparing the speediest humans alive—exceptional athletes who have trained for years with the help of coaches and others for the sole purpose of sprinting prescribed distances on tracks as fast as possible—with average, untrained mammals. While elite human runners can briefly attain speeds of twenty miles per hour or more, the majority of fit humans can rarely exceed fifteen miles per hour, which is probably a more reasonable estimate of maximum sprinting speed for most of human evolution. Unless you are a world-class sprinter, you have little chance of outrunning a squirrel. Why are humans so comparatively slow?

The Trouble with Two Legs

If, as some religions teach, we are made in God's image, then God must be a slow runner. The alternative evolutionary explanation for our relative slowness is that our ancestors were selected about seven million years ago to become habitually upright. Although bipedalism has some benefits, it also came with drawbacks. In addition, to making us clumsy in trees, prone to tripping and falling, and more vulnerable to lower back pain, becoming two-legged made us perilously slow.

To appreciate how bipedalism doomed us to being lead-footed, consider that to walk or run, your legs must generate force against the ground. The harder your legs push down against the ground, the faster you can run. And herein lies the basis for why upright humans are comparatively slow: while a dog or chimpanzee has four legs with which to push on the ground to generate power (power is the rate of doing work), we have only two. In fact, when we run, only one leg is on the ground at any given moment to lift and push us forward. Less power means less speed. Just as my humble car's four-cylinder engine can attain half the speed of a V-8 Ferrari, two-legged humans can run only half as fast as similar-sized four-legged animals. Greyhounds are about twice as fast as elite sprinters.

If you just thought of ostriches, which can attain speeds of forty-five miles per hour, then you've also realized that having two legs is not an insurmountable detriment to speed.[10] Sadly, when our ancestors got off all fours, we never evolved the adaptations that help these flightless birds run fast. Hindrances we inherited from our primate ancestors include big, cumbersome legs and feet. Compare your legs with the hind limbs of a horse, dog, or even an ostrich as shown in figure 10. Because our clunky feet are oriented horizontally along the ground, our ankles are just a few inches above the floor, but these species run on their toes, giving their long legs three major segments. Primates' cumbersome feet are ideal for grasping onto branches and climbing trees, but our stumpy legs diminish speed by shortening each stride. The relatively stubby legs of primates are also relatively thick all the way down: we have fat ankles and big feet. Because dogs, horses, and ostriches have highly tapered hind limbs with small feet, their legs' center of mass is closer to their hips, making their legs easier to swing. Finally, our primate feet lack claws, which act as natural cleats, or hooves, which act as natural shoes.

Being upright has another disadvantage: when running, we lost the use of our spines as stride-extending springs. Watch a slow-motion video of a greyhound or a cheetah galloping. When it lands on its back legs, its hind paws land below the shoulders as its long, flexible spine curves like a powerful bow, storing elastic energy. Then as the animal's hind limbs push off, the spine rapidly unbends, releasing elastic energy to help catapult it into the air and increase its stride length.[11] Our short, little upright spines do nothing to help us run faster, but instead struggle to keep our inherently tippy upper bodies stable while also dampening the shock wave that travels from the foot up to the head every time we hit the ground.[12]

In sum, humans have been slowpokes ever since that fateful transition seven million years ago when our ancestors became bipeds. If I were a hungry saber-toothed tiger in Africa back then, I would have relished hunting down early humans because they would have been much easier to overcome than antelopes and other fast four-legged prey. Yet even if our ancient ancestors were easy pickings, they must

FIGURE 10 Lateral views of running dog, human, and ostrich. Note that the human has a thicker, less tapered leg with only two major segments and big, clunky feet.

have sometimes sprinted for their lives. After all, to avoid being a tiger's dinner, you need only to run slightly faster than the next guy. That would be no problem for Usain Bolt, who can sprint more than twice as fast as I can. But, like most distance runners, I can probably run much farther than Bolt. To what extent does my greater endurance come at the expense of speed?

Fast or Far?

My friend and colleague Professor Jenny Hoffman likes to run preposterously long distances. In one race, she ran 142 miles in twenty-four hours. I'd rather pluck my toenails out one by one than try such a feat, but she claims it's fun and doable by settling into a slow pace and refueling regularly with ginger snaps (which, I note, can be enjoyed without running for twenty-four hours nonstop). Jenny's leisurely ten-minute-per-mile speed during that race needed to be four times slower than the blistering velocities elite sprinters attain. Indeed, had she sprinted flat out to escape a hyena, she'd be gasping for breath within a minute, forcing her to stop or slow down. Had Usain Bolt also been chased, he would have easily outrun Jenny, but he too would have quickly run out of gas. The differences and similarities between Jenny and Usain raise two important questions. What constrains maximum speed over short distances? And why can't we run both fast and far?

When it comes to very short sprints, speed is largely a function of strength and skill. Since sprinters' legs work sort of like hammers that forcefully and rapidly hit the ground, and (as Newton showed) for every action there is an equal and opposite reaction, the harder the legs push downward and backward against the ground, the harder the ground pushes the body upward and forward. For this reason, maximum speeds for hundred- and two-hundred-meter sprints are largely limited by how effectively a runner's leg muscles can produce force during the fleeting period of time—as little as a tenth of a second in elite sprinters—each foot is on the ground.[13] Usain can thus sprint short distances much faster than Jenny primarily because he can strike the ground much harder.

Over longer distances, however, both Usain and Jenny run out of gas if they don't moderate their pace because of the way all organisms struggle to transform fuel rapidly into usable energy. For this reason, it is common to analogize bodies with combustion engines: just as a car's engine burns gas, bodies burn food, and if we run too fast, our bodies run out of fuel like a car that is driven too fast. The problem with that analogy, however, is that bodies are more like battery-powered cars, and instead of having one enormous battery that we recharge occasionally, our cells use millions of tiny organic batteries that we have to recharge constantly.

These ubiquitous miniature batteries, which power all life on earth, are called ATPs (adenosine triphosphates). As the name implies, each ATP consists of a tiny molecule (an adenosine) attached to three molecules of phosphate (a phosphorus atom surrounded by oxygen atoms). These three phosphates are bound to each other in a chain, one on top of the other, storing energy in the chemical bonds between each phosphate. When the last of these phosphates is broken off using water, the tiny quantity of energy that binds it to the second phosphate is liberated along with one hydrogen ion (H^+), leaving behind an ADP (adenosine *di*phosphate). This liberated energy powers almost everything done by every cell in the body like firing nerves, making proteins, and contracting muscles. And, critically, ATPs are rechargeable. By breaking down chemical bonds in sugar and fat molecules, cells acquire the energy to restore ADPs to ATPs by adding back the lost phosphate.[14] The problem is, however, that regardless of whether we are hyenas or humans, the faster we run, the more our bodies struggle to recharge these ATPs, thus curtailing our speed after a short while.

To understand better this fascinating but speed-limiting system that only evolution could have devised, let's imagine that Usain Bolt and I are simultaneously running from that hyena in Kenya. Although Bolt will initially sprint much faster than I, he, too, will gasp for breath after about thirty seconds because we recharge ATPs using the same three processes (schematized in figure 11) that work one after the next on different timescales—immediate, short-term, and long-term—but that have to compromise between speed and stamina.

The first process (called the phosphagen system) provides energy fastest but most fleetingly. As Bolt and I start running, our muscle cells contain barely enough ATPs to power a few steps. It seems inadvisable to stockpile so little ATP, but these organic batteries, despite being minuscule, store only one charge each, and they are too bulky and heavy for cells to manufacture and store in large quantities. You use more than thirty pounds of ATP during a one-hour walk and more than your entire body weight of ATP over the course of a typical day—an obviously impossible amount to lug around in reserve.[15] Consequently, a human body stores in toto only about a hundred grams of ATPs at any given moment.[16] Fortunately, before our first few steps deplete the leg muscles' scant supply of ATPs, they quickly tap into another ATP-like molecule known as creatine phosphate that also binds to phosphates and stores energy.[17] Unfortunately, those creatine phosphate reserves are also limited, becoming 60 percent depleted after ten seconds of sprinting and exhausted after thirty seconds.[18] Even so, the precious short burst of fuel they provide gives muscles time to fire up a second energy recharging process: breaking down sugar.

Sugar is synonymous with sweetness, but it's first and foremost

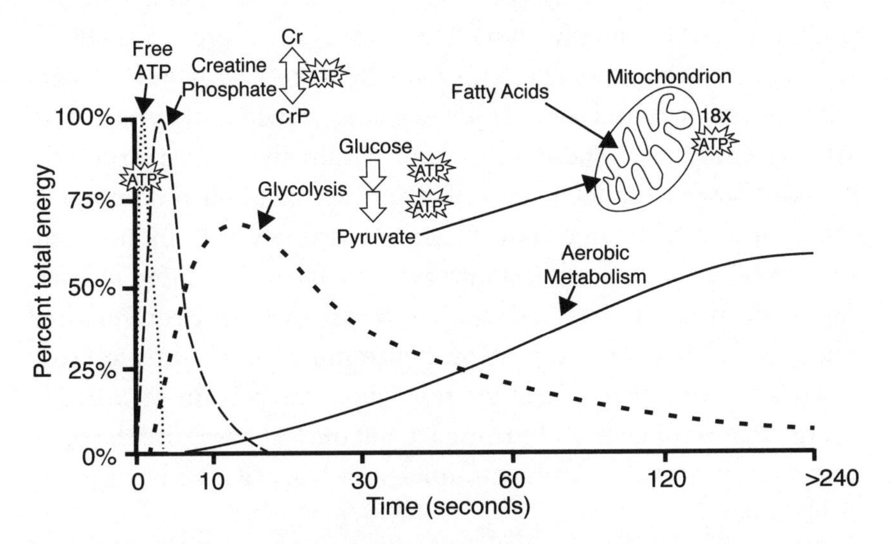

FIGURE 11 Different processes by which muscles recharge ATPs over time. At first, the energy comes nearly instantly from stored ATP and creatine phosphate (CrP); later, energy comes relatively rapidly from glycolysis; eventually, energy must come from slowly aerobic metabolism. Aerobic metabolism occurs in mitochondria by liberating energy either from pyruvate (an end product of glycolysis) or fatty acids.

a fuel used to recharge ATPs through a process termed glycolysis (from *glyco* for "sugar" and *lysis* for "break down"). During glycolysis, enzymes swiftly snip sugar molecules in half, liberating the energy from those bonds to charge two ATPs.[19] Restoring ATPs from sugar doesn't require oxygen and is rapid enough to provide almost half the energy used during a thirty-second sprint.[20] In fact, a fit human can store enough sugar to run nearly fifteen miles. But there is a consequential catch: during glycolysis the leftover halves of each sugar, molecules known as pyruvates, accumulate faster than cells can handle. As pyruvates pile up to intolerable levels, enzymes convert each pyruvate into a molecule called lactate along with a hydrogen ion (H^+). Although lactate is harmless and eventually used to recharge ATPs, those hydrogen ions make muscle cells increasingly acidic, causing fatigue, pain, and decreased function.[21] Within about thirty seconds, a sprinter's legs feel as if they are burning. It then takes a lengthy period of time to slowly neutralize the acid and shuttle the surplus lactate into the third, final, but long-term *aerobic* energy process.

Life demands oxygen, especially if you want to run far. In fact, using oxygen to burn a molecule of sugar yields a whopping eighteen times more ATP than glycolysis. But, once again, there is a trade-off: aerobic metabolism provides substantially more energy but substantially more slowly because it requires a long sequence of steps and an army of enzymes.[22] These steps occur within specialized structures in cells known as mitochondria that are capable of burning not only the pyruvates that come from sugar but also fats and, in emergencies, proteins. Sugars and fats, however, are burned at different rates. Although my body stores enough fat to run about thirteen hundred miles, fat takes many more steps, hence much more time, to break down and burn than sugar. At rest, about 70 percent of a body's energy comes from slowly burning fat, but the faster we run, the more sugar we must burn. At maximum aerobic capacity we burn exclusively sugar.

We can now understand why some people can run faster than others *over long distances*. While Bolt can sprint ridiculously faster than I can by literally hitting the ground harder, the farther we go, the more of an advantage I would probably have over him if we ever raced.

That's because everyone's aerobic system kicks in when they start exercising, but the maximum level of energy obtained this way varies highly from person to person. This important limit, illustrated in figure 12, is termed maximal oxygen uptake, or VO_2 max. Having your VO_2 max measured can be a little scary. Typically, you are fitted with a mask connected to a machine that measures oxygen consumption (as described in chapter 2) while you run on a treadmill. As the treadmill goes faster and faster, you use more and more oxygen until your ability to use any additional oxygen plateaus and you start to gasp. At this limit, your VO_2 max, you need glycolysis to supply additional fuel to your muscles. Speeds above this range cannot be sustained, because muscles become acidic. Fortunately, your VO_2 max has little effect on speed during short bursts of maximum intensity, such as a thirty-second sprint, but the longer the distance, the more it matters. For a hundred-meter dash, only 10 percent of your energy comes from aerobic respiration, but that percentage increases to 30 percent over four hundred meters, 60 percent for eight hundred meters, and 80 percent for a mile.[23] The farther you go, the more your maximum speed benefits from a high VO_2 max (which, as we will see, you can increase by training).[24]

At long last, we can now appreciate why Bolt and I, not to mention that hyena, inevitably trade off energy to go either fast or far.

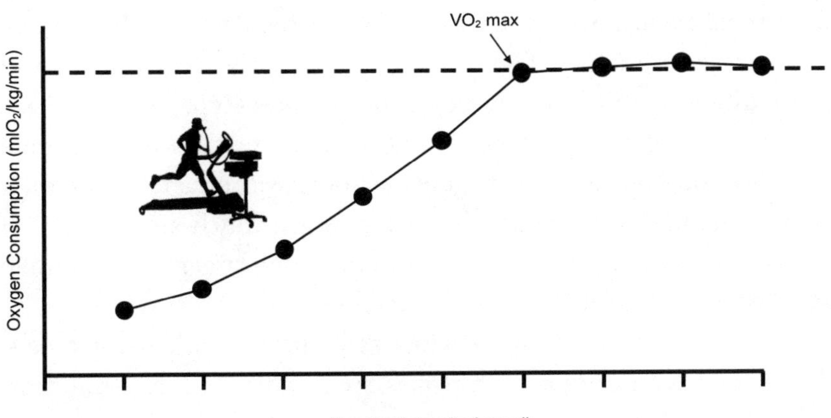

FIGURE 12 How VO_2 max is measured. As one speeds up, one eventually reaches a maximum level of oxygen uptake, which is one's VO_2 max.

Although Bolt leaves me in the dust when sprinting, we both quickly burn through our scanty ATP and creatine phosphate stores while we ramp up our glycolysis rate to maximum levels. Because we both sprint above our VO_2 max, we both have to stop after about thirty seconds and pant to recharge our molecular batteries and clear the acid from our muscles.[25] If, heaven forbid, the hyena keeps chasing us before we recover, we would need to rely more heavily on our aerobic systems and thus run more slowly. The longer the chase, the better I might do relative to Bolt because I probably have more endurance.

Fortunately for Bolt, that wouldn't happen, because if he turned around after thirty seconds to catch his breath, he would see the hyena eating me for breakfast. And therein lies an important reminder: even though the fastest humans have little chance of outrunning hyenas, to survive you sometimes need only be least slow.

Do You Want the Genes for Red or White Meat?

Even if you are scrawny like me, your muscles make up a little over a third of your weight and consume roughly one-fifth of your daily calories. They are probably worth every calorie because they hold you up, keep you warm, and allow you to move, but I suspect you rarely give much thought to how they work or what they look like. Unless you are a surgeon, butcher, or anatomist, most of the muscles you see probably arrive cooked on a plate. Interestingly, while the flesh of fish, chicken, and beef taste unalike, if you were to compare the basic structure of these different animals' muscles under a microscope, you'd be hard pressed to detect differences. That's because muscles evolved more than 600 million years ago to generate force by contracting, and their basic structure and function haven't changed much since then.[26] If so, why do Bolt's muscles—apart from his being larger—allow him to run faster while mine help me run farther?

To address this question, let's look at a muscle in microscopic view. As you can see in figure 13, muscles are bundles of long, thin cells, called fibers. Each fiber, in turn, is made up of thousands of strands, fibrils, that in turn contain thousands of banded structures called sar-

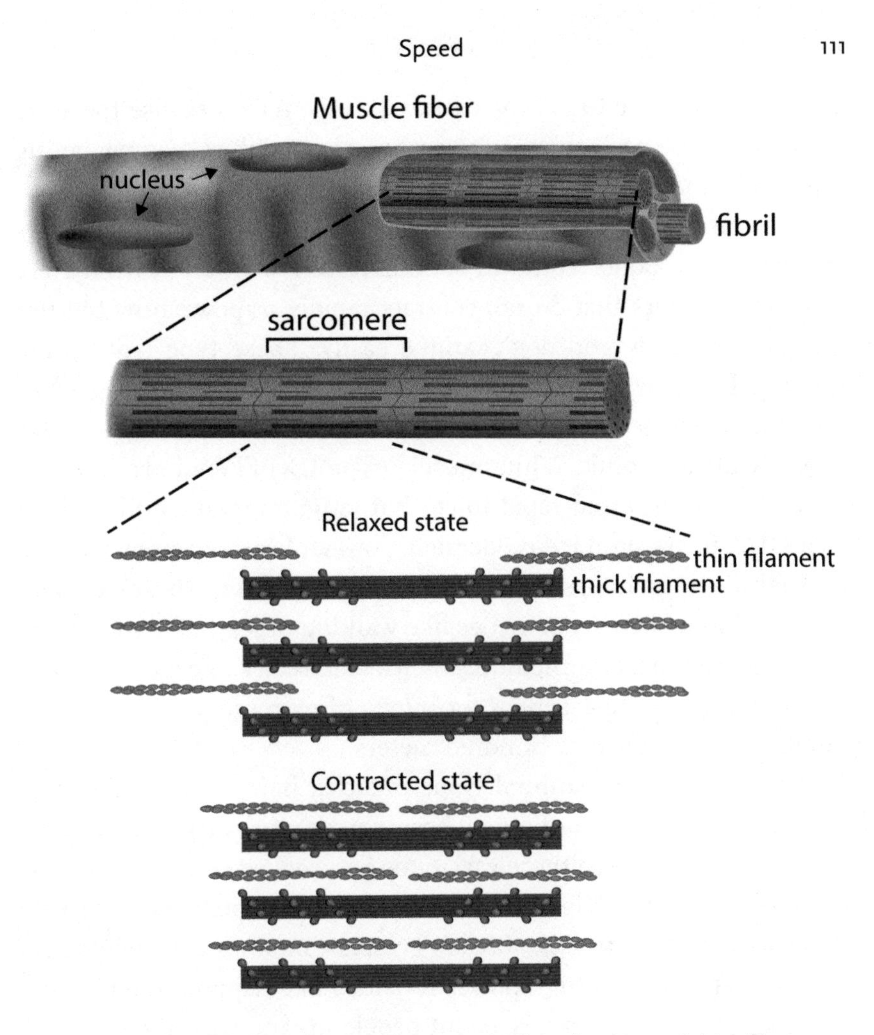

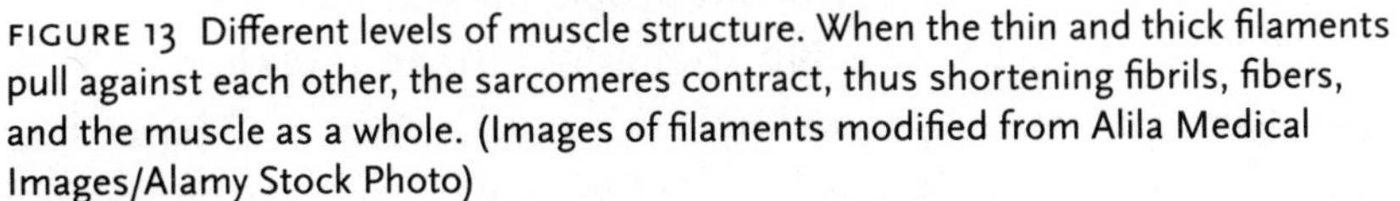
FIGURE 13 Different levels of muscle structure. When the thin and thick filaments pull against each other, the sarcomeres contract, thus shortening fibrils, fibers, and the muscle as a whole. (Images of filaments modified from Alila Medical Images/Alamy Stock Photo)

comeres (Greek for “flesh component”). Sarcomeres generate pulling forces because they are made of two key proteins—one thin, the other thick—that try to slide past each other like interlacing the fingers of your two hands. This contractile action occurs whenever a nerve sends an electrical signal to the muscle, causing tiny projections on the thick filaments to pull against the thin filaments much like a tug-of-war team pulls on a rope. The ratcheting action of each projection exerts a minuscule tug at the cost of a single ATP. Because there are billions of these projections in every muscle, and they keep ratcheting repeatedly, the many little tugs swiftly add up.[27]

All muscle cells work similarly, but the fibers of skeletal muscles that move our bones come in several varieties. At one extreme are slow-twitch fibers that do not contract rapidly or powerfully but use energy aerobically and don't fatigue easily. These type I fibers are colloquially known as red muscle because of their darker tinge.[28] At the other extreme are fast-twitch (type II) fibers, which come in two types: white and pink. White muscle (type IIX) fibers burn sugar to generate powerful and rapid forces but fatigue rapidly. Pink muscle (type IIA) fibers produce moderately powerful forces aerobically and thus fatigue at an intermediate rate. Altogether, red fibers are ideal for sustained low-intensity activities like walking or jogging a marathon, pink fibers are best for medium-intensity activities like racing a mile, and white fibers are essential for bursts of extreme power but short duration like sprinting a hundred meters.

Like those of any animal, your muscles have a mixture of red, pink, and white muscle fibers whose percentages vary from muscle to muscle. You can see these variations in a cooked chicken. Whereas the bird's legs and thighs have more red slow-twitch fibers to help them strut around all day, chicken breasts contain mostly white fast-twitch fibers for brief high-power activities like flapping their wings. It is thankfully unnecessary to cut people up and cook their flesh to see how their muscles vary, but studies of human muscle fiber variation do require people to act as living pincushions and have their muscles biopsied. This procedure is like having an injection except the hollow needle is thicker, and instead of injecting something, the needle sucks up a tiny chunk of muscle. It hurts a little. Yet having your muscles biopsied would reveal extraordinary variation within your body. Many of your muscles have a roughly fifty-fifty mixture of slow- and fast-twitch fibers, but the muscles you use mostly for generating power like your triceps are about 70 percent fast-twitch fibers, and those you use primarily for walking or other non-forceful activities like the deep muscles of your calf (the soleus) are roughly 85 percent slow-twitch fibers.[29]

Percentages of muscle fiber types also vary from person to person, which brings us back to why Bolt can run so much faster than I can. In 1976, a pioneering but somewhat painful study that biopsied the outer calf muscles (gastrocnemius) of forty people found that ordinary nonathletes tend to have equal percentages of fast- and slow-twitch fibers, elite sprinters have about 73 percent fast-twitch fibers, and professional distance runners average 70 percent slow-twitch fibers.[30] Thousands of additional biopsies from an assortment of muscles have since confirmed these results: most of us have slightly more slow- than fast-twitch fibers, but athletes who excel at speed and power sports like Usain Bolt are dominated by fast-twitch fibers, and those who specialize in endurance sports such as the legendary marathoner Frank Shorter tend to have a preponderance of slow-twitch fibers.[31] In addition to being more fast-twitch dominated, sprinters have larger muscles than distance runners.[32]

Similar variations are also evident in the leg muscles of species specialized for speed or endurance. Whereas speedsters like greyhounds and cheetahs have highly muscular legs with mostly fast-twitch fibers, animals evolved for endurance like fox terriers and skunks have less powerfully built legs dominated by slow-twitch fibers.[33] So, if like Usain Bolt you have a preponderance of fast-twitch fibers, you have the potential to be fast without much endurance, and if like Frank Shorter you have mostly slow-twitch fibers, then you, too, can be a great marathoner but have no chance to win the hundred-meter dash. And if you are like most of us, you'll be so-so at both. Perhaps I can blame my mediocre sprinting ability on the genes I inherited from my parents that made me a tortoise instead of a hare.

Nature Versus Nurture

Or can I? As with many simplistic notions regarding nature versus nurture, a closer look reveals the need for caution before leaping to conclusions. Every aspect of our bodies is the product of innumerable interactions among the roughly twenty-five thousand genes we inherited from our parents and the environments in which we have lived, starting with the womb. Few traits have a simple genetic basis, and disentangling the effects of genes, environmental factors, and their interactions affecting speed versus endurance is no exception.

Whenever biologists want to study the heritability of a trait like running speed, the best data come from twins. The most common kind of twin study is to compare a trait such as the hundred-meter sprint time of identical twins who share 100 percent of their genes with that of fraternal twins who share only 50 percent. If the identical twins have hundred-meter times more similar than the fraternal twins', then genes must strongly influence speed; if not, then environmental factors likely predominate. The difference between the two groups of twins provides a numerical estimate of heritability. Despite being plagued by error (how do you put a single, accurate number on athletic ability?), numerous such studies have found that genes explain about half of people's athletic talent.[34] That said, take this percentage with a giant grain of salt because heritability estimates of athletic performance vary widely from study to study. Heritability estimates of speed range from 30 to 90 percent, and those of aerobic capacity range from 40 to 70 percent.[35] These two- to threefold differences in heritability estimates are a valuable reminder that individual studies typically fail to capture the messy complexity of the real world. There is no doubt that all of us inherit anatomical, physiological, and behavioral characteristics that help us excel at specific athletic skills, but the development of these skills is influenced at least as strongly but variably by the environments in which we develop and live. Great athletes like Bolt are both born and made.[36]

Another way to assess contributions to speed versus endurance from nature versus nurture is to look for the specific genes that explain these variations. Bad news here, too, for Team Nature. Without exception, genetic studies (and there have been many) have failed to identify a single gene with a big effect. So far, the best candidate gene associated with athletic talent goes by the insipid name of ACTN3. This gene codes for a protein that helps muscles remain stiff under high forces. Crucially, it has two different versions: a normal R, and a mutant X, which functions poorly, thus causing the muscle to be more elastic. A highly publicized 2003 study of Australian athletes found that the X version of ACTN3 was common among nonathletes and endurance athletes but was almost nonexistent among elite sprinters, weight lifters, and other athletes whose sports require lots of force and power.[37] This discovery caused some parents to pay geneticists to test their children to determine what kinds of sports they should encourage. If their kid had two copies of the X gene, they were told to dissuade them from sprinting and push long-distance running or swimming. However, as researchers collected more data, excitement over ACTN3 fizzled. One study of Greek sprinters showed the gene explained at best 2.3 percent of the variation in forty-meter sprinting times,[38] and other studies have found the gene has no predictive value at all among Africans and other non-Europeans.[39]

To make matters worse, despite explaining very little, ACTN3 is the most potent of the more than two hundred genes so far associated with athletic performance.[40] That doesn't mean genes aren't important. Within any population you can find people whose leg muscles are predominantly fast-twitch or slow-twitch, and genes seem to account for about 40 percent of this variation.[41] In addition, there is limited evidence that people whose ancestry is from West Africa may have slightly higher percentages (about 8 percent) of fast-twitch fibers in some muscles than people of European descent.[42] However, we have yet to find single genes that account for major differences in running performance within or between populations. We must therefore conclude that the contrasting abilities of elite sprinters, marathoners, and average folks are not caused by just a few influential genes. Instead, athletic capabilities such as sprinting speed resemble other complex traits such as height. Height, for example, is highly heritable but influenced by more than four hundred genes, each with small effects that add up.[43] I am slightly short because of the combined effects of several hundred genes that I inherited from both sides of my family that summed up to five feet nine inches. My height, moreover, might also have been influenced by what I ate, how much I slept, and the stresses and illnesses I experienced during my childhood. Why would even more complex traits like speed or endurance be any different?

Last but not least, the evidence that many hundreds of genes of small effect only partly influence athletic talent also challenges common beliefs about the trade-off between speed and endurance. Among the tens of thousands of genes we inherit, some help us run slightly faster, and others help us run farther. There cannot be any simple genetic basis for making people either tortoises or hares. Most of us are a little of both. So why does it seem so obvious that some people are destined to run far and others fast? And even if you prefer jogging to sprinting, should you sometimes pick up the pace?

In Praise of High Intensity

At the age of thirty-four, the journalist Asher Price decided to train for a year to see if he could jump high enough to dunk a basketball. Although he was moderately tall (six feet two), the deck was otherwise stacked against him: in addition to his advancing age, Asher was not particularly athletic, slightly overweight, and recovering from testicular cancer. When he called me beforehand to ask if I thought his dream was unattainable, I am ashamed to admit I threw water on his aspirations, telling him, "Look, if you're not able to dunk into your mid-thirties, you're probably not going to be able to dunk now."[44] You'll have to read Price's charming account of his effort, *Year of the Dunk,* to find out who was right, but my reflexive prediction reflects the consensus view that outstanding athletic abilities arise from a mix of natural talent and training, and that feats like dunking that require producing high, rapid forces are at odds with feats that demand endurance.

Yet consensus isn't truth. Although the notion that most of us are condemned to being either mediocre tortoises or hares contains a kernel of veracity, I think our perceptions about speed versus endurance have been warped by paying too much attention to elite, professional athletes. To be sure, world-class sprinters have no chance when competing against top marathoners and vice versa, but these extraordinary athletes, whose abilities lie at the extreme ends of human performance, have little relevance to the rest of us pedestrians. Consider that the fastest marathoners run 26.2 miles in roughly two hours, running mile after mile at about a 4:40 pace. Can you run that fast for even a mile? If you can, I'm impressed because very few people can run a single mile at that speed. Compared with 99 percent of humanity, these marathoners evince no trade-off between speed and endurance. Instead, they are evidence that you can run *both* fast and far.

Other kinds of athletes also illustrate how endurance and speed can coexist. Professional soccer players run an average of eleven to twelve kilometers per match, combining about twenty-two minutes of fast and explosive sprinting with sixty-eight minutes of slower running and walking.[45] Are they tortoises or hares? Obviously, they are both. This is not to say that some trade-offs between endurance and speed don't exist, but they are largely masked by individual-level variation in overall athletic talent. For example, among elite decathletes, who must excel at both endurance and power, those who do better at events like the hundred-meter dash and shot put that require speed or explosive bursts of force also do better at endurance events like the fifteen-hundred-meter run.[46] And just as the best human athletes tend to be best at everything, individual frogs, snakes, lizards, and salamanders that produce the most rapid bursts of power also have the most endurance.[47]

In the grand scheme of things, our evolutionary history doomed humans as a whole to being slow compared with most quadrupeds, but doesn't it make sense that when we compete against other humans, most of us need to be both tortoises and hares? Also our hunter-gatherer ancestors engaged in many diverse activities like walking, carrying, digging, fighting, preparing food, and possibly even a little swimming every now and then.[48] And apart from sometimes running long distances, they occasionally sprinted like the dickens to escape lions or each other. Today's hunter-gatherers excel at endurance, but measurements of their top speed indicate they also run reasonably (but not blisteringly) fast—about twelve to seventeen miles per hour.[49] At the other end of the spectrum, I know football players who spent years training to be fast and strong but decided later in life to run marathons. We can train our bodies to do an astonishing range of things, and while some workouts must favor our inner hare or tortoise, why can't we do both?

The answer is we can. And, what's more, it's highly effective. For those of us who think we are better suited for endurance than extreme speed, abundant evidence shows that occasional, regular bouts of high-intensity exercise make us not only stronger and faster but also fitter and healthier. This form of training, known as high-intensity interval training (HIIT), involves alternating short bouts of intense anaerobic exercise such as sprinting with less intense periods of recovery. To be clear, HIIT isn't weight training; it is basically intense cardio. So let's finish our exploration of speed by examining how HIIT can help make us sprint faster without compromising our endurance.

We'll begin with plyometric drills, also known as jumping-training exercises. A typical plyometric exercise might be a sequence of ten or so exaggerated skips in which you jump as high and fast as possible on one leg at a time raising the opposite knee along with both arms. With every landing, your hips, knees, and ankles flex, thus stretching your leg muscles and making it extremely challenging for them to contract explosively.[50] These jumps rapidly fatigue your fast-twitch fibers. Next, do an equal number of butt kicks. Then try to repeatedly sprint a hundred or two hundred meters as fast as you can, thus demanding from your muscles rapid, forceful contractions and depleting their ATP and phosphate stores. Such HIIT workouts are hard and can make you sore for days.

But they work. If you keep up a regimen of two sessions a week of HIIT, your muscles will gradually improve their ability to produce high, rapid forces in part by augmenting how many fibers contract simultaneously when stimulated by nerves. In addition, your muscles will change composition. Although HIIT cannot stimulate your body to produce more fast-twitch muscle fibers, the ones you have will thicken, making you stronger and hence faster.[51] On average, sprinters' muscles are more than 20 percent thicker than distance runners'.[52] HIIT can also modify slower, more fatigue-resistant pink fibers into faster, more fatigable white fibers; lengthen fibers slightly, thus boosting their shortening speed; and increase the percentage of fibers in a muscle that contracts, thereby increasing force.[53] But these and other changes don't happen on their own, and require constant effort to maintain. If you *want* to run faster, you have to *try* to run faster.

The benefits of regular HIIT go well beyond its effect on muscles. Among other payoffs, HIIT increases the heart's ability to pump blood efficiently by making its chambers larger and more elastic. HIIT also augments the number, size, and elasticity of arteries and increases the number of tiny capillaries that infuse muscles. HIIT further improves muscles' ability to transport glucose from the bloodstream and increases the number of mitochondria within each muscle, thus supplying more energy.[54] These and other adaptations lower blood pressure and help prevent heart disease, diabetes, and more. The more we study the effects of HIIT, the more it appears that HIIT should be part of any fitness regimen, regardless of whether you are an Olympian or an average person struggling to get fit.

These benefits highlight a final lesson to draw from our exploration of speed. Although elite athletes like Usain Bolt teach us about the limits of human performance, we mustn't forget that average, everyday humans are capable of remarkable physical feats that we should celebrate and that played far more important roles in the evolutionary history of our species. None of our Stone Age ancestors had the opportunity to spend years training to sprint precisely a hundred meters in a straight line as fast as possible in front of a crowd of spectators. Instead, they evolved to be jacks-of-all-trades, good at a wide range of athletic challenges worthy of both tortoises and hares. Hopefully, they were not chased too often by hyenas, let alone lions or saber-toothed tigers, but those moments must have been consequential, otherwise you and I might not be here. . . .

SIX

Strength: From Brawny to Scrawny

MYTH #6 We Evolved to Be Extremely Strong

Nobody picks on a strong man.

—Charles Atlas

Soon after my grandparents came to America, they settled in Brooklyn, New York. Among my childhood memories of visiting them is being intrigued by the bodies on Brooklyn's famous Coney Island Beach on hot summer days. You can't hide your physique in a bathing suit. Everyone in my family was short and unmuscular, but there on New York's most popular beach every imaginable body type was on nearly full display: short to tall, underweight to obese, smooth to hairy, repulsive to attractive, scrawny to brawny. I remember thinking that when I grew up, I wanted to have one of those impressive, attractive muscled bodies.

Little did I know that, according to legend, it was on that very same beach that America's attitude toward muscular physiques took a dramatic turn. The story begins in 1903 when Angelo Siciliano, aged ten, stepped off the boat on Ellis Island. Angelo was just another poor Italian newcomer to New York who spoke no English. Abandoned by his father, Angelo and his mother settled with his uncle in Brooklyn, where they struggled to achieve the American dream. As a child,

Angelo was apparently sickly, weak, and the victim of regular beatings from an abusive uncle and gangs of bullies. And then, in his own words: "One day I went to Coney Island and I had a very pretty girl with me. We were sitting on the sand. A big, husky lifeguard, maybe there were two of them, kicked sand in my face. I couldn't do anything and the girl felt funny. I told her that someday, if I meet this guy, I will lick him."[1]

A few days later, still feeling the sting of humiliation, Angelo had an epiphany during a school trip to the Brooklyn Museum. Impressed by the bulging muscles on statues of Greek gods, he realized he could restore his pride and achieve manhood by bulking up. Angelo claims he spent several months sweating away to little effect in his bedroom using weights, ropes, and elastic grips, but then had a second epiphany watching lions stretch at the Bronx Zoo. Wondering how lions could develop such strength without using weights, Angelo figured they must get strong by "pitting one muscle against another." He started to experiment with what he called "dynamic tension," which today we call isometric training. It worked. When he showed off his new physique a few months later on storied Coney Island Beach, a friend reportedly said, "You look like the statue of Atlas on top of the Atlas Hotel!" Soon thereafter, Angelo Siciliano changed his named to Charles Atlas.[2]

Charles Atlas was hardly the first bodybuilder to make money off his bulging muscles, but he became the most successful muscleman of his era and helped launch the modern physical culture movement. After making a few bucks as a strongman in Coney Island (people paid to walk on his stomach), Atlas became a model, won a contest as the "World's Most Perfectly Developed Man," and started a mail-order course that promised to help every scrawny kid and flabby man in America achieve a glorious macho frame. In comic strips, pamphlets, and other advertisements, Atlas endlessly retold his story, capitalizing on archetypal insecurities: losing your girl, not being manly, fear of weakness and decrepitude. To be sure, the desire to be virile was hardly new, but Atlas's promises were especially potent to legions of men whose pride had been dashed by the Great Depression and whose insecurities were being stoked by the Industrial Revolution as machines replaced human labor. Bulking up was a way to restore millions of wounded egos, and Atlas became the new high priest of the manly physique.

Ever since Charles Atlas, countless youngsters have grown up enticed by advertisements that appeal to deep-seated desires to be Herculean. Atlas helped inspire generations of fitness gurus and celebrity musclemen including Jack LaLanne and Arnold Schwarzenegger. Gyms sprang up throughout America and elsewhere, first with dumbbells, barbells, and other weights and then with newfangled contraptions. A big leap forward was the Nautilus machine, which uses weights on pulleys to place an adjustable but constant level of resistance on muscles throughout their entire range of motion. As Nautiluses and other devices made resistance training more effective and efficient, physical culture evolved from a subculture to a mainstream multibillion-dollar industry.[3]

It was inevitable that, amid the many efforts to extol and market the benefits of resistance exercise, weight lifting and notions about human evolution would collide. A major spark was the 1988 bestselling book *The Paleolithic Prescription,* which argued that most "diseases of civilization" were caused by our bodies not being adapted to modern lifestyles.[4] The book mostly focused on food, spawning the Paleo Diet, but its way of thinking soon extended to exercise, kindling the primal fitness movement.[5] Just as paleo dieters believe (illogically) it is healthiest to eat like cavemen, primal fitness enthusiasts believe it is best to work out like our muscled ancestors of yore.[6]

To learn more about the primal fitness movement, I once exercised with one of its superstars, Erwan Le Corre. The occasion was the New York City Barefoot Run. On a sparkling weekend day in September, I joined several hundred paleo-primal enthusiasts on the ferry to picturesque Governors Island in New York City's harbor, not far from the Statue of Liberty. Our main event was to run barefoot around and around the island's 2.1-mile path, but the participants were just as intent to drink beer, eat barbecue, and socialize. Erwan—a tall, athletic Frenchman who runs a fitness camp in Santa Fe, New Mexico—immediately became the focus of attention thanks to his movie-star looks, impressive physicality, and exuberant approach to exercise. He turned the path around Governors Island into a barefoot playground. Whenever the fancy took him, he leaped off the trail to climb trees, jump over benches, and chase squirrels. In the same vein, attendees of Erwan's camp are encouraged to do everything the human body was apparently supposed to do including running, walking, leaping, crawling, climbing, swimming, lifting, carrying, throwing, catching, and fighting. Picture brawny, shirtless men carrying enormous logs and swinging through trees.[7]

I've since tried to learn more about the primal fitness movement from books, websites, conferences, and talking to enthusiasts. As best as I can tell, the majority of these modern-day cavemen and cave-women believe our ancestors had extremely muscular, lean bodies thanks to a lifetime of "natural movement." By this, they mean moderate levels of endurance exercise interspersed with tasks that require enormous strength such as lifting boulders or fighting lions. Weight training is thus a bedrock of the primal fitness movement. One of its major advocates, Mark Sisson, prescribes an entire lifestyle exemplified by an imaginary but supposedly typical caveman named Grok. According to Sisson, "Grok didn't engage in a chronic pattern of sustained moderate-to-difficult-intensity efforts like today's devoted fitness enthusiasts tend to do." Instead, "Grok's life demanded frequent bursts of intense physical effort—returning gathered items (firewood, shelter supplies, tool materials, and animal carcasses) to camp, climbing rocks and trees to scout and forage, and arranging boulders and logs to build shelter."[8]

The most successful intersection between primal fitness and physical culture is CrossFit. Started in 2000 by Greg Glassman in Santa Cruz, California, CrossFit has grown into a worldwide cultlike movement. The first time I walked into a CrossFit gym, I was underwhelmed. The gym was an old garage; instead of gleaming machines, TVs, and mirrors, I saw only weights, climbing ropes, and stationary bikes. But a CrossFit workout quickly dispels any notion this is an unserious endeavor. The strategy is to alternate intense cardio with equally intense resistance exercises such as lifting kettlebells, doing handstand push-ups, and climbing ropes. By following each day's unique prescribed "WOD" (Workout of the Day) in teams modeled after a platoon of marines, CrossFitters encourage each other to keep up a relentless, total-effort exercise session. At the end, everyone is physically spent but ecstatic. Apart from the impressive fitness effects of such routines, most CrossFitters believe they are participating in an ancient tradition of total-body athleticism based on the considerable strength that was supposedly necessary for human survival. As a devoted CrossFitter friend of mine told me, "Being strong is primal."

As we have repeatedly discussed, our ancestors rarely if ever lifted weights or did other kinds of exercise for the sake of health and fitness, but does this kind of workout even remotely resemble the sorts of physical activities they used to do? Were they really that strong, or would the intense, exhausting WODs of CrossFitters be as alien to most hunter-gatherers as paying taxes or reading a book?

Strength Through the Ages

In 1967 and 1968, two doctors, Stewart Truswell and John Hansen, journeyed deep into the Kalahari Desert of Botswana to document the health of the San hunter-gatherers living there. Their detailed, careful analysis dispassionately assessed people's weight, height, nutritional condition, cholesterol levels, blood pressure, and more, but one line stands out: "Bizarre results of old injuries were noted such as one man . . . [who] had survived an unarmed fight with a leopard. He is left with a facial paralysis, weakness of the extensor tendon of his bowstring finger and chronic osteitis of the humerus. But he *had* killed the leopard with his bare hands."[9]

Wow! Taking out a leopard with your bare hands seems worthy of Hercules, but how many more hunter-gatherers over the last few million years were less fortunate and thus unable to recount their encounters with fearsome predators? Regardless, these stories fuel stereotypes. Primal fitness websites and books, not to mention a few scientific papers, sometimes include descriptions of hunter-gatherers and other nonindustrial peoples as resembling CrossFit devotees thanks to their natural cross training.[10] To quote one book that recommends we "go wild" to free our bodies from the afflictions of civilization: "Imagine for a second a group of Maasai men—the storied herders of Kenya—making their way across the Serengeti, an effortless trot of lithe, formed bodies, perfect conditioning, and a beauty of economy and motion that is the envy of every dedicated gym rat? Where are their personal trainers?"[11]

Having met many Maasai, not to mention a few hunter-gatherers, I am sorry to report that these characterizations are overblown, and I cannot help but notice they almost always focus on young men. Careful measurements of strength in hunter-gatherer men and women are limited but indicate that hunter-gatherers, young men included, are lean and modestly strong but not brawny. In general, tropical hunter-gatherers also tend to be more slight than strapping. The average Hadza man, for example, is five feet four inches tall and weighs 117 pounds; Hadza women average four feet eleven inches and 103 pounds. Body fat percentages are about 10 percent in men and 20 percent in women, just at the margin of being classified as underweight.[12] Measurements of Hadza grip strength as well as estimates of overall upper-body strength and muscle size put them squarely within Western norms for their age and below those of most highly trained athletes.[13] Although not musclemen, the Hadza are lean and fit and appear to have the sort of overall strength one expects of people whose livelihood depends on a diverse range of regular physical activities from digging to running to climbing trees.

The Hadza are just one population, but they appear to be about the same in terms of size, muscle mass, and strength as other hunter-gatherers such as the San of the Kalahari, the Mbuti of central Africa, the Batek of Malaysia, and the Aché of Paraguay. The Aché, for example, live in the Amazon and are about as small and wiry as the Hadza, with nearly identical average grip strength values.[14] When asked to do as many push-ups, pull-ups, and chin-ups as they could, the Aché were fit but similar to Westerners with one important difference: as they aged, their strength declined at lower rates, presumably reflecting how they stay physically active throughout the life span including middle age.[15] Men reach peak strength in their twenties; women achieve peak strength slightly later but barely lose muscle as they get older. As a result, there is little difference in strength between older Aché men and women.

Musclemen and -women today and CrossFitters, in particular, bulk up with weights and machines, but how would hunter-gatherers or Maasai herders get that strong without gyms? According to legend, the ancient Greek wrestler Milo of Croton developed his physique by lifting a calf over his head every day until it grew into a full-sized cow. The story seems about as likely as the legend of his death (Milo was supposedly torn apart by wolves after he unintentionally trapped himself in a tree that he was trying to rip apart with his bare hands). As we will see, weight lifters have perfected the art and science of bulking up with just the right number of repetitions using dumbbells or weight machines that are more effective, faster, and easier to use than hoisting wriggling cows or ripping apart trees. Although hunter-gatherers lift heavy things from time to time, the resistance they apply to their muscles comes mostly from carrying things, from digging, or from lifting their own body weight. Equipment-free, body-weight-based actions such as push-ups, pull-ups, squats, and lunges help develop strength, but they have one major drawback: as you gain strength, the weight you lift remains unchanged. Without gyms, hunter-gatherers simply cannot do what is necessary to get superstrong, let alone achieve a body worthy of Charles Atlas or Milo.

And for good reason. If I were a hunter-gatherer, beyond lacking gym facilities, I'd want to be reasonably but not extremely strong. One potential drawback of bulking up too much is sacrificing power. Strength is how much force I can produce; power is how rapidly I produce it. Strength and power are not independent, but there is some trade-off between the two: a strong woman may be able to lift a cow above her head, but not rapidly. In contrast, a powerful woman cannot lift as substantial a load, but she can hoist less hefty things more swiftly and repeatedly. Some tasks like jumping high or scampering quickly up a tree depend more on power than strength. The ability to lift above your head something twice or more your body weight is a bizarre, dangerous feat that probably had little practical value in the Stone Age. Even today, average sedentary human beings benefit more from power than strength. Many activities of daily living such as lifting a bag of groceries and rising from a chair require rapid bursts of force. As we will see later, maintaining these power capabilities is especially vital as we age.[16]

Another drawback of being superstrong that mattered in the Stone Age is its caloric cost. Bodybuilders who can lift a cow must also eat as much as a cow. Well, almost. Recall that muscle is an expensive tissue, accounting for about one-third of a typical person's body mass and one-fifth of her or his energy budget. I need about three hundred calories a day to sustain my unmuscular frame. Beefed-up weight lifters, however, can be more than 40 percent muscle mass, which means they carry as much as twenty added kilograms of costly flesh.[17] If I ever decide to bulk up like that, I'll have to eat two hundred to three hundred more calories a day to pay for my new physique. While obtaining an extra three hundred calories is a trivial task today (accomplished by wolfing down a milk shake), the challenge of foraging daily for those additional calories in the Stone Age would have compromised one's reproductive success.[18]

All in all, musclemen and gym rats have physiques you would not have encountered often or perhaps ever among our hunter-gatherer ancestors who lacked the wherewithal to get that ripped, wouldn't have needed such strength, and would have been taxed by the caloric costs. But what about our more distant forebears? How do modern human hunter-gatherers compare with apes and extinct cavemen who so often are depicted as being powerful and strong?

Brawny Apes and Cavemen?

In 1855, the Academy of Natural Sciences in Philadelphia commissioned Paul du Chaillu, a young French American adventurer, to explore West Africa.[19] Du Chaillu had traveled to the region as a child with his father, who owned a trading depot off the coast of Gabon, and he was intensely curious to return. So, at the age of twenty, he journeyed up the Ogooué River of Gabon deep into the jungle, an area of Africa then unknown to Europeans. There he spent the next three years documenting the region's different cultures and peoples and slaughtering thousands of birds and mammals—all to be sent back to Philadelphia and stuffed. Du Chaillu also claimed to be the first European to observe gorillas in the wild. He describes these apes in his first book, *Explorations and Adventures in Equatorial Africa* (1861), as terrifying "half man, half beast" and the "King of the African forest." Figure 14 shows a typical illustration. Du Chaillu's narrative was so popular he returned a second time and then penned a series of adventure yarns, including *Stories of the Gorilla Country*.[20]

This cruel and racist book is disturbing to read. In one heartbreaking chapter, du Chaillu recounts how he captured his first live gorilla, "Fighting Joe." The poor creature, only three years old, was bagged and collared after du Chaillu's men shot its mother. Unsurprisingly, Fighting Joe did not take well to captivity and being orphaned, reacted violently, and repeatedly tried to escape what can only be described as brutal, inhumane conditions in a makeshift jail. After a few weeks, including one nearly successful dash for freedom, poor Fighting Joe died.

One of the many children who read du Chaillu's book was Merian Cooper, who claims it inspired his 1933 movie, *King Kong*. This truly monstrous gorilla epitomizes popular culture's outsized view of ape strength. In English, to go ape still means to express wild anger. Even though du Chaillu's accounts of gorillas were ridiculed by scientists of the time (Darwin described him as a "malignant old fool"), there is little question his writings and drawings along with subsequent portrayals by later explorers influenced how early scientists viewed our ape ancestry as brutish.[21] Thanks to the pioneering work of Jane Goodall, Dian Fossey, and others, we now appreciate that apes in the wild, though far from angels, are mostly peaceable.[22] However, the assumption that our ape ancestors from the forests of Africa were extremely muscular remains deeply entrenched. Supposedly, a 120-pound male chimpanzee is strong enough to rip your arm from its socket.

The first scientific test of chimpanzee strength was in the 1920s by John Bauman, an American biology teacher who also coached a college football team. Bauman noted that "while all authorities on the anthropoid apes judged them to be greatly the superior of the human being in strength, no exact tests of their strength were cited, so that it seemed . . . that even a few definite strength tests would be of interest and value." Accordingly, he rigged a crude apparatus to measure the strength of chimpanzees and orangutans in zoos. Bauman was initially unable to entice any ape to pull more than halfheartedly on his contraption until he tried a female chimp named Suzette whose "tricky and malicious disposition" inspired her to pull "viciously" to destroy the device. After a few trials, Suzette apparently managed to use her entire body to pull 1,260 pounds, about three to four times more than any burly football player could manage.[23]

Bauman's amateurish estimate of Suzette's strength using an untested, uncalibrated, and probably inaccurate instrument is still cited regularly despite repeated failed efforts to replicate her feat. In 1943, a Yale primatologist named Glen Finch carefully replicated Bauman's experiments on eight adult chimpanzees, none of whom could muster more strength than adult male humans.[24] A generation later, U.S. Air Force scientists devised a bizarre contraption—resembling a cross between a metal cage and an electric chair—to measure how much force chimps and humans could generate when flexing their elbows. The only adult chimpanzee they managed to train to use this device was about 30 percent stronger than the strongest human also measured.[25] More recently, Belgian researchers showed that seventy-five-pound bonobos can jump twice as high as humans who weigh twice as much, indicating that both species jump the same height per pound.[26] Finally, and perhaps most definitively, a laboratory analysis of muscle fibers demonstrated that a chimp's muscles can produce at most 30 percent more force and power than a typical human's.[27] Although these studies differ in terms of methods, they collectively reveal that adult chimps are no more than a third stronger than humans. Contrary to the old meme, a chimp couldn't rip your arm from its socket in an arm-wrestling contest. That said, you'd still probably lose.

According to popular belief, you should also think twice about wrestling a burly caveman. The apotheosis of this stereotype is the quintessential troglodyte, the Neanderthal. Ever since the nineteenth century, when the first Neanderthal fossils were excavated from European caves, these Ice Age cousins have ignited imaginations. It has always been clear from their anatomy that Neanderthals differed from people today: they had enormous browridges, sloping foreheads, chinless jaws, and generally robust skeletons. When the first of these fossils were discovered, some experts mistook them for idiots, criminals, or bowlegged Cossacks who had somehow stumbled into caves, died, and got buried.[28] More sober scholars recognized them as an extinct species of human, but couldn't restrain their prejudice. When the Irish geologist William King formally defined the species *Homo neanderthalensis* in 1864, just five years after Darwin published *The Origin of Species* and three years after du Chaillu's accounts, he left little doubt about his distaste for these savage antecedents "whose thoughts and desires . . . never soared above those of a brute."[29]

The Neanderthals' unfortunate caveman stereotype was amplified unforgettably in the early twentieth century by Marcellin Boule, an influential French paleontologist who described in detail the first nearly complete skeleton from the cave of La Chapelle aux Saints. Sadly, Boule's reconstruction of the "Old Man of La Chapelle" was disastrously off the mark. He mischaracterized Neanderthals as savage, stupid, amoral, and burly with a stooped posture. Boule's most lasting contribution to the caveman cliché can be seen in widely viewed reconstructions from the time such as the one in figure 14.

FIGURE 14 *Left*, Du Chaillu's depiction of "my first gorilla" from his 1861 book, *Explorations and Adventures in Equatorial Africa. Right*, illustration by the Czech artist F. Kupka based on Boule's reconstruction of the Neanderthal from La Chapelle aux Saints. The illustration was published in the *Illustrated London News* in 1909.

These and many subsequent depictions of Neanderthals have menacing, hunched, apelike poses, and I suspect it is hardly coincidental they are as muscular, hairy, and bent-kneed as du Chaillu's gorillas.

Fortunately, Neanderthal stereotypes have undergone a much-needed rectification. Scholars now recognize Neanderthals as intelligent, highly skilled close cousins whose brains were as big as ours and with whom we share more than 99 percent of the same genes. But were they stronger than us? One line of evidence comes from estimates of their body size. If you had access to my bones, you could reconstruct my height and approximate my weight from measurements of certain skeletal dimensions. The same methods suggest that Neanderthal males averaged 5 feet 5 inches (166 centimeters) and 172 pounds (78 kilograms) while females were 5 feet 2 inches (157 centimeters) and 145 pounds (66 kilograms). Neanderthals were thus shorter but heavier than most humans today. If Neanderthals had similar percentages of body fat as Inuit hunter-gatherers from the Arctic, then they must have been extremely muscular. According to the anthropologist Steven Churchill, Neanderthal males and females averaged a total of thirty-two and twenty-seven kilograms of muscle, respectively, suggesting their muscles were 10 to 15 percent larger, hence stronger.[30]

Another reason to suspect that Neanderthals and other so-called archaic humans from the Ice Age were more muscular than people today is their robust bones. In general, the more we load our bones, especially when we are young, the thicker they become.[31] However, the most intriguing evidence comes from their skulls. True to stereotype, male Neanderthals have massive faces with bigger, more menacing browridges and thicker cranial vaults than females. These robust features might be a consequence of more testosterone.[32] Testosterone famously stimulates sex drive and aggression, but it also enhances secondary sexual features including masculinized features of the upper face such as browridges.[33] Higher testosterone levels may be responsible for the bigger upper faces and larger browridges of male chimpanzees than those of their gentler cousins, bonobos, and perhaps the same was also true of male Neanderthals and other archaic humans.[34] This hypothesis is relevant to our discussion of strength because testosterone also helps build muscle, which is why some athletes use it illegally. In one (legal) experiment, researchers gave high doses of testosterone for ten weeks to twenty normal men, half of whom also lifted weights, and compared them with controls who received a placebo and didn't work out.[35] Compared with undoped controls, the men given just testosterone added about six pounds of muscle and got 10 percent stronger; those doped with testosterone who also lifted weights added thirteen pounds of muscle and got about 30 percent stronger.

From these various lines of evidence, it seems that Neanderthals and other archaic humans from the Ice Age, like chimpanzees, were moderately more muscled than the average Joe, including contemporary hunter-gatherers. How, then, do some people like Charles Atlas achieve great strength, worthy of, if not more impressive than, a Neanderthal?

Crush the Resistance!

On November 28, 2015, Eric Heffelmire was in his garage in Vienna, Virginia, underneath his jacked-up truck, repairing a corroded brake line. All of a sudden the jack slipped, pinning Mr. Heffelmire and spilling gasoline, which instantly ignited. Fortunately, his daughter Charlotte—five feet six and weighing just 120 pounds—saw what happened and rushed to the scene. "I lifted it [the truck] the first time, he said 'OK, you almost got it,'" Charlotte told reporters. "Finally managed to get it out, it was some crazy strength, pulled him out." And as if that weren't enough, Charlotte then climbed into the flaming truck and drove it on three wheels out of the burning garage, rescued her sister's baby, and then called the fire department.[36]

Charlotte's heroic deed is an example of hysterical strength, the ability of everyday people to muster superhuman feats of muscle in life-or-death situations. In such emergencies, the body releases massive amounts of adrenaline and cortisol, allowing the heroine to maximally contract every muscle fiber in her body. Some scientists are skeptical of these inevitably anecdotal reports because such feats cannot be replicated in the laboratory and they involve more strength than the body can theoretically produce. Regardless, most of us are stronger than we think and never achieve our full potential because the nervous system sensibly inhibits us from going all out, thus tearing muscles, breaking bones, and possibly killing ourselves.[37] One especially lethal concern when we lift heavy weights is that we need to unceasingly push blood through clenched muscles to our brains. Since any interruption in blood supply might cause us to faint, perhaps fatally, high-resistance exercise requires the heart to generate high pressures that have to be withstood, especially by the heart itself and by the aorta. For this reason, as blood pressure shoots up, we instinctively inflate our chests and briefly hold our breath. This vital reflex, known as the Valsalva maneuver, lessens stress on the heart, and it also helps rigidify the trunk and stabilize the spine.[38]

Even if I am capable of more brawn than I think, I am not very muscular. I cannot count how many times I have resolved to work out in the gym on a more regular basis to get stronger. My first foray into the weight room was in high school, but I quickly retreated. Friends in college and graduate school occasionally enticed me to lift weights, but I never tried working out regularly until I hit my late thirties, looked in the mirror, and realized I was entering middle age a weakling. So I joined a health club a few blocks from my house, hired a trainer, and set to work.

I hated it. After assessing my lack of strength, my trainer prescribed me a conventional routine that involved several sets of repetitions on a dozen machines along with some free weights, sit-ups, push-ups, lunges, and tormenting exercises involving large rubber balls. No pain, no gain meant being constantly sore. Getting strong also interfered with my running. Even when my legs didn't hurt, they felt dead. The gym was also a cheerless basement that reeked of stale sweat and had no natural light. No one seemed to be having any fun as they progressed from one machine to the next under fluorescent lights doing their repetitions with an air of grim determination. Despite getting stronger, I quit the gym after six months.

Since then, I have attempted several times to reboot my efforts to gain strength. I've hired other trainers and tried different gyms, but I just can't seem to make it stick. Instead, I have developed a gym-free routine doing push-ups, squats, and a few other exercises in the privacy of my own home. I am by no means strong, but is this enough? If Charles Atlas managed to bulk up in his bedroom, why can't I?

The basic principle behind resistance exercise is to make your muscles generate force against an opposing force such as your body's own mass or an external load like a dumbbell, a stack of weights, or a cow. In essence, you use something heavy to *resist* your muscles' efforts to contract. Not all physical activities involve much resistance. Swimming involves minimal resistance because water is fluid; walking and running are also low-resistance activities because the ground pushes back on just a few of your leg muscles for only a portion of each stride cycle. Gyms are therefore effective for gaining strength because they are filled with weights and fiendish machines designed to keep your muscles constantly battling resistance over a broad range of motions.[39] Yet not all resistance exercises are the same or have the same effect.

Imagine you are holding a heavy weight in your hand to do biceps curls. If you are curling the weight upward by flexing your elbow, your biceps muscle is generating force while shortening, technically known as a *concentric* muscle action. Concentric contractions are the primary means by which muscles move us. Muscles, however, don't always shorten. If you hold the weight steady without moving it up or down, your biceps will still try to shorten but won't actually change its length, an *isometric* muscle action. Isometric muscle actions can be challenging, but it is even harder to lower the weight very slowly by extending your elbow. This sort of *eccentric* muscle action requires your biceps to fire as it lengthens. Will your biceps get stronger if you focus on concentric, isometric, or eccentric muscle actions?

Concentric contractions are critical for movement, but as Charles Atlas supposedly intuited in the Brooklyn Zoo, they are generally less potent for building muscle than eccentric and isometric muscle actions.[40] Athletes, trainers, bodybuilders, and others interested in getting stronger therefore tend to incorporate plenty of eccentric and isometric kinds of resistance training. Most infamously, they tend to follow the maxim "No pain, no gain." To explore why, imagine you are in the gym right now doing three sets of ten to twelve biceps curls using a challenging weight. At first you will feel fatigued, but hours later you'll become sore. This happens because you are making your biceps generate more force than it can easily handle against a resisting force (the weight), literally tearing it apart at the microscopic level. Filaments snap, membranes rip, connective tissues split.[41] This so-called microdamage triggers short-term inflammation, accounting for the swelling and soreness. More important, by intentionally shredding the muscle a little, you stimulate growth because the microdamage stimulates affected muscle cells to turn on a cascade of genes. Among other things, these genes augment the total number and thickness of muscle fibers, thus expanding the muscle's diameter, making the muscle stronger.[42]

Although "no pain, no gain" is a bedrock principle and mantra for serious weight lifters, you may be relieved to know you can strengthen your muscles without getting horridly sore and walking around like a mummy. You do have to repeatedly stress your muscles beyond their customary capacity, but shredding them isn't always necessary to turn on the genes that promote growth.[43] If you primarily want to gain strength, you'll get the most bang for your buck by slowly doing a few demanding repetitions of weights that require eccentric or isometric contractions.[44] That said, if you are more interested in power and endurance, you'll derive more benefit from multiple sets of fifteen to twenty rapid concentric repetitions on less demanding weights with only brief rests between sets.[45] Lifting weights a few times a week, moreover, is especially helpful to stay healthy and vigorous as we age.

Aging Muscles

Wouldn't it be nice to sleep like a bear through the cold, miserable winter of my native New England and wake up with the first buds of spring? Not being a bear, however, I'd wake up seriously weakened. When bears hibernate, they preserve their muscle mass despite months of starvation and physical inactivity.[46] In contrast, humans confined to bed for far shorter periods lose muscle at an alarming rate.[47] After three weeks of bed rest, leg muscles can shrink up to

10 percent.[48] Even worse, astronauts in the gravity-free environment of space can lose 20 percent of their muscle mass in just a week or two.[49]

Fortunately, the process of aging is not as ruinous for muscles as bed rest or spaceflight, but muscular atrophy—the gruesome technical term is "sarcopenia," Greek for "loss of flesh"—is a major cause of disability and disease among the elderly. As we age, muscle fibers typically dwindle in size and number, and nerves degenerate.[50] The result is a loss of strength and power. On average, grip strength in industrialized countries like the United States and the U.K. declines about 25 percent from the age of twenty-five to seventy-five.[51] In the town of Framingham, Massachusetts, a few miles from my house, the percentage of women unable to lift just ten pounds has been shown to increase from 40 percent among fifty-five- to sixty-four-year-olds to 65 percent among seventy-five- to eighty-four-year-olds.[52] This trend is worrying. As people lose strength, they become less able to perform basic tasks such as rising from a chair, climbing the stairs, and walking normally. Increasing feebleness in turn makes people even less active, leading to a vicious cycle of deterioration.

Sarcopenia is a silent epidemic of aging that needs more attention, especially because debilitating declines in muscle function and capacity are largely preventable. Studies of aging demonstrate that hunter-gatherers, like postindustrial Westerners, lose strength as the years pass, but at a considerably slower rate.[53] As figure 15 illustrates, the average seventy-year-old Aché woman in the Amazonian rain forest has a grip strength typical of a fifty-year-old woman in England.

Elderly hunter-gatherers and others who remain active throughout their life span testify to the welcome news that using muscles retards muscle loss as we age. Indeed, aging does not put an end to muscles' capacity to respond to resistance exercise; instead, modest levels of resistance exercise slow and sometimes reverse sarcopenia regardless of age thanks to the mechanisms we have already reviewed. Dozens of randomized control trials have found that prescribing moderate, non-strenuous levels of weight training helps older individuals increase their muscle mass and strength and thus improve their ability to function normally and stay active without requiring assistance.[54] One

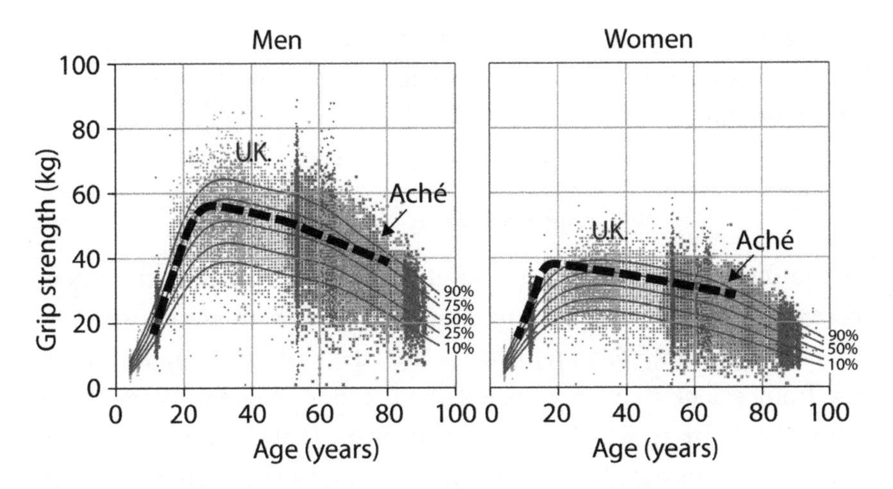

FIGURE 15 Comparison of male and female grip strength at different ages in the U.K. and among Aché foragers. (Modified with permission from Dodds, R. M., et al. [2014], Grip strength across the life course: Normative data from twelve British studies, *PLOS ONE* 9:e113637; and Walker, R., and Hill, K. [2003], Modeling growth and senescence in physical performance among the Aché of eastern Paraguay, *American Journal of Human Biology* 15:196–208)

study even demonstrated marked improvements in strength among eighty-seven- to ninety-six-year-old men and women following eight weeks of resistance training.[55] Critically, by halting and reversing sarcopenia, these interventions decrease injury risk and enhance quality of life for the elderly.

Sarcopenia is concerning for its own sake but, frighteningly, is also associated with other diseases we will consider toward the end of this book. Most obviously, as muscle mass declines, people load their bones less, contributing to osteoporosis. This furtive disease occurs when bones become too frail to sustain the loads they incur, causing them to snap or collapse. Because weakened muscles lead to less physical activity, sarcopenia is also a risk factor for other conditions associated with inactivity, including heart disease and type 2 diabetes. Happily, numerous studies confirm that non-extreme levels of resistance exercise confer significant metabolic and cardiovascular benefits, including improving muscles' ability to use sugar and lowering levels of harmful cholesterol.[56] Done properly and not in excess, strength training also helps prevent injuries.[57] Finally and importantly, warding off sarcopenia in old age helps prevent depression and other mental health conditions.

Weighing How Much Weights to Do

If you dislike spending time and effort in the gym, you are in good company: neither do most superheroes. Spider-Man got his strength and other powers from a radioactive bite, the Hulk and Captain America from scientists who mutated their genes, Wonder Woman and Thor from parents who happened to be gods. Only Batman works out, but he is a fantastically wealthy philanthropist whose parents' murder inspired a lifelong compulsion to rid the world of crime.[58] As a mere mortal who needs to work for a living, avoids gyms, and prefers cardio over weights, I'd like to know how much resistance training is enough to accomplish my goals.

A good place to get carefully reviewed, consensus advice is the American College of Sports Medicine. Its most recent expert panel's review of the evidence suggests I supplement my weekly quota of aerobic exercise with twice-weekly bouts of strength training that involve eight to ten different resistance exercises with ten to twelve repetitions each.[59] Once I hit sixty-five years old, they recommend I increase my weight training to ten to fifteen repetitions.

I wonder how these recommendations would strike our ancestors hundreds of generations ago. Beyond marveling at a machine-filled world that doesn't require much strength to survive, they would probably be confused by how we spend money to needlessly lift things whose sole purpose is to be lifted. While some of us get aerobic physical activity in our "normal" environments by walking to work instead of driving or taking the stairs instead of the elevator, few jobs today require much resistance physical activity. Shopping carts, baby carriages, wheels on suitcases, forklifts, and other devices emancipate us from having to lift or carry anything anymore. Thus to get resistance exercise, we do bizarre things like repeatedly lift weights in the gym. Fortunately, the biological response to such actions appears to be the same as having to carry children and food, dig holes, lift rocks, and do whatever resistance activities people did in the Stone Age.

While modern weight lifting might amuse our hunter-gatherer ancestors, they'd probably be relieved to know that, like them, we don't need to be more than moderately muscular to ward off sarcopenia and other associated diseases like osteoporosis. It bears repeating that for most of human evolutionary history too much muscle was more of a cost than a benefit. Were I a hunter-gatherer struggling to get enough food, the benefits of too much strength would probably be overshadowed by its added cost, making less energy available for other needs. Both then and today, I'd mostly want to be just strong enough for my normal activities of daily living.

That said, we have yet to consider two brawny activities as old as the hills that sometimes demand strength as well as speed and power and which can have major effects on reproductive success: fighting and sports.

SEVEN

Fighting and Sports: From Fangs to Football

MYTH #7 Sports = Exercise

Probably the battle of Waterloo was won on the playing-fields of Eton, but the opening battles of all subsequent wars have been lost there.

—George Orwell, *The Lion and the Unicorn* (1941)

About fifty baboons surround us near a grove of acacia trees on a typically hot and dusty day in Tarangire National Park, Tanzania. My eyes are drawn to two scrawny infants playing rough-and-tumble. The frisky little monkeys grab each other's tail and roll around in the dirt like wrestlers as they nip and paw each other. Seemingly oblivious to the high-spirited youngsters, a nearby adult female is grooming a male twice her size. She concentrates intently as her nimble fingers search deftly in the dense fur along his back for ticks. When she grasps one, she pops it in her mouth. The big male seems blissfully calm. Elsewhere, other baboons are feeding, nursing, or just hanging out. Then another slightly smaller male approaches the grooming couple. As the infants scamper to safety, the big male rises, barks, and exposes his daggerlike canines. In a trice, the two males meld into a whirling, snarling ball of gnashing fangs, fur, and tails. Everyone—baboon and human—stops what they are doing to watch the two males fight viciously for the next ten seconds. Then, just as rapidly, the combat ends when the smaller male runs away screaming. Judging from the way he licks his forelimb, he got a bite on his upper arm. Eventually, playing, grooming, and other tranquil activities resume.

If you ever watch a troop of baboons, you'll see plenty of scenes like that. Baboons live in big groups with scores of males and females. Both sexes have dominance hierarchies that begin during infancy through play and then transform into recurring acts of aggression. Violence affects both males and females, but males are usually the aggressors. Young adult males must fight to become top-ranking males. In turn, dominant males spend much of their time vigorously defending their status and preventing other males from mating. Tempers are on edge, stress levels are high, fights are frequent. Attitude and strategy matter, but victory is largely influenced by speed, strength, size, and agility. Baboons, moreover, are typical among primates. If you spend a week with a troop of chimpanzees, you'll observe numerous fights, some disturbingly brutal. Male chimpanzees frequently attack other males as well as females to gain dominance and control mating opportunities. Occasionally they kill.[1]

We humans are nicer. Visit a park in any town to observe a group of fellow humans, and you'll see children playing, but it is highly improbable any adults are fighting. Instead, the adults are peacefully monitoring the children, hanging out, or participating in sports like soccer and basketball. Grown-up humans play more than adults of other species, and we fight far less often than other primates like baboons and chimpanzees. Even the most belligerent human groups ever studied engage in violence about 250 to 600 times less frequently than chimpanzees.[2] Are adult humans so nonaggressive because, as we have seen, we evolved to be slow and weak? And have we exchanged fighting for playing, especially in the context of sports?

According to consensus, the answer is yes. We traded brawn for brains. Instead of relying on speed, power, and strength, humans evolved to cooperate, use tools, and solve problems creatively.

I think this widely held view is only partly true. Although the last two chapters highlighted how humans are comparatively slow and feeble, speed and strength have hardly ceased to matter. Instead, these brawny characteristics, albeit diminished, played vital roles in the saga of human activity in large part because of the special way humans compete physically with each other and our prey. Yes, we are less violent on a day-to-day basis and less able to use raw power and strength than chimpanzees and other primates, but humans haven't stopped fighting altogether. Instead, we have changed the way we fight and how often. Some of us also hunt. It follows that regardless of whether you fight or hunt infrequently with fangs or fisticuffs, speed and strength can still have pivotal evolutionary consequences, especially if you get life-threatening injuries. Because dead people can't have babies or help existing children survive, any heritable advantages or disadvantages that affect fighting and hunting ability should have strong effects on selection.[3]

Speed and strength also remain important components of exercise, including sports. All mammals play when they are young to develop athletic skills useful for fighting, yet in every culture humans young and old engage in sports and other forms of play. It is not coincidental, moreover, that most games and sports emphasize qualities like speed and strength that blur the line between play, fighting, and sometimes hunting. If you think about it, the athletes we most admire and reward tend to be those who outcompete others according to the brawny precepts of the Olympic motto, *Citius, Altius, Fortius* (faster, higher, stronger).

Therefore, let's conclude our evolutionary and anthropological exploration of speed, strength, and power by exploring how fighting and sports (and to a lesser extent hunting) influenced the evolution of human physicality. Please note some topics in this chapter necessarily focus more on males than females for the simple reason that males fight more than females (largely due to testosterone), but as we will see, females also play a role in human fighting and sometimes hunting, and they certainly engage in plenty of sports. Regardless of sex, however, our first step is to consider the behavior that often underlies the reason we fight in the first place: aggression.

Are Humans Naturally Aggressive?

Had I grown up in a war-torn country or a violent neighborhood, perhaps I would be less inclined to opine that humans fight less than chimpanzees or baboons. Indeed, working on this chapter made me intensely aware of how ignorant and inexperienced I am about violence. I have never tried martial arts or aggressive sports like wrestling. Nor am I one of those Hemingwayesque types who enjoys watching boxing and bullfighting. For that matter, I've witnessed only a handful of fights in my entire life, none serious. Curious, I decided to seek out a fight. Because I was too chicken to get myself into a brawl at a seedy bar, I attended a cage fight in a town outside Boston armed only with my postdoctoral student Ian.

Arriving at the fight club filled me with doubt. Having watched some Hollywood fight movies, I expected the venue to be an abandoned factory in a disreputable part of town, but Ian and I found ourselves walking to the back of a bowling alley in a dilapidated shopping mall. My skepticism evaporated, however, as we pushed through a dense throng of hundreds of inebriated, bellowing young men plus a handful of women who surrounded an octagonal cage made out of chain-link fence. Heavy metal music was blaring at the highest possible volume, and I could almost smell the testosterone. Over the next few hours Ian and I saw half a dozen mixed martial arts (MMA) fights. Compared with chimpanzee and baboon fights, these matches seemed to be in slow motion. Every contest started with the opponents circling each other warily, mostly boxing and sometimes kicking, but inevitably the fighters ended up crashing to the floor and wrestling intensely as they tried to protect their heads, get on top, and pummel their opponent with punches, elbow jabs, kicks, and body slams. One fighter won by strangling the other guy with his legs, leaving the loser gasping desperately on the cage floor.

To some, martial arts are an intense kind of physical, masculine poetry. To quote Joyce Carol Oates: "I have no difficulty justifying boxing as a sport because I have never thought of it as a sport. There is nothing fundamentally playful about it, nothing that seems to belong to daylight, to pleasure. At its moments of greatest intensity it seems to contain so complete and powerful an image of life—life's beauty, vulnerability, despair, incalculable and often self-destructive courage—that boxing is life and hardly a mere game."[4] I myself see little beauty in boxing, wrestling, MMA, and other violent sports, but I do appreciate the challenge of fighting well. Watching cage fighters drove home the extraordinary strength necessary to survive, especially as those guys writhed on the cage's floor, straining almost every muscle to avoid serious injury and to hurt their opponent as much as possible. I was amazed no one's neck was snapped—a thought amplified by the posters on the wall of MMA fighters who had died, one of them the previous year following a fight in that very same cage. In fact, skill and attitude appear to be more important determinants of who wins than strength. As the combatants struggle, they must fight with their minds as much as their bodies to overcome pain and fatigue and figure out how to win.

These qualities aside, there is no whitewashing the violence and raw aggression. Nearly one in four MMA fights results in an athlete getting hurt, making it the most injurious sport for which there are data.[5] The last event of the evening was between two fighters we will call Slippery Steve and Bareknuckle Bob. To say these lightweights fought like hell is an understatement. As they kicked, punched, and wrestled fiercely on the floor, neither missed a chance to wound the other. When standing, they stomped on each other's feet. When rolling on the ground, they used their knees, feet, elbows, and fists to land whatever blows they could as hard as possible. The fight ended when Bareknuckle Bob appeared to break Slippery Steve's arm. As Steve stormed in pain out of the cage, holding his limp arm, he roared, "I ain't fucking done yet!" The crowd roared back almost as aggressively.

Cage fights and other violent sports remind us that humans are capable of enjoying and engaging in extraordinary aggression. But does that mean we are as aggressive as chimpanzees or baboons? After all, Bareknuckle Bob and Slippery Steve are professional entertainers paid to fight. Though motivated to beat the crap out of each other, they were subject to official regulations (no biting or blows to the genitals). Bareknuckle Bob and Slippery Steve are not all that different from boxers or wrestlers whose sports are legitimate enough to be part of the Olympics. For that matter, how different are MMA fighters from American football players who are also paid to risk injury as they battle each other on the field with padding and helmets?

Pugilistic sports and other forms of human aggression raise an age-old debate about human nature. Deep down are we naturally peaceful and cooperative creatures who become aggressive when corrupted by civilization? Or are we naturally aggressive and civilized by culture? Who is more aberrant: wimpy me or aggressive Slippery Steve?

To be honest about my biases, I was raised to be aware of humanity's tendencies and capacities for violence but to believe that humans evolved to be primarily moral, peaceful, and cooperative. I am glad to be a generally nonviolent human and not an ape: if I were a chimpanzee, I'd spend an appreciable part of my day trying to avoid being beaten up or killed. Only a human being would risk death by running into a burning building to save the life of an unrelated stranger or a pet. Even rough sports like cage fighting have rules and umpires to protect the participants from too much harm. In this regard, I am drawn to the philosophy of Jean-Jacques Rousseau and his followers who believe that our natural tendency is to behave morally and that many acts of human violence can be traced to corrupting cultural attitudes and conditions.[6]

Humans may be highly cooperative, but we do sometimes fight each other. Males especially. Furthermore, humans alone in the animal world have invented arrows, darts, guns, bombs, drones, and other weapons that make us frighteningly lethal. Even a feeble, unskilled human can maim or kill thousands with a trigger or button. Violence is woven into every culture, including hunter-gatherer societies, calling into question assumptions that we are naturally benign and unaggressive.[7] I thus also give credit to Thomas Hobbes and his followers who see human tendencies toward aggression as ancient, intrinsic, and sometimes adaptive.[8] As detailed comprehensively by Steven Pinker, our species has become exponentially less violent only very recently thanks to social and cultural constraints, many fostered by the Enlightenment.[9]

How, then, do we reconcile our extraordinary capacities for cooperation and conflict avoidance (Rousseau) with our capacities for aggression (Hobbes)?

A persuasive resolution to this age-old debate was proposed by Richard Wrangham, who points out that we wrongly conflate two profoundly different kinds of aggression: proactive and reactive.[10] According to Wrangham, humans differ from other animals, especially our ape cousins, in having exceedingly low levels of reactive aggression but much higher levels of proactive aggression. We correspond to Rousseau in terms of reactive aggression and to Hobbes in terms of proactive aggression.

To illustrate this difference, imagine I just now rudely snatched this book from your hands. You might shout indignantly and try to grab it back, but it is unlikely you will attack me. Your brain would immediately inhibit any major act of reactive aggression. If you were a chimpanzee, however, you'd probably respond to my theft with instantaneous, uninhibited violence. Unless I were the dominant male in the troop, without pausing to think, you'd give me a thumping and retrieve your book. One widely reported case of this sort of reactive aggression that is only too common among chimpanzees involved an adult chimp named Travis who had spent his entire life peacefully as part of Sandra and Jerome Herold's family. Then, in February 2009, at the age of fifteen, he flew off the handle after one of Sandra's friends, Charla, picked up one of his favorite toys. Travis's immediate and savage attack left Charla with no hands and without much of her face including her nose, eyes, and lips.[11]

Road rage is one example of how humans sometimes aggress reactively like Travis, but such incidents are rare and shocking because as children we rapidly learn to suppress these reactive instincts. Yet nonreactive adult humans can excel at purposeful, planned forms of hostility. This kind of proactive aggression is characterized by predetermined goals, premeditated plans of action, attention to the target, and lack of emotional arousal. Chimpanzees sometimes engage in proactive aggression, but humans have taken planned, intentional forms of fighting to new heights such as ambushing, kidnapping, premeditated homicide, and, of course, war. Arguably, hunting and combative sports like boxing are also forms of proactive aggression. And, importantly, hunting and other forms of planned aggression are utterly different psychologically from reactive aggression. Violent criminals, ruthless dictators, torturers, and other proactive aggressors can simultaneously be loving spouses and parents, reliable friends, and patriotic fellow citizens who remain utterly calm and pleasant in situations that would send a chimpanzee or a toddler into a rage. They also don't need to be as physically powerful.

How do we evolve from strong, dangerous apelike animals with high reactive and low proactive aggression to wimpy, cooperative, playful humans with low reactive and high proactive aggression? One long-standing argument, still debated, is that this transition occurred early in our evolutionary history, just after we diverged from the apes and became upright.

Stand Up and Fight?

It is not hard to find evidence for homicide in the Stone Age. As examples, one Neanderthal from a site in Iraq died sixty thousand years ago from being lanced by a spear whose point was left embedded in his spine, and Ötzi the "Iceman," whose body lay frozen for five thousand years in an Alpine glacier, was shot in the back with an arrow.[12] The site of Nataruk, Kenya, is especially shocking. Today, Nataruk is a hot, dusty scrubland, but ten thousand years ago it was a lagoon where an entire band of hunter-gatherers—twenty-seven men, women, and children—were killed. By studying these bones, Marta Lahr and her team reconstructed a grisly scene.[13] Some of the skeletons have broken hands, suggesting they were bound, and all of them bear traces of traumatic deaths: fragmented cheekbones, bashed-in skullcaps, fractured knees and ribs, puncture wounds from projectiles. These and other clues, including the fact that some of the victims were infants and pregnant women, suggest that these hunter-gatherers were massacred through proactive aggression and then dumped without burial.[14]

Sites like Nataruk incite controversy because many anthropologists believe that intergroup violence of that scale postdates the origins of farming. When I first learned about hunter-gatherers, I was taught they are generally peaceful because they are egalitarian, own no property to fight over, and are highly mobile. When intragroup conflicts arise, hunter-gatherers can just move. Elevated levels of interpersonal violence and large-scale aggression were attributed to the corrupting effects of contact with farmers and Westerners.[15] However, evidence for violence among preagricultural societies was always there if you looked.[16] As Richard Wrangham has argued, instead of asking when humans became less aggressive, we need to ask when humans became less reactively and more proactively aggressive.

One long-standing idea, which traces back to Darwin, is that the human lineage long ago became fundamentally less brutish and violent than apes. Unlike Rousseau, Darwin was no romantic, but he had a benevolent view of human nature. In his 1871 masterpiece, *The Descent of Man,* he reasoned (somewhat long-windedly) that reduced aggression was a key driving force early in human evolution:

> In regard to bodily size or strength . . . we cannot say whether man has become larger and stronger, or smaller and weaker, in comparison with his progenitors. We should, however, bear in mind that an animal possessing great size, strength, and ferocity, and which, like the gorilla, could defend itself from all enemies, would probably, though not necessarily, have failed to become social; and this would most effectually have checked the acquirement by man of his higher mental qualities, such as sympathy and the love of his fellow-creatures. . . .
>
> The slight corporeal strength of man, his little speed, his want of natural weapons, &c., are more than counterbalanced, firstly by his intellectual powers, through which he has, whilst still remaining in a barbarous state, formed for himself weapons, tools, &c., and secondly by his social qualities which lead him to give aid to his fellow-men and to receive it in return.[17]

Darwin's view that human cooperation, intelligence, diminished strength, and reduced aggression evolved as an ensemble since our divergence from the apes has been popular ever since he penned those words. But the horrors of several world wars have inspired more

Hobbesian interpretations of human evolution. The strongest proponent of the early-humans-are-killers camp was Raymond Dart. Dart was an Australian who reluctantly moved in 1922 to Johannesburg, South Africa, to teach anatomy. His move turned out to be fortuitous thanks to the nearly complete skull of a juvenile *Australopithecus*, nicknamed the Taung Baby, that landed in his lap two years later. Within a year, Dart gained worldwide fame when he argued correctly that the skull indicated humans evolved from small-brained apelike creatures in Africa rather than from large-brained ancestors in Europe. Dart concluded mistakenly, however, that the many other broken bones also present in the limestone pits in which the Taung Baby and other fossils were found were hunted by early hominins. Dart initially echoed Darwin's theory that bipedalism freed the hands of early hominins to make and use hunting tools, which in turn selected for big brains, hence better hunting abilities.

Then, in a famous 1953 paper, clearly influenced by his war experiences, Dart proposed that the first humans were not just hunters but also murderous predators.[18] Dart's words are so astonishing, you have to read them:

> The loathsome cruelty of mankind to man forms one of his inescapable characteristics and differentiative features; and it is explicable only in terms of his carnivorous, and cannibalistic origin. The blood-bespattered, slaughter-gutted archives of human history from the earliest Egyptian and Sumerian records to the most recent atrocities of the Second World War accord with early universal cannibalism, with animal and human sacrificial practices of their substitutes in formalized religions and with the world-wide scalping, head-hunting, body-mutilating and necrophilic practices of mankind in proclaiming this common bloodlust differentiator, this predaceous habit, this mark of Cain that separates man dietetically from his anthropoidal relatives and allies him rather with the deadliest of Carnivora.

Dart's killer-ape hypothesis, as it came to be known, was popularized by the journalist Robert Ardrey in a best-selling book, *Afri-*

can Genesis, that found a ready audience in a generation disillusioned by two world wars, the Cold War, the Korean and Vietnam Wars, political assassinations, and widespread political unrest.[19] The killer-ape hypothesis left an indelible stamp on popular culture including movies like *Planet of the Apes, 2001: A Space Odyssey,* and *A Clockwork Orange.*

But the Rousseauians weren't dead yet. Reanalyses of bones in the limestone pits from which fossils like the Taung Baby came showed they were killed by leopards, not early humans.[20] Further studies revealed these early hominins were mostly vegetarians. And as a reaction to decades of bellicosity, many scientists in the 1970s embraced evidence for humans' nicer side, especially gathering, food sharing, and women's roles. The most widely discussed and audacious hypothesis, proposed by Owen Lovejoy, was that the first hominins were selected to become bipeds to be more cooperative and less aggressive.[21] According to Lovejoy, early hominin females favored males who were better at walking upright and thus better able to carry food with which to provision them. To entice these tottering males to keep coming back with food, females encouraged exclusive long-term monogamous relationships by concealing their menstrual cycles and having permanently large breasts (female chimps advertise when they ovulate with eye-catching swellings, and their breasts shrink when they are not nursing). Put crudely, females selected for cooperative males by exchanging sex for food. If so, then selection against reactive aggression and frequent fighting is as old as the hominin lineage.[22]

Anthropologists, many from the Rousseau camp, have spent the last forty years disputing the sex-for-food hypothesis. The biggest problem is that early hominin males appear to be at least 50 percent bigger than females.[23] Lucy, the famous *Australopithecus afarensis* female from 3.2 million years ago, weighed slightly less than thirty kilograms, but males of her species weighed about fifty kilograms. This difference in body size, termed sexual dimorphism, is a reliable indicator of competition between males within species. If I had to fight other guys without weapons to get a girlfriend or wife, I would have a strong advantage if I were as big as possible, and I'd have little hope of passing on my genes if I were tiny. Unsurprisingly, whenever species have high levels of male-male competition, selection drives up body size in just males. Among gorillas and baboons, species in which males fight to control harems of many females, males are twice as big as females, but among pair-bonded gibbons that fight much less, males are just 10 percent bigger than females.[24] Chimpanzees are intermediate, with males being about 30 percent bigger.[25]

Because our australopith ancestors—especially males—probably fought each other as much as if not more than chimpanzees do, human aggression must have declined at some point in the last two million years within the human genus, *Homo*. The question is when.

Better Angels of Our Genus

The beginnings of *Homo* are murky, but by two million years ago *Homo erectus* had evolved. Compared with earlier hominins, this pivotal species had bigger brains, smaller teeth, and nearly humanlike bodies. Also *Homo erectus* males were probably about 20 percent bigger than females.[26] Because reduced dimorphism suggests reduced male-male conflict, maybe our lineage has been kinder and gentler ever since *Homo erectus*. Coincidentally, the archaeological record also indicates that *Homo erectus* were bona fide hunter-gatherers who hunted large animals, gathered many kinds of plants, made sophisticated stone tools, and shared food in camps.

Hunting and gathering matter. Despite evidence that hunter-gatherers are no angels (by some estimates, almost one-third of male deaths in such societies arise from violence[27]), you simply cannot survive as a hunter-gatherer without being highly cooperative—far more than chimpanzees. One form of cooperation is the division of labor between sexes in which females forage more for plants and men do the lion's share of hunting and honey gathering. Although gathered plants usually provide the bulk of dependable calories, meat and honey are high-status, calorie- and nutrient-rich foods necessary to sustain the group's needs, especially those of nursing mothers. Indeed, hunter-gatherer mothers cannot acquire enough calories for themselves and their offspring without some provisioning from males and grandmothers.[28] Hunter-gatherer males also must cooperate more than males in other species. Men often hunt in small groups and frequently come home empty-handed. By sharing meat from successful hunts, hunters ensure there is enough food to go around every day. Hunter-gatherers also collaborate to take care of children and fend off predators. Altogether, decreased size dimorphism, increased cooperation between and among the sexes, and the importance of women's roles in hunter-gatherer societies have led anthropologists to speculate that humans have been less aggressive since the origins of the genus *Homo*.[29]

Homo erectus hunter-gatherers likely cooperated extensively, but that doesn't mean they didn't fight. There are several reasons to think that if we had a time machine to observe them one or two million years ago, we would see more interpersonal violence than today. Aside from evidence for proactive violence among contemporary hunter-gatherers, two thorny facts don't entirely square with the view that we stopped fighting ever since we became hunter-gatherers.

The first fact is muscle. The average adult man today is 12 to 15 percent heavier than the average adult woman, but women have much higher percentages of body fat masking underlying differences in muscle mass. Whole-body scans show that males average 61 percent more muscle mass then females, with most of that difference in the upper body.[30] Men's extra brawn, moreover, is added during puberty, when testosterone levels shoot up, accelerating muscle growth in the arms, shoulders, and neck.[31] In this regard, human men resemble male kangaroos, whose upper bodies also enlarge during adolescence to help them fight.[32] Enhanced upper-body muscularity in male humans might also have been selected for hunting, but we cannot rule out aggression.

The second fact is literally staring us in the face. Consider the faces of assorted males in the genus *Homo* lined up for you in figure 16. Note that until about 100,000 years ago, even in some of the earliest *Homo sapiens,* males tend to have massive, heavily built faces and menacingly large browridges. The earliest *H. sapiens* males have smaller, less robust faces than Neanderthals and other non-modern humans, but truly lightly built, "feminized" faces don't appear until less than 100,000 years ago.[33] It is intriguing to hypothesize that

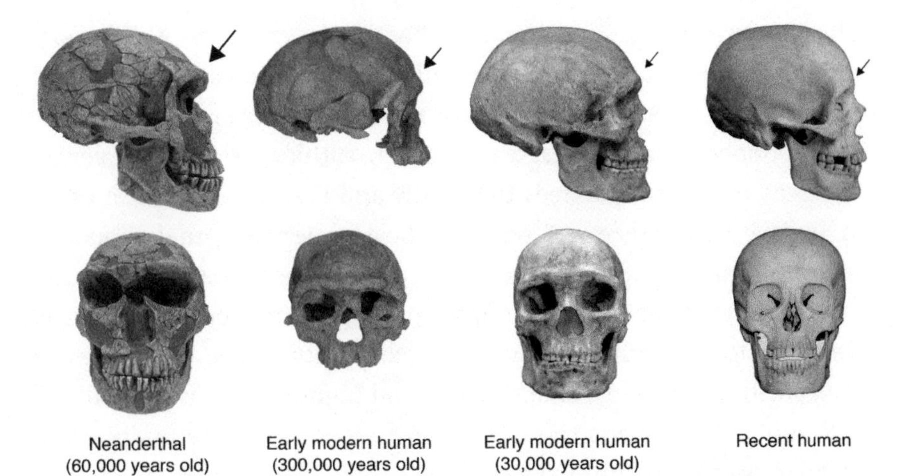

FIGURE 16 Side and front views of skulls from a male Neanderthal and male *H. sapiens* from different time periods. Notice how the face has recently become smaller (more gracile). The arrows point to the browridges, which are less built up in smaller-faced humans. (Photos from Lieberman, D. E. [2011], *The Evolution of the Human Head* [Cambridge, Mass.: Harvard University Press])

these big faces reflect higher levels of testosterone during adolescence. In males today, elevated testosterone contributes to not only higher libidos, more impulsivity, and more reactive aggression but also bigger browridges and larger faces.[34] Another molecule that possibly affects facial masculinization is the neurotransmitter serotonin, which reduces aggression; less masculinized faces are associated with higher levels of serotonin.[35]

Reductions of masculine features associated with aggression have caught the attention of biologists because they mirror many changes seen in other animals, especially domesticated species. I have little fear walking up to a pig on a farm or my neighbor's dog, but I wouldn't dream of approaching a wild boar or a wolf in the same way. Over generations of breeding, farmers have reduced the aggressiveness of these and other animals by selecting for lower levels of testosterone and higher levels of serotonin.[36] Correspondingly, many domesticated species have smaller faces. Intriguingly, some wild species also evolved reduced aggression, less territoriality, and more tolerance on their own through another kind of selection known as self-domestication.

The best example are bonobos. Bonobos are the rarer, less well-known cousins of chimpanzees that live only in remote forests south of the Congo River in Africa. But unlike male chimpanzees and gorillas, male bonobos rarely engage in regular, ruthless, reactive violence. Whereas male chimpanzees frequently and fiercely attack each other to achieve dominance and regularly beat up females, male bonobos seldom fight.[37] Bonobos also engage in much less proactive violence. Experts hypothesize that bonobos self-domesticated because females were able to form alliances that selected for cooperative, unaggressive males with lower levels of androgens and higher levels of serotonin.[38] Tellingly, like humans, bonobos also have smaller browridges and smaller faces than chimpanzees.[39]

Many scientists are testing the idea that humans also self-domesticated.[40] If so, I'd speculate this process involved two stages. The first reduction occurred early in the genus *Homo* through selection for increased cooperation with the origins of hunting and gathering. The second reduction might have occurred within our own species, *Homo sapiens,* as females selected for less reactively aggressive males.

Let's now turn back to the issue of strength and fighting. As you might have noticed, the story I have told about the last two million years has two conflicting threads. On the one hand, our ancestors became hunters and thus must have benefited from plenty of brawn, especially among males; on the other hand, we became less reactively aggressive and more cooperative, which presumably reduced selection for being big and strong. Among the solutions to this contradiction are that humans fight and hunt upright and with weapons.

Fighting Before Weapons

The last time I was in a fistfight I was eleven years old and it didn't go well. Since then, I've had a pleasantly peaceful life, but if, heaven forbid, I must fight again, I'd benefit from a weapon. All human cultures, including hunter-gatherers, rely frequently on weapons. In the three years he spent with San hunter-gatherers in the Kalahari, the anthropologist Richard Lee documented thirty-four incidences of fights without weapons and thirty-seven with weapons.[41] Whereas many of the fights using spears, arrows, or clubs were apparently premeditated and proactive, all the weaponless fights he described were short, sudden, and reactive. Judging from other such accounts, I suspect this pattern is universal. If you are plotting to attack someone, you'd be foolish not to use a weapon, but in an unplanned fight you can use a weapon only if you happen to be carrying one. Reactively aggressive fights are thus more often weaponless and consequently less often lethal.[42]

But once upon a time, all fights—reactive and proactive—were weaponless. Slippery Steve and Bareknuckle Bob demonstrate how trained martial arts fighters can inflict serious damage, but I suspect the best human martial artists would be torn to shreds if they had to confront a chimpanzee. Chimpanzee combat can be lightning fast, and they attack not just with powerful arms and legs but also with large, razor-sharp canines. Chimpanzees sometimes stand up to kick, slap, claw, and punch (open-handed or with fists), but they also maneuver deftly on all fours. All in all, their fights are fast, furious, and full-bodied.

Human fights are different. There is an entire field of research, hoplology (from the Greek word *hoplos* for a plate-armored animal), that studies martial arts, stage combat, and the use of weapons.[43] One can also watch on YouTube hundreds of disturbing videos of street fights recorded by bystanders on cell phones. These and other lines of evidence show that human fighting is distinctive largely because we battle on just two legs. As noted by the biologist David Carrier, one advantage of fighting upright is that it enables animals to use their arms as weapons or shields and to hit downward with maximum force. Although apes, bears, and kangaroos sometimes stand up to fight, most animals, including chimpanzees, prefer to attack and retreat on all four legs, which are faster and more stable than two. Because humans are necessarily slow and unstable on two legs but even less maneuverable on all fours, human combatants are trained to fight bipedally in a crouched posture, almost dancing, with their arms out and in front of their heads. Upright human fighters hit, parry, and grapple, and they sometimes also kick, which can generate a lot of power but carries a greater risk of falling. Once grounded, however, wrestling humans are especially vulnerable because it becomes harder to flee or protect oneself. Whoever is on top literally has the upper hand.

Another distinctive aspect of unarmed human fights is the focus on the head. Headbutts are a Hollywood staple, but real-life human brawlers rarely attack with their heads. For one, human teeth are ineffective weapons. Lacking fangs and a snout, we can at best chomp on an opponent's finger or ear. More crucially, our big, vulnerable brains need to be protected. Whereas chimpanzees and other animals use their heads when fighting to snap and tear with razor-sharp fangs, trained human fighters shield their heads behind their hands and arms. In addition, chimpanzees attack each other everywhere on the body, but humans most commonly aim for the head, hoping for a knockout or a fractured jaw.[44] David Carrier and Michael Morgan controversially proposed that early humans evolved a long thumb and short fingers partly to make a compact fist for punching and that our big jaws and cheekbones are similarly adapted to withstand being punched.[45]

While bipedalism handicaps human fighting by making us ungainly and puts a premium on using our forelimbs to land blows and protect our heads, unarmed human fighters are otherwise like the rest of the animal kingdom in benefiting from a combination of size, strength, skill, and attitude. Obviously, bigger individuals are more likely to win because they are stronger and heavier and have longer arms with bigger fists.[46] However, as in other species, size and strength are not deterministic. One point of unanimity among experts is that fighting is largely a learned skill.[47] Every form of martial arts emphasizes balance, posture, developing protective reflexes, and generating force effectively with proper technique.[48] Additionally, one cannot overestimate a combatant's willingness to take risks and persevere.[49] I would never fight over a sandwich, but defending my family would be a different matter. The degree to which combatants are motivated and formidable is so influential for determining a fight's outcome that humans, like other animals, spend much effort advertising and sizing up these qualities in potential opponents.[50] Like snarling dogs, rival humans often strut, shout, and expand their chests before deciding whether to fight. From an evolutionary perspective, such posturing makes sense. For both winner and loser, it is almost always best to back down if the outcome is predetermined.

The odds and outcomes of victory or defeat were revolutionized, however, when weapons were invented.

Fighting with Weapons

In a memorable scene from *Raiders of the Lost Ark,* Harrison Ford sprints frantically through a crowded market but finds his way blocked by a gargantuan assassin brandishing a fearsome scimitar. As both combatants and the audience gear up for an epic duel, Ford grins sheepishly and then shoots the swordsman.[51] The scene is violent but funny and on target. Ever since spears, arrows, and other projectile weapons were invented, puny Davids have been better able to vanquish massive Goliaths. How did humans become weaponized, and how did weapons affect human athleticism, especially strength?

When Jane Goodall first published evidence in the 1960s that wild chimpanzees made tools, her observations astonished the world. Subsequent research has documented that chimpanzees make a variety of simple tools, including sticks modified into tiny spears to lance small mammals hiding in holes in trees.[52] Chimpanzees also hurl rocks, branches, and other objects when displaying and fighting. But these weapons are not very lethal by human standards, especially in the hands of chimpanzees, whose overhand throwing aim is dreadful.

Like chimpanzees, early hominins must have employed simple wooden tools, but a seismic shift happened sometime between 3.3 and 2.6 million years ago with the invention of stone tools, about the same time as the oldest evidence for meat eating.[53] We know from wear traces on tools and cut marks on ancient bones that these early tools helped early hominins butcher animals.[54] I can vouch for their effectiveness because every year the students and faculty in my department have a goat roast in which we make and use simple stone tools to butcher a goat (and, no, we don't hunt or kill the goat). We also know from microscopic wear on the tools' edges that some were used to cut plants including wood.[55] It is not a great leap of imagination to suppose that by two million years ago early members of the human genus, *Homo,* had crude wooden spears, which after all are just long, sharpened sticks. How useful would these spears have been for fighting or hunting? Did the evolution of spears and other projectiles change the human body?

Having a spear is better than not having one, but they are not easy to use. Throwing spears accurately and forcefully from all but the shortest distances takes skill born from hours of practice. Further, untipped spears do not create as much tissue damage as spears with stone points, an innovation that dates to only 500,000 years ago.[56] And if you have only one spear, missing your target can leave you unarmed and vulnerable. A more controlled way to kill with a spear is thrusting, but this alternative comes with a serious disadvantage: you must get up close to your prey or victim, putting yourself at risk. There is evidence that Neanderthals in Europe hunted with thrusting spears, but the frequency and pattern of injuries on their skeletons suggest they paid a heavy price for getting up close and personal with their prey.[57] If you ever go on safari, please don't leap out of your vehicle, charge a wildebeest, and try to plunge a spear into its flank. . . .

Maybe throwing that spear doesn't seem like such a bad idea after all? Most of us today throw solely when we play games and sports, but in almost all cases we throw overhand. When chimpanzees or other primates want to throw things accurately, they do so underhand.[58] If you ever spy an ape or monkey in the zoo preparing to lob underhand feces in your direction, run! If, however, the animal is trying to hurl overhand with more force, you can relax because they lose any ability to aim with overhand throws. Humans are the only species capable of throwing overhand fast and on target. Actually, only humans (male and female) who practice. When Neil Roach and I were doing experiments on the biomechanics of throwing, I distinguished myself as the least able to throw either hard or accurately. Even so, I was better than any ape thanks to a series of adaptations that first appear around two million years ago and that underlie human throwing capabilities.

To appreciate how evolution made us good at throwing projectiles

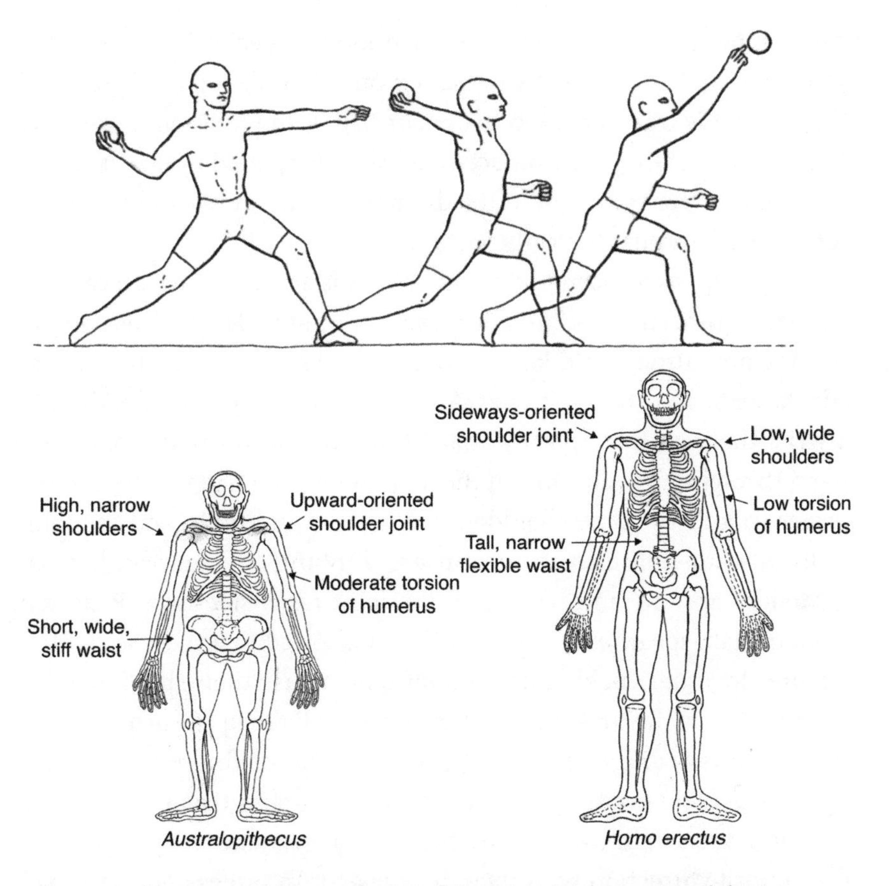

FIGURE 17 Anatomy and biomechanics of throwing. As the top panel depicts, throwing is a whiplike motion, in which energy is sequentially transferred and added from the legs to the hips and then to the torso, the shoulders, the elbow, and finally the wrist. *Homo erectus* (*bottom right*) but not *Australopithecus* (*bottom left*) has a number of features that make this motion possible. (Figures modified from Roach, N. T., and Lieberman, D. E. [2014], Upper body contributions to power generation during rapid, overhand throwing in humans, *Journal of Experimental Biology* 217:2139–49; Bramble, D. M., and Lieberman, D. E. [2004], Endurance running and the evolution of *Homo*, *Nature* 432:345–52)

such as spears, let's begin with the two key elements of a first-class throw: velocity and accuracy. Stand up and try throwing something light and harmless like a crumpled piece of paper as hard as possible at a target. Note that the throw's velocity comes from using your body like a whip as illustrated in figure 17. If you tried hard, you stepped into the throw and then sequentially rotated your hips, back, shoulders, elbow, and finally wrist. At each joint, especially the shoulder, you generated energy that you passed on to the next joint.[59] Some of this energy is transferred to the paper ball at the moment of release. In turn, your ability to throw accurately is determined by how well you are able to move your arm in the direction of the throw and to release the projectile at just the right instant.

Throwing hard, accurately, and reliably is a unique human capability that requires hours of practice. Part of this skill derives from neural control, but humans unlike apes also evolved some special anatomical adaptations, shown in figure 17. Apart from a highly mobile waist and a wrist that can flex upward, many human features that make overhand throwing possible are in the shoulders, which generate half the power in a throw. Ape shoulders are narrow and high, and the joint faces upward—all features useful for climbing. In contrast, human shoulders are low and wide, and the joint faces sideways. Research from my lab, spearheaded by Neil Roach, showed that these and other features together enable throwing humans to use their shoulders like a catapult.[60] In the first part of a throw, we cock the upper arm by holding it sideways and rotating it backward. This cocking motion stores up considerable elastic energy in the muscles and tissues that cross the shoulder. Then, when we unwind, the arm rotates like a spring in the opposite direction with incredible speed. In professional baseball pitchers, this rotation can be nine thousand degrees per second, the fastest motion recorded in the human body.[61] To finish the throw, we then extend the elbow, flex the wrist, and release the projectile.

When Neil and I looked at the fossil record, we noticed that all the features that enable us to throw well show up by two million years ago in the species *Homo erectus*.[62] Given that humans also started hunting around then, throwing capability was probably selected to help put meat on the table. It would be naive, however, not to suppose that early hominins sometimes also threw spears or rocks at each other as well. I suspect *H. erectus* children spent hour upon hour practicing their throwing skills and developing upper-body strength.

All over the planet, millions of children are keeping up this tradition. On many afternoons, my younger neighbors are in the street throwing baseballs and footballs, fantasizing about becoming great athletes. It was not too many generations ago that similar kids practiced the same throwing skills and dreamed of becoming great hunters or warriors. The reason we no longer associate throwing with fighting and hunting is that technology has moved on. While simple untipped spears and slings were once the only lethal projectiles available, ever-accelerating innovations have transformed how we kill from a distance. The first big breakthrough was about 500,000 years ago with the invention of stone points that could be hafted onto spears. Within the last 100,000 years, humans then devised the bow and arrow, the spear-thrower (atlatl), harpoons, nets, blowguns, hunting dogs, poison-tipped arrows, and traps.[63] Think of the weapons invented since that make it even easier and safer to kill from a distance.

As cultural evolution has dissociated physical activities like throwing from their combative and lethal origins, have our bodies also changed? Probably they have because cultural and biological evolution are not independent. Consider fire and clothing. With these inventions, hominins were able to move into new, colder environments that then permitted selection for features like lighter skin away from the tropics.[64] Since cooking became common, human digestive physiology evolved to make us now dependent on cooking to survive.[65] Weapons invented since the Iron Age probably haven't been around long enough to influence human evolution, but what about spears and other projectiles?

First spears. Beyond our adaptations for throwing, recall that males in the human genus shrank from being about 50 percent to just 15 percent bigger than females. Much of that size reduction is probably explained by less male-male competition, but we cannot rule out the possibility that spears diminished the benefits of having a big body when hunting or fighting. That said, upper-body muscle mass is on average 75 percent greater in human males than females.[66] As we have seen, strength in the shoulders, arms, and torso matters for throwing, not to mention wrestling and other competitions.

A more speculative hypothesis is that the invention of the bow and arrow and other cutting-edge projectile technologies revolutionized the costs and benefits of reactive aggression. For the first time, slight Davids like me could take down Goliaths, and females could defend themselves more effectively against male aggressors. Weapons like the bow and arrow also helped less brawny individuals hunt effectively and with reduced risk at a distance. Since the bow and arrow was invented 100,000 years ago, it has probably been less advantageous to be big and reactively aggressive.[67] Wouldn't it be ironic if the evolution of projectile weapons helped domesticate humans? I suspect less reactive aggression also helped spur the evolution of another human universal: sports.

Be a Good Sport?

Sports are organized forms of play, and some like fencing and boxing make no effort to hide they are ritualized forms of fighting, but if you ever want to witness a truly blatant confirmation of the association between combat and sport, visit Florence, Italy, in June to see the Calcio Storico Fiorentino (shown in figure 18).

I stumbled upon this violent spectacle several decades ago on a visit to Florence. I had arrived by plane that morning and took a walk to recover from jet lag. As I wandered toward one of Florence's grand squares, the Piazza Santa Croce, I saw people streaming into stands surrounding a sand-covered arena the size of a football field. Intrigued, I managed to get a seat in a sea of rowdy Florentines all dressed in green. As I later learned, this competition, which dates to the fifteenth century, involves a series of matches between teams representing the city's four quarters. Most of the competitors were bare-chested and wearing Renaissance-looking pants. The referees had swords. For about an hour, the teams participated in what can only be described as a cross between rugby and a giant, ruthless cage fight. Each team of twenty-seven men was ostensibly battling to throw a soccer-sized ball over narrow slits at each end of the field, but that involved serious brawling to help their teammates advance the ball toward the goal or prevent their adversaries from doing the same. Apart from a few no-no's like kicking someone in the head, these guys did anything they could to beat the crap out of each other including boxing, wrestling, headbutting, tripping, and choking. With each goal scored and with every broken nose and cracked rib, everyone around me was on their feet cheering, "Verdi! Verdi!" By the match's end, many of the players had blood streaming down their faces, and quite a few had already been carried out on stretchers. It was a battle.

FIGURE 18 Scenes from Calcio Storico Fiorentino. (Photo by Jin Yu/Xinhua)

Sports, even those as extreme as Calcio Storico Fiorentino, evolved from play. Almost all mammal infants play to develop the skills and physical capacities needed to hunt or fight as adults.[68] Additional benefits of play include helping youngsters to learn or change their place in social hierarchies, to forge cooperative bonds, and to defuse tensions. Humans are no different except our play often uses tools like balls and sticks, and like dogs and a few other domesticated species we continue to play as adults.[69] In every culture, games and sports emphasize skills useful for fighting and hunting such as chasing, tackling, and throwing projectiles. It is widely acknowledged, however, that sports differ from play in one key respect: whereas play is unorganized and unstructured with no particular rules or outcomes, sports are competitive physical activities between opponents according to established rules and criteria for winning.[70] By this definition, some pastimes that require little strength or fitness are classified as sports including darts and bowling.

Honestly, I have nothing against darts and bowling, but the traditional definition of sports excludes one fundamental and crucial characteristic revealed by an evolutionary anthropological perspective: the control of reactive aggression. Even in violent sports like hockey and football, it is against the rules to lash out violently at an opponent.

Homer illustrates this point dramatically in *The Iliad*. For most of the poem, the Greeks squabble and fight bloody skirmishes under the walls of Troy. Like male chimpanzees, the Greeks feud incessantly, jockeying for power, status, and females. But in the penultimate book, one of the most important moments of the epic, they stop for a sort of mini-Olympics. The impetus for these competitions is the slaying of Patroclus, the beloved companion of the Greek hero, Achilles. Beside himself with grief, Achilles sponsors a day of funeral games to honor the corpse and to please the gods, who apparently enjoy watching sports. The games include boxing, footraces, chariot races, and throwing competitions, but the wrestling match between Ajax and Odysseus stands out. It's a classic matchup: Ajax is a strongman who relies on brute force; Odysseus is smaller, wiry, and cunning. Predictably, they fight to a stalemate. After many exciting rounds of dramatic lifts, rib-cracking falls, and clever moves, Achilles steps in and calls a draw: "No more struggling—don't kill yourselves in sport! Victory goes to both. Share the prizes."[71]

Generations of readers have wondered why Homer interrupts the siege of Troy with wrestling and other sports, but Achilles's message exemplifies Richard Wrangham's argument: the Greeks need to stop fighting among themselves and instead cooperate if they are to end their ten-year siege. They should stop being reactively aggressive with each other and be only proactively aggressive toward the Trojans. As with war, suppressing reactive aggression and following rules are fundamental to most sports. Indeed, sports might have evolved as a way to teach impulse control along with skills useful for hunting and controlled proactive fighting. What is more unsportsmanlike than punching an opponent who scores a goal or, even worse, punching a teammate who scores instead of you? Professional tennis players aren't even allowed to say rude things on court.

Surely other hominins including Neanderthals engaged in play, but I hypothesize that sports evolved when humans became self-domesticated. As noted above, it is primarily among domesticated species that adults play, and among the many reasons humans in every culture play sports, one is to teach cooperation and learn to restrain reactive aggression. Regardless of whether you are trying to beat your opponent to a pulp in a cage or impress the judges of a synchronized swimming competition, to be a "good sport" you have to play by the rules, control your temper, and get along with others. Sports also foster habits like discipline and courage that are crucial for proactive aggression such as warfare. Perhaps the Battle of Waterloo really was won on the playing fields of Eton.

There are additional, powerful reasons sport is not just universal but also wildly popular. Sports can be fun to do and entertaining to watch, they foster community spirit, and they are extremely lucrative. Few other human activities regularly draw more than 100,000 live spectators, not to mention billions of television viewers. From an evolutionary perspective, individuals may also be drawn to sports because they can improve their reproductive success. Just as good hunters and fighters in small-scale societies have more offspring, good athletes—both male and female—get to show off their physical prowess, achieve high status, and attract mates.[72]

Last but not least, sports have recently become an excellent way to get exercise and thus promote physical and mental health. In spite of occasional Christian biases against the pleasures of the flesh (Calvin and his Puritan followers had especially dim views of sports), centuries of educators and philosophers have advocated sports for the nobility and other elites who otherwise might never need to be physically active. To quote Rousseau: "Do you, then, want to cultivate your pupil's intelligence? Cultivate the strengths it ought to govern. Exercise his body continually; make him robust and healthy in order to make him wise and reasonable. Let him work, be active, run, yell, and always be in motion. Let him be a man in his vigor, and soon he will be one in his reason."[73] My university follows this tradition, but thankfully for both women and men. Harvard's Department of Athletics sponsors forty varsity teams, involving almost 20 percent of students. Its official mission to promote "education through athletics" states that sports "help our students grow, learn, and enjoy themselves while they use and develop their personal, physical, and intellectual skills."[74]

In the final analysis, humans are physically weaker than our ancestors not because we evolved to fight *less* but because we evolved to fight *differently:* more proactively, with weapons, and often in the context of sports. Along the same lines, we didn't evolve to do sports to get exercise. As a form of organized, regulated play, sports were developed by each culture to teach skills useful to kill and avoid being killed as well as to teach each other to be cooperative and nonreactive. Sports took on the role of providing exercise only when aristocrats and then white-collar workers stopped being physically active on the job. Now in the modern, industrial world we market sports as a means of exercising to stay healthy (I'm still not convinced about darts). Yet true to their evolutionary roots, many sports still emphasize skills useful for fighting and hunting that involve strength, speed, power, and throwing projectiles.

As a closing thought, consider the world's most popular sport, soccer. Soccer requires most of the same behavioral skills useful in other team sports including cooperation and reduced reactive aggression. But soccer also demands another characteristic that is especially important for health, that we humans excel at, and that helps separate us from the rest of the animal world: endurance.

PART III

Endurance

EIGHT

Walking: All in a Day's Walk

MYTH #8 You Can't Lose Weight by Walking

The pay is good and I can walk to work.

—John F. Kennedy

To provide a glimpse of a normal day's walk way back in time, let me tell you about one occasion when I asked two Hadza hunter-gatherers, Hasani and Bagayo, if a colleague and I could follow them on a hunt. They graciously agreed provided we were as quiet as possible, obeyed their requests to stay back when necessary, and did not slow them down.

We started just after dawn, when it was still deliciously cool and there was dew on the grass. Hasani wore a colorful cloth wrapped around his waist along with a striped yellow and black shirt; Bagayo was wearing shorts and a well-used Manchester United soccer shirt. Both hunters were shod in homemade sandals, and the only things they carried were a bow, a quiver of arrows, and a short knife. In contrast, I was equipped for adventure. I wore a big-brimmed hat, lightweight boots, a high-tech shirt that wicks away moisture and blocks UV rays, and trail pants. I also had my cell phone, a GPS watch, and in my backpack two bottles of water, sunscreen, insect repellent, spare glasses, some apples and energy bars, my Swiss Army knife, and, because you never know, a flashlight and a tiny first-aid kit.

As soon as we left camp, I had to focus. Hasani and Bagayo walked briskly, but there was no trail, and the footing was treacherous. Big rocks lurked everywhere under the lush grass (it was the rainy season), threatening to turn any misstep into a twisted ankle. As we hiked down a steep escarpment toward a wooded valley with Lake Eyasi shimmering in the distance, Hasani and Bagayo stopped frequently to look for footprints and other signs of game. They were almost entirely silent, communicating rarely, briefly, and softly. At first, we searched in the cracks of giant boulders for hyraxes, cat-sized creatures that look like rodents but are actually relatives of elephants. Then we followed the trail of a kudu, whose prints were fresh from that morning. We never saw the kudu, but around midmorning we came across some impala antelopes. Hasani motioned us to get down low, stripped off his shirt, removed his sandals, and slowly crept barefoot through the brush toward the antelopes. At the same time, Bagayo circled around to the other side. My colleague and I sat quietly, hoping not to ruin the hunters' chances. About fifteen minutes later I heard the sound of an arrow loosed, and soon thereafter Hasani returned looking irritated. No words were needed to explain that he had missed. So off we went again as it got hotter and hotter.

And then our walk changed thanks to a honeyguide bird. These little brown birds have been collaborating with humans in Africa for thousands, possibly millions of years.[1] In typical fashion, the honeyguide loudly tweeted out its distinctive, insistent, chattering song—*tch, tch, tch tch, tch, tch!*—and then flew from tree to tree singing periodically to make sure we followed. Within ten minutes our little friend delivered us to a beehive. Delighted, Hasani and Bagayo made a fire, smoked out the bees by stuffing the hole with smoldering grass (with a few stings for their efforts), and then robbed the bees of a large portion of the nest's honeycombs. They devoured these on the spot, spitting out the beeswax, thus rewarding our avian guide.

The day's walk soon became a honey-collecting expedition. As we headed back in the general direction of camp, Hasani and Bagayo went from one beehive to the next, practicing the same routine: they made a fire, smoked out the bees, and chomped on sweet, waxy honeycombs. After we visited five hives, it was afternoon and hot, and Hasani and Bagayo, now chatting away and enjoying their sugar high, had clearly given up any intention of hunting. We got back to camp around 1:30 p.m., more than five hours after we had left, with nothing but honey in our bellies. According to my GPS, I had taken 18,720 steps covering a distance of 7.4 miles.

Later, at dinner, I talked to Bagayo and some other hunters. Everyone agreed that times had gotten tough because encroaching Datoga herders and their cows were depleting the region of game. The hunting isn't as good as it used to be, they said. More often than not, the men come back without meat, and they rely increasingly on the plants women gather as well as honey and traded food. Still, just about every day, the men do what we did: they venture forth from camp to hunt, collect honey, and get their hands (and teeth) on anything edible. The women also walk a lot every day, but having accompanied them, I think they have more fun than the men. On a typical day, a group of women and children hike several miles until they find a good place to dig for tubers. Everyone then just plunks down and digs while sitting, chatting, nursing, and sometimes singing as they extract tuber after tuber from the hard, rocky soil. Some of the tubers are consumed raw on the spot, some are cooked for lunch, and others disappear into slings to carry home. On the way there and back, the women and children sometimes pause to collect berries and other food.

But mostly they walk. If there is one physical activity that most fundamentally illustrates the central point of this book—that we didn't evolve to exercise but instead to be physically active when necessary—it is walking. Average hunter-gatherer men and women (Hadza included) walk about nine and six miles a day, respectively, not for health or fitness but to survive.[2] Every year, the average hunter-gatherer walks the distance from New York to Los Angeles. Humans are endurance walkers.

For most people in the postindustrial world, walking is still a necessity but has ceased to bear any resemblance to endurance. Unless you are disabled, you probably walk a little to get to work, even if it's from your car to your office and back. You also walk to the bathroom, to lunch, to shop, and to do countless other small but necessary tasks. Perhaps you took a stroll to relax or (even more bizarrely) you walked on a treadmill going nowhere, but the majority of steps you took today probably seemed essential rather than optional. The big difference between you and Bagayo and Hasani is that their survival demands up to 20,000 steps per day, whereas data culled from millions of cell phones indicate the average American takes 4,774 steps (about 1.7 miles), the average Englishman takes 5,444 steps, and the average Japanese 6,010.[3] Consider also that these numbers are averages. That means millions of Americans take fewer than 4,774 steps per day. Beyond numbers, there are other differences in how we walk. Hunter-gatherers walk in minimal sandals or barefoot, they usually carry food and babies, and they either bushwhack or tread on the simplest of trails on varied terrains. Until a few generations ago, no one ever walked in cushioned, supportive shoes on hard, flat sidewalks, let alone treadmills.

These changes raise a host of questions, including how, why, and how much we evolved to walk and how walking affects aging and health. If there is any single exercise prescription we repeatedly hear, it is to walk about ten thousand steps a day. As one popular book put it, "Walking on a regular basis—whether going for a brisk, structured walk, or just fitting in more steps every day—can help you shed pounds and inches and, most importantly, keep them off."[4] Or in the cautious parlance of two exercise scientists: "10,000 steps/day appears to be a reasonable estimate of daily activity for apparently healthy adults and studies are emerging documenting the health benefits of attaining similar levels."[5] Yet, while almost all weight-loss programs prescribe daily walking, some experts claim it is impossible to lose weight by walking because even a lengthy walk burns an insignificant number of calories and simply makes people hungry. A widely read cover article in *Time* magazine in 2009 titled "The Myth About Exercise" stated, "Of course it's good for you, but it won't make you lose weight."[6]

As a first step to explore these confusing, contradictory claims, let's consider the peculiar manner in which humans walk in the first place, tottering about on only two legs.

How We Walk

Most people take for granted the ability to walk, but every day, just a mile from my house, the Spaulding Rehabilitation Hospital welcomes patients who struggle to retrain their bodies to walk. The gait clinic is a big, brightly lit room that looks more like a gym than a medical facility because it is filled with equipment like walkways, treadmills, and weights. When I last visited, there were about ten patients in the clinic, each working with a physical therapist. Some of the patients were battling neurodegenerative diseases; others had suffered strokes or accidents. One of them, let's call her Mary, a woman in her thirties, had injured her spine in a car accident and kindly allowed me to observe her physical therapy session.

What struck me most was Mary's focus. After easing herself out of her wheelchair onto a walkway with handrails on both sides, Mary concentrated on the simple challenge of putting one leg in front of the other while bearing her weight. Her left leg had trouble, but her right leg simply wouldn't behave. With every step she had to consciously command uncooperative muscles to perform basic motions that used to be instinctive: first flex her hip, then bend her knee, then extend her knee, and so on. At her side, a physical therapist provided encouragement and advice with every step. As Mary's gait improved and she gained confidence, the therapist added incremental challenges like tiny three-inch-high hurdles to help Mary relearn basic movements. For the rest of the session, Mary and her therapist progressed through a series of exercises to strengthen and regain control of specific muscles, and they reviewed the exercises she would do at home. "You are making progress," her therapist said, encouragingly, but both of them knew that Mary had a long way to go before she might again walk unassisted.

Unless you are like Mary, you have probably given little thought to the act of walking since you started toddling at about one year old.[7] That effortlessness is a remarkable achievement of your amazing nervous system, which dynamically controls the many dozens of muscles needed to put one foot in front of the other in varied and sometimes treacherous conditions including rocky mountain paths and icy sidewalks. Sadly, it often takes an accident or a stroke for you to appreciate these patterned movements and reflexes, which must accomplish two major things: move you efficiently and keep you from falling over.

Moving your body efficiently when walking is not special to humans. Whether you walk on two or four legs, the dominant function of a leg is to be a pendulum. This is illustrated in figure 19, but if a picture is worth a thousand words, then action is worth even more, so take a few steps around the room and focus on what your right leg is doing. Notice when it isn't on the ground, it swings forward like the pendulum on a grandfather clock with its center of rotation at the hip. This "swing phase" of a stride is primarily powered by your hip muscles. Your leg's pendular action flips, however, at the end of the swing phase when your foot collides with the ground. At this instant, your leg becomes an upside-down pendulum whose center of rotation is the ankle. In essence, your leg becomes a stilt during this "stance phase" of the stride.

The stilt-like behavior of legs during stance is key to understanding how you use energy when you walk. During the first half of the stance phase, muscles vault your body up and over that leg, elevating your center of mass about two inches (five centimeters). That upward lift expends calories but stores potential energy, just as if you were to raise this book. Then during the second half of stance, your body converts that potential energy to kinetic energy by falling downward and forward, as if you were to drop the book. Eventually, your swing leg collides with the ground, halting your body's fall and starting a new cycle. Walking thus costs calories to raise the body's center of mass in the first half of stance, then redirect it upward and forward from one step to the next, and to swing the arms and legs.[8] While at least one foot is on the ground at all times during a normal walk, the key energetic principle that moves you forward is using your legs like pendulums to exchange potential and kinetic energy. Quadrupeds like dogs and chimpanzees use their four legs in just the same way.[9]

Sequencing, coordinating, and powering the pendular movements that make up each stride is important, but as Mary and others who have lost the ability to walk attest, the biggest hurdle when walking is to not fall over. Unlike quadrupeds, which always have at least two

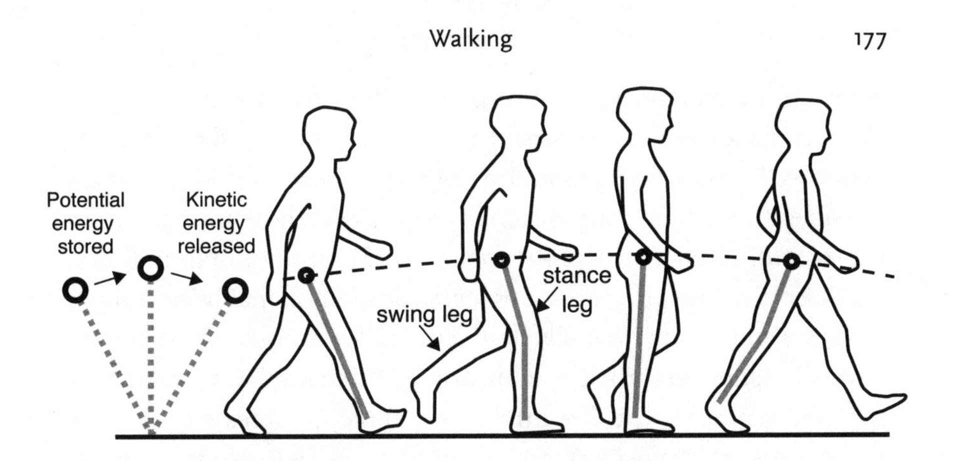

FIGURE 19 Walking mechanics. While each leg is on the ground, it acts like an upside-down pendulum storing up potential energy in the first half of stance, which is then partially recovered as kinetic energy in the second half of stance.

legs on the ground when walking, bipedal humans have only one foot on the ground for most of each stride. As we walk, our bodies usually want to fall sideways. Because we have vertical torsos, our tippy upper bodies also sway forward and backward and from side to side. And with only two legs, humans topple over easily if perturbed. Have you ever seen a dog or cat trip and fall while walking? If you want added evidence of how unstable bipeds can be, especially on uneven or slippery surfaces, watch a quadruped trying to walk on its hind legs. Even chimpanzees and gorillas, which regularly walk upright, lurch awkwardly. Their hips and knees are constantly bent like Groucho Marx's, their arms swing vigorously, and their entire torsos rotate exaggeratedly in concert with their hips, causing them to look as if they were drunk.[10]

Fortunately, natural selection bequeathed humans many ingenious features to keep us from toppling over when we walk on two legs. One of the most critical adaptations is the unique shape of our pelvis, shown in figure 20. Whereas quadrupeds like apes and dogs have tall, flat pelvic bones that face backward, the bowl-shaped human pelvis is short and wide and curves to the side. This curvature repositions muscles that are *behind* the hip in quadrupeds to run along the *side* of the hip in humans. Because of that lateral orientation, when only one leg is on the ground, those muscles can contract to keep the pelvis and upper body from falling sideways toward the swing leg. You can do a simple experiment to test this function (known as hip abduction): stand on just one leg with your hips level for as long as possible. After about thirty seconds you will feel those muscles on the side of your hip burning as they get tired of keeping you from falling.

The other conspicuous and essential adaptation that helps humans walk upright is our uniquely long, curved lower back. Chimpanzees have stiff, short lower backs with usually three lumbar vertebrae, but humans typically have five lumbar vertebrae that create a backward curve. This curve positions the upper body above the hips; without it the torso would always be falling forward, requiring one to use muscles in the hip and back to keep it upright.

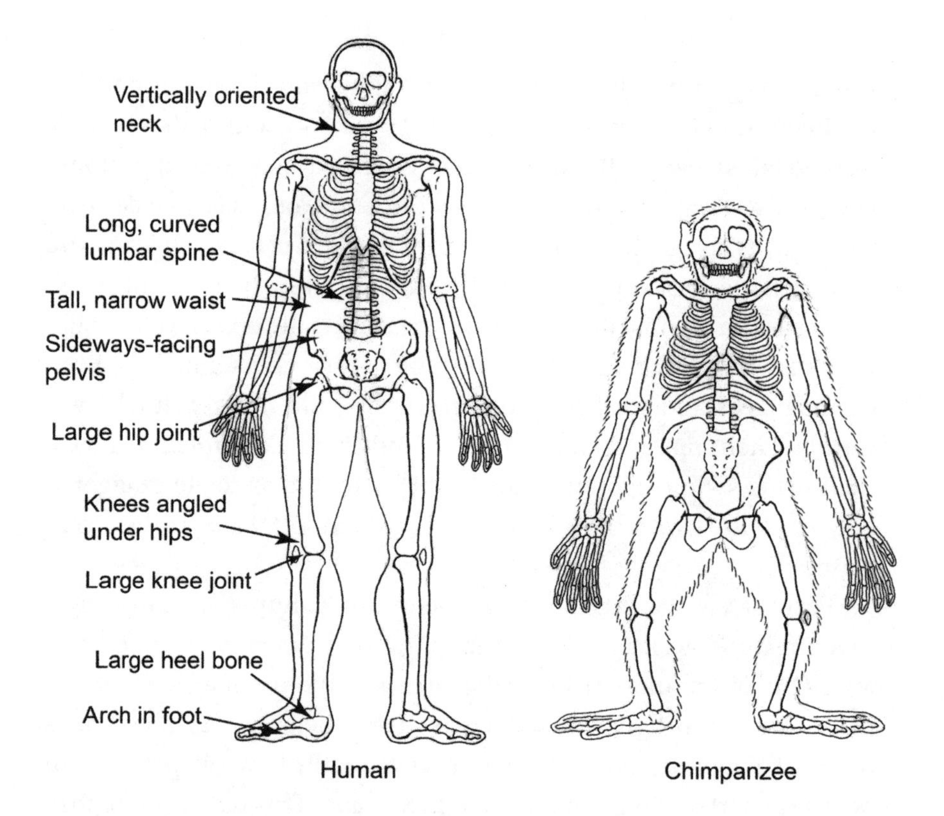

FIGURE 20 Adaptations for efficient, effective bipedal walking in humans (*left*) that are not present in chimpanzees (*right*). (Modified from Bramble, D. M., and Lieberman, D. E. [2004], Endurance running and the evolution of *Homo, Nature* 432:345–52)

Humans inherited many other adaptations to walk bipedally including expanded heel bones, arches in the feet, a big toe that points forward, stabilized ankles, long legs, buttressed knees, inwardly angled thighs, expanded hip joints, and a downwardly oriented foramen magnum. My colleagues and I have also shown that by frequently walking barefoot, evolutionarily normal humans develop thick calluses that protect our feet like shoes, but unlike shoes don't blunt sensory perception from the ground.[11] Almost no one thinks of these and other features when they walk, but they are nonetheless operating silently and efficiently in the background. We realize their importance only when an injury or illness interferes with their function. Even stubbing your little toe can turn the simple act of walking into a hellish struggle. Our many adaptations for maintaining stability as we wobble precariously on two legs raise an interesting, ancient question: Why walk on just two legs when four are obviously better?

Four Legs Good, Two Legs Bad?

In 2006, millions of people learned about an unfortunate family in Turkey with a genetic mutation that causes them to walk on all fours. Videos, including a BBC documentary, showed them moving slowly and awkwardly on their hands and feet in their homes, on the street, and in fields, their butts in the air and their necks craned upward so they could see where they were going. The researcher who initially studied them, Dr. Uner Tan, named the genetic syndrome after himself, claiming the family's apelike gait was an example of human "devolution" that held new clues to how and why humans became bipedal.[12] In actual fact, their gait was unlike any primate's, and they walked on all fours only because the mutation they carried impairs a region of the brain, the cerebellum, that controls balance.[13] If you or I couldn't balance on two legs but needed to get somewhere, we'd walk like this family not because it is atavistic but for simple, urgent biomechanical reasons.

Uner Tan syndrome is not evolutionarily informative, but the widespread interest it evoked illustrates how speculation about the origin of our unusual two-legged gait has been unceasing since Darwin's day. Among other theories, bipedalism is thought to have evolved as an adaptation for carrying food, foraging upright, saving energy, making and using tools, keeping cool, seeing over tall grasses, swimming, and showing off genitalia. These hypotheses range from sensible to dubious, but all of them require knowing what we evolved from: our last common ancestor with chimpanzees. Did this "missing link" knuckle walk like a chimpanzee by resting its weight on the middle digits of its fingers? Did it swing in trees like a gibbon? Or did it climb cautiously above branches on all fours like a monkey?

Sadly, "missing link" is an apt term for this enigmatic ancestor because it is very much missing. The rich, moist, and acidic soils of the African rain forests that apes inhabit quickly destroy bones after animals die, leaving behind almost no fossil record of our closest relatives and their ancestors, including the missing link. Absence of evidence for this species provides fertile ground for speculation and bickering, but many lines of evidence point in the same direction. If we were to travel in a time machine seven to nine million years ago to Africa, it is more likely than not that our last common ancestor with chimpanzees would look something like a chimpanzee and would be knuckle walking and sometimes climbing in a forested habitat.[14] That is important because when scientists have measured the cost of knuckle walking, they find it is very energetically inefficient. Like a gas-guzzling car, knuckle-walking chimpanzees burn through calories.

The first evidence for the high cost of chimpanzee walking dates to a 1973 experiment by C. Richard Taylor and Victoria Rowntree, who trained juvenile chimpanzees to walk on a treadmill while wearing oxygen masks so their energy expenditure could be measured.[15] In addition to finding that the chimpanzees spent as many calories walking on two legs as on four legs, Taylor and Rowntree found that walking was almost three times more costly in chimpanzees than in humans and other mammals of the same size. A generation later, these results were confirmed in adult chimpanzees using more modern methods by Michael Sockol, Herman Pontzer, and David Raichlen.[16] Pound for pound, average humans spend the same amount of energy to walk a given distance as dogs and most other quadrupeds, but chimpanzees spend slightly more than twice as many calories.[17] Chimpanzees waste energy when knuckle walking because their lurching, Groucho Marx–like gait with constantly bent knees and hips requires their leg muscles to work extra hard to hold up their bodies.[18]

To appreciate why the high cost of chimpanzee knuckle walking helps explain the origins of bipedalism, consider that most chimpanzees live in fruit-filled rain forests. If they walk a typical two to three miles a day, their inefficient gait costs them about 170 calories daily. That expense is evidently worth it to allow them to be adept at climbing trees and helps explain why chimpanzees typically walk only as much as sedentary Americans. According to Richard Wrangham, the farthest he ever saw chimpanzees walk was an unusually long patrol, nearly seven miles, by a group of males. Apparently, these guys were totally exhausted by their long trek and barely moved the next day.

The woeful inefficiency of knuckle walking is rarely a problem for chimpanzees deep in the forest, but it must have been a serious challenge for our missing-link ancestors about seven to nine million years ago. During this period of rapid climate change, the rain forest that covered much of Africa shrank and split into thousands of fragments interspersed with drier, open woodlands. For apes living in the depth of the rain forests, life went on as usual, but those at the margins of the forest must have faced a crisis. As woodlands replaced the forest, the fruits that dominated their diet became less abundant and more dispersed. They had to travel farther to get the same amount of food. Because life is fundamentally about acquiring and using scarce energy to make more life, those better able to conserve energy would have had a reproductive advantage. However, because these apes still benefited from using their long arms, fingers, and toes to climb trees, natural selection evidently favored those who walked efficiently without compromising their ability to still climb effectively. The solution was bipedalism. Individuals who could still scamper nimbly up and down trees but also had hips, spines, and feet that helped them save hundreds of calories a day by walking upright probably had higher reproductive success. Despite being slower and less stable on two legs, over many generations these apes became gradually better at walking upright until eventually they were a new species. We are their descendants.[19]

To fathom just how advantageous it is to walk upright instead of knuckle walk like an ape, let's return to that morning I spent with Bagayo and Hasani. That 7.4-mile walk likely cost me a respectable 325 calories. However, if I were as inefficient as a chimpanzee, the walk would have cost me roughly 700 calories. By walking upright instead of knuckle walking, hunter-gatherers like Bagayo and Hasani save more than 2,400 calories per week, adding up to 125,000 calories a year. That's roughly enough energy to run about forty-five marathons.[20]

But what of the other theories accounting for bipedalism? Although bipedalism helps us carry things, forage upright, use tools, and keep cool, none of these offer a compelling explanation for why bipedalism first evolved. Chimpanzees have no problem walking upright to carry things; they just do it inefficiently. Moreover, there is no evidence that apes can't forage effectively upright; the oldest stone tools appear millions of years after bipedalism; and walking upright helps us keep cool only in open habitats that hominins didn't initially inhabit.

The forces that drove our ancestors to walk upright eons ago may seem irrelevant today, but they aren't. For millions of years until the postindustrial era, our ancestors had to walk something like five to nine miles every day to survive. We evolved to be endurance walkers. Yet, like our ancestors, most of us retain a deep-seated drive to spend as little energy as possible by walking only when necessary. That instinct to conserve calories points to another key difference between walking today and in the past: how much we carry things like babies, food, fuel, and water.

Beasts of Burden

Of all our necessities, water is one of the most vital. But if you are like me, you rarely think about how you get it. When I want water, I find a faucet, which I effortlessly turn and, presto, out comes clean water. Our distant ancestors would have considered this magical. For millions of years, people who weren't camped beside a lake, stream, or spring had to lug water long distances every day. Even during the early Industrial Revolution, people in cities and towns fetched water daily from communal pumps.

To help appreciate what it's like to lack running water, let's return to the rural community of Pemja, Kenya, where my students and I do research. This beautiful region has rolling hills interspersed with granite outcrops and is dotted with tiny farms that grow mostly corn. Water flows in the valleys, but there are no wells, pumps, or other means of supplying water to people's homes or crops. Streams and springs are communal sites where people bathe, wash clothes, and obtain water to cook and drink. Once a day, women fill enormous plastic drums of water, which they hoist on their heads and carry home up steep, rocky trails. I can barely carry one of these containers for a hundred yards, but the women of Pemja are so strong and adept at carrying these drums they make it seem easy.

It isn't. Carrying forty to sixty pounds of water is hard work and requires skill and practice. To get a sense of what it entails, one of my former students, Andrew Yegian, who studies the biomechanics of carrying (and is much stronger than I am), once tried to carry on his head a brimming ten-gallon water container from a stream in the valley up a long, steep hill to the school at the center of the community. Laughing at the absurdity of a young, foreign man offering to carry her water, a woman of about thirty happily gave Andrew her recently filled yellow barrel, and a small crowd set off behind Andrew to watch him try to keep pace with another woman, twice his age, also lugging water up the twisty, rock-strewn trail. As you can sort of see in figure 21, she used just one hand to steady the barrel on her head as she took short, graceful steps; Andrew lurched awkwardly with both hands, trying to keep the barrel from falling over. He stumbled frequently, sweated profusely, and groaned quietly as the hill got steeper and steeper and the load felt heavier and heavier despite all the water that sloshed out. I am happy to report, however, that Andrew made it and was rewarded with a raucous ovation as he staggered into the schoolyard.

Imagine carrying water like this day after day, year after year. In a world without beasts of burden or wheels, people also had to lug firewood, children, and everything they gathered and hunted. I'll bet that schlepping a dead kudu for five miles is tiring. Further, whenever hunter-gatherers move camp, which they do every month or two, they must carry all their belongings. Carrying is thus another important,

FIGURE 21 Water carrying at Pemja. *Left*, an experienced water carrier using just one hand to balance the water barrel; *right*, Andrew, who is struggling with two hands. (Photo by Daniel E. Lieberman)

quotidian form of endurance physical activity associated with walking. Beyond the strength and skill required, it costs extra energy.

In theory, the cost of carrying something should be approximately proportional to its weight. Carrying an infant who weighs 10 percent of your body weight should be like being 10 percent heavier and thus cost you 10 percent more calories when you walk. If only it were that easy. Dozens of studies have found that carrying loads less than half one's body weight typically costs an extra 20 percent of the added weight, and when loads get really heavy, the costs increase exponentially.[21] Carrying stuff while walking is generally expensive because we not only spend more calories to elevate more weight during the first half of stance but also have to spend more energy to redirect our body as a whole upward and forward at the end of each step. In addition, when we carry things, our muscles have to work harder to keep us and the load stable.

Energy is so precious and carrying used to be so frequent and necessary that humans have devised many ingenious ways to carry loads as economically as possible. All these ways, however, require strength, practice, and skill. One method is to carry stuff on your head. Neophytes like Andrew and me do this ineptly, but we and other researchers have found that women in Africa who regularly carry water and other heavy loads on their heads learn to carry up to 20 percent of their body weight without incurring additional cost.[22] The trick is they balance the load stably and stiffen each leg as they vault over it, saving up more potential energy, which they get back as kinetic energy. Another cost-saving method is to carry things with a tumpline, a strap attached to the load that goes around the top of the head. Tumplines demand strong neck muscles and a forwardly bent neck and back. I've seen women in Mexico and Ethiopia carrying massive bundles of firewood with tumplines, which are also used by Himalayan porters who have been shown to carry heavy loads with 20 percent more efficiency than Westerners using backpacks.[23] Another clever technique is to balance heavy objects on your shoulders using poles made of flexible materials like bamboo. Eric Castillo and I found that Chinese porters save energy this way by timing their steps so the body rises when the pole falls and vice versa, thus reducing vertical oscillations.[24] And just so you know, weights carried higher up in a backpack cost slightly less energy to hoist than those carried closer to the hips as long as you bend forward slightly.[25]

In every culture, people carry different things and in different ways, but it's my impression that women in many cultures do the lion's share of carrying. For example, in Pemja, women carry almost all the water and firewood. This is doubly unfair when they are pregnant, highlighting another difference between walking today and in ancient times. Women in the United States tend to become less physically active when pregnant, but taking it easy was not feasible for most mothers-to-be until recently.[26] According to the anthropologist Marjorie Shostak, hunter-gatherer women in the Kalahari consider pregnancy "women's work" and travel their usual distances carrying normal loads right until they give birth.[27]

Pregnancy poses a special carrying challenge for bipedal mothers. The abdomen in pregnant quadrupeds can expand sideways and downward to accommodate the extra size and weight of the fetus and placenta. As figure 22 shows, quadrupeds also position that extra mass stably within the rectangle of support provided by their four legs. Pregnant bipeds, however, have a space and balance problem. In addition to pressing downward on the pelvic floor, the growing fetus and placenta lie in front of the body's center of mass. Pregnant women are thus constantly in danger of toppling forward. Staying upright while standing and walking demands extra work by the back and hip muscles as the baby gets bigger. To compensate, pregnant mothers sometimes lean back, but this characteristic posture places extra stress on the curve of the lower back that can lead to back pain. Lower back pain is bad enough today, but imagine having a debilitating backache if you must walk long distances while carrying things. Evidently the problem was serious enough to lead to selection on the female spine. As Katherine Whitcome, Liza Shapiro, and I showed,

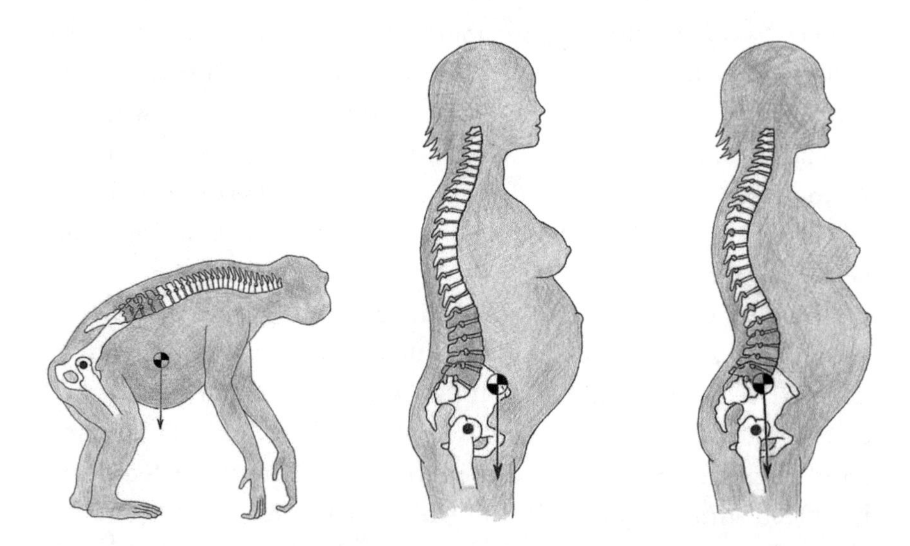

FIGURE 22 Comparison of pregnant chimpanzee (*left*) and human (*center and right*). The chimpanzee's center of mass (*circle*) is supported by her four legs, but an upright human mother's center of mass when pregnant causes her upper body to pitch forward when she stands normally (*center*). If she leans back (*right*), she stabilizes her center of mass but has to curve her lower back more, which places more stress on the lumbar vertebrae. The joints on these vertebrae, however, are extra reinforced, and the lumbar curve is spread over three vertebrae, not two vertebrae, which is the case in males. (Figure from Whitcome, K. K., Shapiro, L. J., and Lieberman, D. E. [2007], Fetal load and the evolution of lumbar lordosis in bipedal hominins, *Nature* 450:1075–78)

two vertebrae create the curvature in the lower back in males, but by three million years ago australopith females had evolved to spread that curve more gently over three vertebrae and to have larger, more effectively oriented joints.[28]

And that's just pregnancy. Once a baby is born, our ancestors and many people today still have to carry it everywhere they go without strollers, car seats, and other modern gizmos. When hunter-gatherer women set out from camp, they put their infants on their backs in slings and hoist toddlers on their hips. Then, on the way home, they also carry food in slings on the back or in baskets on the head. Hunter-gatherer women toting infants plus food often carry as much as 30 percent of their body weight.[29]

All in all, we not only walk less today than we used to but also carry less stuff when we walk. Given how many calories people today must be saving compared with our ancestors by walking and carrying so little, how is it possible that walking—as some claim—is an ineffective way to lose weight?

Can You Walk Off Extra Pounds?

If you want to start a fight in a room of exercise scientists, shout loudly, "Exercise doesn't help you lose weight!" Then run. Until recently, it was considered a universal truth that moderate exercise like walking is essential for losing weight. But as the obesity epidemic has mushroomed and billions of people struggle and fail to shed unwanted pounds, two opposing camps have emerged on this issue. Some experts vigorously defend walking and other kinds of exercise as indispensable for any weight-loss program, but others have come to view these efforts as ineffective. As is so often the case, the debate oversimplifies a complex issue that defies a simple yes or no answer.

On the face of it, it seems preposterous to think that walking doesn't help with weight loss. Recall that energy balance is the difference between the calories one ingests and the calories one spends. You probably burn roughly 50 calories more by walking a two-thousand-step mile than driving the same distance. So trudging ten thousand additional steps a day (five miles) will expend a respectable extra 250 calories per day.[30] To be sure, those ten thousand added steps might make you hungrier, but if you snack sensibly and consume 100 calories less than you walked off, those supplementary steps will eventually amount to a deficit of about 3,000 calories a month. That amount is just shy of 3,500 calories, the supposed number of calories in a pound of fat according to a much-cited, overly simplistic, and inaccurate 1958 study.[31] Further, low- to moderate-intensity activities like walking burn relatively more fat than carbohydrates (hence the "fat-burning zones" on some exercise machines).[32] As a result, lots of people try to trudge away extra pounds.

Biological systems such as bodies are messy, and anyone who has struggled to lose weight knows that simple theories rarely apply to the convoluted realities of weight loss. What works for one person fails for another, and while many people successfully shed pounds when they start a new weight-loss plan, satisfaction often turns to frustration as the initial rate of weight loss diminishes and then reverses. Study after study has shown that overweight or obese people prescribed standard doses of exercise for a few months usually lose at most a few pounds. For example, one experiment with the clever acronym DREW (Dose Response to Exercise in Women) assigned 464 women to 0, 70, 140, and 210 minutes of slow walking a week (140 minutes is about five added miles). Apart from their prescribed exercise, the women took about five thousand additional steps per day as they went about their normal activities. After six months, those prescribed the standard 140 minutes a week lost only five pounds, while those assigned 210 minutes lost a paltry three pounds (more on this unexpected result below).[33] Other controlled studies on overweight men and women report similarly modest losses.[34]

For someone who is fifty pounds overweight, losing three to five pounds over half a year is a frustrating drop in the bucket. Accordingly, a stock response to these studies has been to declare exercise futile for trimming your waist. Before we entirely dismiss the weight control benefits of walking, the most fundamental type of endurance physical activity, let's examine the major arguments behind this contention through the lens of evolutionary anthropology.

The first is the specter of *compensatory mechanisms,* notably fatigue and hunger. If I walk ten thousand extra steps, I'll be more tired and hungry, so I'll rest and eat more to recoup lost calories. From an evolutionary perspective, these urges make sense. Because natural selection ultimately favors those who can allocate as much energy as possible to reproduction, our physiology has been tuned over millions of generations to hoard energy, especially fat. Further, because almost no one until recently was able to become overweight or obese, our bodies primarily sense if we are gaining or losing weight rather than how much excess fat we have. Whether you are skinny or stout, negative energy balance—including dieting—causes a starvation response that helps us restore energetic equilibrium or, better yet, gain weight so we can shunt more energy toward reproduction.[35] It's unfair, but losing ten pounds elicits food cravings and the desire to be inactive regardless of whether one is skinny or obese.

And therein lies another key difference between walking today and in ancient times. If I walk ten thousand extra steps to place my body in negative energy balance, it is literally a piece of cake for me to wipe out the extra cost of such a walk. The ease of refueling with a donut or a Gatorade or just by sitting at my desk for the rest of the day helps explain the counterintuitive result we just saw from the DREW study in which the women who exercised the most lost less weight than predicted: they ate more.[36] Happily, more than a dozen studies on the effects of exercise, food intake, and non-exercise physical activity on weight loss found that modest doses of prescribed exercise rarely cause people to spend the rest of the day as couch potatoes erasing the benefits of their exertions.[37] However, several experiments that required large doses of exercise (one involved training for a half marathon) did cause exercisers to eat more.[38] When the body regulates energy balance like a thermostat, it apparently does so more through diet than through physical activity.

This leads us to the next common argument against walking to lose weight, that we need to walk a ridiculous number of miles to lose just a few pounds. As we have already seen, this critique is true thanks to our evolutionary heritage as efficient, long-distance walkers. If I follow a standard prescription of briskly walking thirty minutes a day, almost two extra miles, I'll spend about a hundred extra calories per day, theoretically allowing me to shed approximately five pounds in half a year—about the reductions most studies report. If a skinny hunter-gatherer mother loses five pounds in six months, she's in trouble, but many obese American dieters aim to lose about fifty-five pounds.[39] Losing that many pounds that quickly through exercise alone would theoretically require Herculean efforts like running eight miles a day. Although far from easy, dieting is unquestionably more effective for shedding many pounds.

While walking 30 minutes a day won't lead to rapid, spectacular weight losses, an evolutionary and anthropological perspective puts a different spin on the argument that walking expends too few calories to shed excess pounds. While the commonly prescribed two-mile daily walk expends a pittance—just 4 percent—of the average person's daily energy budget of twenty-seven hundred calories, that pittance is partly attributable to setting the exercise bar so low. It bears repeating that the standard public health recommendation is 150 minutes of moderate exercise every week. This amounts to a paltry 21 minutes a day, one-sixth the level of physical activity among nonindustrial people like the Hadza.[40] Although jobs, commuting, and other obligations fill our days with necessarily sedentary activities, the average American still spends at least eight times as much time (170 minutes per day) watching television.[41] It is no wonder that studies using modest exercise doses report modest weight losses.

Lo and behold, studies that prescribe higher, more evolutionarily normal levels of exercise, including walking, have the potential to be more effective for weight loss. One intriguing study asked fourteen overweight and unfit men and women to hew to the standard 150 minutes a week by walking briskly five times a week, but assigned another sixteen individuals the task of walking twice as much. Apart from the prescribed exercise, both groups were otherwise free to eat and sit as much as they wished. After twelve weeks, the 150-minute-a-weekers barely lost any weight, but the 300-minute-a-weekers lost an average of six pounds.[42] At this rate they potentially could lose twenty-six pounds in a year. An even more demanding study compared obese men prescribed seven hundred calories of exercise a day (about five miles of jogging) with men asked simply to cut back their diets by the same number of calories. Over three months, both groups lost almost seventeen pounds (seven and a half kilograms), but the ones who exercised lost more unhealthy organ fat, even though they also ate more.[43]

Another issue is time. Just as most dieters want to lose weight fast, researchers who study the effect of exercise on weight loss are also pressed for time. For practical reasons, they have to conduct the experiment relatively quickly without too many participants dropping out, then analyze and publish the results. Consequently, few studies measure the effects of more than a few months of exercise. Short-term studies pose a problem, however: because walking is so energy efficient, it takes months or years for small doses of exercise to add up to substantial weight losses. But it's possible. Just as that daily four-dollar cup of coffee at Starbucks adds up to nearly fifteen hundred dollars a year, someone who manages to walk an hour a day without compensating by eating more calories could theoretically lose an impressive forty pounds in two years.

Nothing about metabolism, however, is simple, and one last and important complication regarding efforts to walk off weight is a still poorly understood phenomenon known as metabolic compensation. Once again, studies of the Hadza play a role in how we understand this phenomenon. When Herman Pontzer and his colleagues measured daily energy expenditures in the Hadza, they were surprised to find that the highly active Hadza spend about the same total number of calories per day as sedentary industrialized people with the same lean body weight.[44] In addition, when Pontzer and colleagues collected energetic data from adults in many countries including the United States, Ghana, Jamaica, and South Africa, they observed that more active people spent only slightly more calories per day than more sedentary people who weighed the same. In addition, individuals who were more physically active didn't have total energy budgets as high as their exertions would predict.[45] How could someone who spent five hundred extra calories a day exercising not have a total energy budget that is five hundred calories higher? The proposed explanation is that people's total energy budgets are constrained: if I use five hundred extra calories walking, I'll spend less energy on my resting metabolism to help pay for my exertions.[46]

This controversial idea (termed the constrained energy expenditure hypothesis) is still being tested, as is its relevance to weight loss. If correct, then contrary to many people's expectations, exercisers might spend almost the same number of total calories per day as similar-sized but more sedentary individuals despite devoting more energy to being active. To appreciate the implications of this phenomenon, consider that the Hadza spend about 15 percent more of their total energy budget on activities like walking, digging, and carrying.[47] In addition, as we will see later, exercise can stimulate repair and maintenance mechanisms that elevate people's resting metabolic rates—an "afterburn"—for a few hours to as much as two days afterward.[48] Yet if very active Hadza hunter-gatherers and exercising industrial people have total energy budgets that are about the same as similar-sized but physically inactive industrials, then they must spend less energy on other things like maintenance or reproduction. This may seem implausible, but we have already seen this phenomenon in people who lose a lot of weight, like the extreme dieters in the Minnesota Starvation Experiment whose resting metabolic rates plummeted.

To what extent and in what circumstances physical activity shifts metabolisms thus offsetting efforts to lose weight remains to be elucidated, but the fact remains that many studies have shown that exercise, including walking, can lead to weight loss. But to do so, one needs to walk considerably more than half an hour per day for many months. In addition, people who exercise more may compensate metabolically, negating some of the effects of added physical activity. Finally, it truly is faster and it's often easier to lose weight by dieting because everyone needs to eat but no one has to exercise, and not eating five hundred calories of energy-rich food (four slices of bacon) requires less time and effort than walking five miles a day. Please, I do not wish to trivialize how hard it is to exercise if one is unfit and overweight: it can be uncomfortable, unpleasant, and disheartening, and disabilities can make it challenging or impossible. But for those unwilling or unable to run, swim, or do other vigorous exercises, walking remains an inexpensive and pleasant way to get a moderate and useful dose of physical activity.

Even more important, regardless of how one initially loses weight, keeping the weight off almost always demands physical activity. The majority of dieters who do not exercise regain about half their lost pounds within a year, and thereafter the rest typically creeps back slowly but surely. Exercise, however, vastly increases the chances of maintaining weight loss.[49] One example of this payoff comes from an experiment conducted here in Boston. When doctors put 160 overweight police officers on low-calorie diets for eight weeks, some with and some without exercise, all the officers lost sixteen to thirty pounds (seven to thirteen kilograms) with the ones who exercised losing slightly more. But once the crash diet was over and the policemen went back to their normal diets, only the officers who continued to exercise avoided weight regain; all the rest regained most or all of the pounds they initially lost.[50] Many other studies confirm that physical activity, including walking, helps keep those lost pounds off.[51]

Maybe those ten thousand steps a day aren't such a bad idea after all. . . .

Ten Thousand Steps?

In the mid-1960s, a Japanese company, Yamasa Tokei, invented a simple, inexpensive pedometer that measures how many steps you take. The company decided to call the gadget *Manpo-kei,* which means "ten-thousand-step meter," because it sounded auspicious and catchy. And it was. The pedometer sold like hotcakes, and ten thousand steps has since been adopted worldwide as a benchmark for minimal daily physical activity.[52] Among its virtues, ten thousand daily steps is easy to remember and a modest challenge for most people. Ten thousand steps also includes both exercise and non-exercise physical activity such as doing chores and walking around the house and office.

By chance, ten thousand steps a day also turns out to be a plausible goal. A veritable who's who of medical organizations agree that adults should get at least 30 minutes of "moderate to vigorous" aerobic exercise at least five days a week for a minimum of 150 minutes per week. Critically, these 150 or more minutes are *in addition* to the normal activities of a generally sedentary lifestyle such as shuffling about the house and walking from the car to the store. There are different ways to define what constitutes "moderate" exercise, but by any measure this includes a brisk walk that involves about a hundred steps a minute. Because a 30-minute walk at this pace is usually three thousand to four thousand steps, and anything less than five thousand steps a day falls under the threshold of being "sedentary," a reasonable daily minimum of steps adds up to about eight thousand to nine thousand. Include a few more steps for good measure and, voilà, you have the magic ten thousand! Perhaps not uncoincidentally, the five or so miles most hunter-gatherer women walk a day translates to roughly ten thousand steps.

Yet one question still nags me. If the roughly ten thousand steps we evolved to walk a day is so reasonable, attainable, and sensible, and walking isn't very costly, why do so many of us walk so little? Wouldn't natural selection have favored those of our ancestors who liked to walk because the benefit of those extra few thousand steps outweighed the relatively insubstantial cost?

The answer, once again, is energy. Table 8.1 summarizes the average number of calories spent walking by chimpanzees, hunter-gatherers, and Westerners. As you can see, sedentary Westerners spend as much daily energy walking as chimpanzees, but hunter-gatherers like the Hadza walk about three times more than an average Westerner, spending nearly twice as many calories despite weighing much less. Altogether, chimpanzees and hunter-gatherers spend about 10 percent of their total energy budget trudging about, but Westerners spend only 4 percent.

Table 8.1: Energy spent walking in chimpanzees, hunter-gatherers, and Westerners (sexes averaged)

Group	Weight (kg)	Distance walked per day (km)	Energy spent walking (calories)	Percentage of total energy budget spent walking
Chimpanzees	37	4.0	125	10%
Hadza	47	11.5	216	10%
Westerners	77	4.1	126	4%

From the perspective of a twenty-first-century American, these numbers seem trivial. In a world of energy abundance and comfort, who cares about 100 calories here or there? It bears repeating that if I power walk five miles, I'll expend roughly 250 extra calories, as many calories as I'll acquire from snacking on the granola bar in my backpack. If I really want to expend a lot of energy, I should run those five miles, and if my goal is to lose weight, I should throw away the granola bar, not to mention all the other high-energy foods in my pantry. Many experts, including some who study the Hadza, thus blame the obesity epidemic squarely on industrial diets, not activity levels.

Without discounting the importance of diet, I think this view undervalues the role of moderate physical activities like walking, especially when viewed through an evolutionary lens.

First, the difference between 5 and 10 percent of one's daily energy budget may seem trivial today, but it is hardly chump (or chimp) change to hunter-gatherers (or chimps). With few exceptions, most organs and functions expend a small percentage of one's total energy budget. But these many vital expenses add up quickly. Skimping on thermoregulation, digestion, circulation, repairing the body's tissues, and sustaining the immune system can quickly land us in hot water. When energy is limited, moreover, saving 100 or so calories a day on unnecessary strolls rapidly tallies up over time to thousands of valuable, scarce calories. If the average Hadza mother managed to walk as few steps as typical industrialized women, she might save between 30,000 and 60,000 calories a year, a titanic sum. If we consider that a nursing mother can expend as much as 600 calories a day to produce milk, those calories would help her have larger, healthier babies and store extra fat to tide her over during lean times.[53] By the same token, if average industrialized people walked as much as the Hadza, they would spend approximately 350 calories a day walking. If they didn't compensate for all those spent calories by eating more, they would slowly but surely shed pounds.

In the eighteenth century, the word "pedestrian" came to denote something dull, commonplace, or uninspired, but I hope you agree that walking is hardly a pedestrian topic. We evolved to walk many miles a day in our strange, ungainly, upright, but efficient manner, and the fact that walking doesn't expend a lot of calories is fundamental, not coincidental. Of the many special qualities that make us human including big brains, language, cooperation, making sophisticated tools, and cooking, efficient bipedal walking was apparently the first and remains one of the most important. We wouldn't be here if our ancestors didn't have to walk at least ten thousand steps a day. But that legacy has not remained a necessity. Until recently, walking wasn't exercise, and despite its being economical, we evolved to do it as little as possible. So in today's topsy-turvy world, many of us must either force ourselves to walk more than necessary or find enjoyable alternatives like gardening, housework, Ping-Pong, cycling, or swimming.[54]

And if you think we struggle today to do moderate activities like walking, consider how much less we engage in vigorous types of endurance physical activity, most notably long-distance running.

NINE

Running and Dancing: Jumping from One Leg to the Other

MYTH #9 Running Is Bad for Your Knees

> No more words—he dashed toward the city,
> heart racing for some great exploit, rushing on
> like a champion stallion drawing a chariot full tilt,
> sweeping across the plain in easy, tearing strides—
> so Achilles hurtled on, driving legs and knees.
>
> —Homer, *The Iliad,* book 22, lines 26–30
> (trans. Robert Fagles)

In 1969, when I was five years old, my mother started running. She was in her thirties, unfit, and struggling with a stressful new job at the University of Connecticut, where she had been told that "a woman has to be twice as good as a man" to get tenure. That year, however, her life changed when she joined a small group of women seeking to end discrimination and unfair treatment of women at the university. One of their objectives was to liberate the university's newly built field house, which admitted women only as spectators at games. She needed to take up a sport and, at a friend's suggestion, decided to try running.

Mind you, 1969 was before the jogging boom had started. Stores didn't sell running shoes, *Runner's World* was little more than a leaflet, and amateur joggers like my mother were basically on their own. So she laced on the only brand of sneakers she could find and slogged as fast as possible around an outdoor track as best she could. At first, my mother was unable to run more than a quarter of a mile. But by alternating running and walking, she slowly built up enough endurance to run an entire mile. And then two. She did not enjoy it, but that wasn't the point, was it? Then, when she and her friends finally ran in the field house, they were unceremoniously evicted. Undeterred, she and her friends kept running and also demanded the university open a women's locker room. They were told this was impossible and that even if space could be made, it would go to waste because women would never use it. Plus, women would demand hair dryers.

I am proud to say that the University of Connecticut opened its field house to women in 1970 thanks to my mother and her co-runners. But just as she changed the university by running, running also changed her. The palpitations she had been experiencing went away, and she gradually became addicted to running. She also hooked my father on running, and for more than four decades she jogged about five miles nearly every day, often with my dad, even in the winter. My mother is now in her eighties, and although she seriously damaged her knee several years ago, she still goes to a gym almost every day.

As a child I had no idea my mother was in the vanguard of the women's and running revolutions. But aside from being inspired to start jogging when I was an anxious, hyperactive, and insecure teenager, I have come to realize that I absorbed several important lessons from her that, as we will see, make total sense from an evolutionary and anthropological perspective. First and foremost, my mother didn't start running for her health; she became a runner because she felt it was necessary. In addition, running for my mother was often social, either with her friends or with my dad. And for her, running was about endurance, not speed. She never raced, but instead trotted along at whatever pace she enjoyed for never more than five miles.

I consider my mother a hero and a pioneer, but some exercists (people who brag and nag about exercise) would derisively label her a "jogger" to distinguish her from real "runners." I object to this distinction. Do we sneer at amateurs who play pickup basketball in the park or those who go for a brisk midday walk at lunch? And what is it with non-runners' disapproval of runners? Sometimes when my father-in-law—a non-runner if there ever was one—drove past a runner, he would needle me by declaring "there goes another jogger running himself to an early grave." So-called runophobes like him think running is a form of torture that will ruin your knees and damage your heart, and they are wont to bring up the legend of Pheidippides, the Greek messenger who supposedly collapsed and died after running from the battlefield of Marathon to Athens to bring tidings of victory. (To set the record straight, Pheidippides's death was invented seven hundred years later and popularized by the nineteenth-century poet Robert Browning to add pathos to his poem's climax; it was never mentioned by Herodotus or other ancient historians who wrote accounts of the event.)

To be fair, runners can be equally biased. Some "runophiles" mistake running for a virtue, and the most insufferable tell yarns to anyone who will listen about their race experiences, describe their injuries in excruciating detail, and humblebrag by confessing they run only sixty miles a week, or begin sentences with "And then at mile eleven . . ." Another irritating extreme are "born-to-runners." These enthusiasts have read about how we evolved to run (this is partly my fault) and preach that running is the key to health and happiness, especially if you run barefoot. Fortunately, the majority of runners are simply passionate about running.

While running engenders passions and controversies, let's keep in mind another form of moderate to vigorous aerobic exercise that involves jumping around for hours: dancing. Dancing is a cultural universal even more popular than running, and it may be nearly as ancient and important to being human. Like running, dancing has its extreme enthusiasts, is potentially injurious, and has its own marathon events.

Why are these endurance physical activities so popular, and why do they arouse such passions? How much should we laud their benefits or worry about the injuries they cause? Most fundamentally, did we really evolve to spend endless hours dancing or running? The most seemingly preposterous claim is that slow, unsteady humans can outrun horses.

Man Against Horse?

In 1984, David Carrier, then a graduate student, published a highly original paper titled "The Energetic Paradox of Human Running and Hominid Evolution," which argued that running humans could outdistance antelopes and other speedy mammals in the heat.[1] In addition to reviewing what was then known about the energetics of sweating and running in humans and other mammals, Carrier described an obscure, ancient method of hunting in which runners pursued animals on foot until they collapsed. Sadly, Carrier's paper had little impact at the time on scholars of human locomotion. Back in the 1980s, the big debate was when and how humans became good at walking, and no one thought of humans as good runners. When I asked one of my professors what he thought of Carrier's paper, he replied incredulously, "It is silly to think humans evolved to run, because we are slow, unsteady, and inefficient." He then pointed me to a paper he had written showing that humans ran as uneconomically as penguins.[2] My tail between my legs, I retreated and tried to banish the idea from my mind.

Thirty-two years later I found myself questioning my sanity in Prescott, Arizona, toeing the line on an October morning at 6:00 with forty other runners and fifty-three horses and riders. Ironically, this "Man Against Horse" race was born in the town's saloon in 1983, the year before Carrier's paper was published, when an avid runner, Gheral Brownlow, wagered an equestrian friend, Steve Rafters, that decent runners could beat horses over a long distance. More modestly, I bet my daughter I could beat just one horse over the twenty-five-mile course, which goes up and over Mingus Mountain (seventy-seven hundred feet). Within minutes of the start, however, I was sure I was going to lose my bet. "See you later!" the riders shouted cheerfully to me and the other plodding human runners as they easily passed us in the first mile of the race. If you've ever been overtaken on foot by a trotting horse at the start of a marathon, you can't help but feel insecure about its massive rippling muscles and long tapered legs. What on earth was I doing?

For the next few hours with no horses in sight, I gave up hope of beating even a single horse. As the sun rose, the trail turned off the riverbed, passed through an open plain, and then up the rocky mountain past cactuses, brush, and pine trees. The hotter and higher it got, the more I struggled to keep running upward. Resigned to my fate, I decided to just try to enjoy this stunningly beautiful place. But then, near the top at about the twentieth mile, I passed my first horse, whose rider had stopped to allow the animal to cool down. My heart leaped and I found new energy as I started to pass more horses at the summit and then elatedly hurtled as fast as I could down the steep switchback trails on the mountain's other side. I was sure no horse could ever catch me on such a steep and winding descent. Then, as the trail leveled off toward the bottom, I heard two horses close behind me. Every thudding hoofbeat and rasping snort seemed amplified. I have never been a competitive athlete, but when these two horses passed me on an old dirt trail, my brain turned on a switch I never knew existed, directing my legs to catch up with those horses. With less than half a mile to go, I somehow passed them as they slowed down in the hot open fields, and I have never crossed the finish line of a race with such a runner's high. If I may brag: I beat forty of the fifty-three horses despite an unremarkable time of 4:20.

If you had told me in 1984 or 1994 that I would be running marathons against horses, I would have laughed incredulously. Apart from my esteemed professor's casting doubt on Carrier's argument, I never considered myself a runner. To be sure, inspired by my mother, I used to jog several times a week for a few miles, but I was never on a track team, and I don't think I had ever run more than five miles in my life. My attitude toward running and my path to Mingus Mountain, however, changed irrevocably one day in graduate school thanks to a brilliant colleague, Dennis Bramble, from the University of Utah.[3] Also present was a pig on a treadmill.

So here's the scene. I was running a pig on a treadmill as part of an experiment on how bones respond to loading when Bramble, who was visiting, wandered into the lab. As I stood there, Bramble crossed his arms, cocked his head to the side, and remarked, "Dan, you know that pig can't hold its head very still." Frankly, I had never noticed how pigs held their heads, but as I looked with new eyes, I could see he had a point. Unlike a running dog or horse, which holds its head still like a missile, that pig's head was flopping about like a fish on a beach. Our conversation quickly turned to the importance of stabilizing one's gaze and the hypothesis that animals adapted for running ("cursors" in biological parlance) have a special rubber-band-like structure, the nuchal ligament, at the back of their heads that acts like a spring to keep their heads still. As soon as that pig was back in its pen, Bramble and I were looking at skulls of pigs, dogs, and other animals that, remarkably, preserve traces of whether a nuchal ligament is present. Soon thereafter we were looking at casts of fossil hominins. He showed me that nuchal ligaments were present in dogs, horses, antelopes, and other cursors, but absent in pigs and other non-running animals. Even more excitingly, we could see that gorillas, chimpanzees, and early hominins lack a nuchal ligament, but humans and fossil species from the genus *Homo* had them. If the nuchal ligament was an adaptation to stabilize the head during running, here was evidence that millions of years ago humans had been selected to run.

Over the next few years, Bramble and I collaborated on a series of experiments to study how humans and other animals stabilize their heads when running, and we started to assemble and analyze a list of features that evolved independently in humans and other animals that run a lot. We also documented when and where these adaptations first showed up in fossils. Eventually, we decided to write a paper reviewing these and other lines of evidence. The paper published in 2004 in *Nature* was titled "Endurance Running and the Evolution of *Homo*," and the magazine's cover, which featured our paper, was emblazoned "Born to Run."[4] Our basic argument was that by two million years ago our ancestor *Homo erectus* had evolved the necessary anatomy to run long distances in the heat in order to scavenge and hunt long before the invention of bows and arrows and other projectile weapons. Following Carrier, we argued that these ancestors sometimes hunted by outrunning fleet-footed animals like wildebeests and kudu as I had horses that hot morning in Prescott, Arizona.

If you are skeptical, rest assured you should be. If you walk out your door, most of the humans you see will be walking, not running, and the fastest among them will be slower and more awkward than most animals. Even if we are the tortoises rather than hares of the animal world, how do humans outrun some of the best runners in nature? Further, how would this kind of long-distance running help anyone hunt, especially when there are easier, less exhausting ways to get dinner? A first step toward answering these questions is to consider how humans and other animals run and what features we have that help us do it.

Jumping from One Foot to the Other

If you can, put this book down, walk a few steps, and then break into a run. Although both gaits employ the same anatomy, running has obviously different, more challenging mechanics. When walking, at least one leg is on the ground at all times functioning like an upside-down pendulum as your body vaults up and over it with each step. But the instant you switch to a run, your legs start to function like pogo sticks as illustrated in figure 23. Instead of your body's center of mass initially *rising* at the start of each step, it *falls* as you bend your hips, knees, and ankles. Flexing these joints stretches tendons (especially the Achilles) in your legs, causing them to store up elastic energy like springs. Then, during the second half of stance, the same tendons recoil while your muscles also contract, straightening your joints and pushing you up and into the air. And all the while, you instinctively lean slightly forward, bend your elbows, flex your knees more as you swing them, and move your arms opposite to your legs. In essence, running is jumping from one leg to another.

If by chance you happen to be running alongside a trotting horse, it, too, is jumping along, using its legs as springs in spite of those legs being longer, bigger, and twice as numerous. In fact, bipedal running is equivalent to quadrupedal trotting. Just as your arm on one side

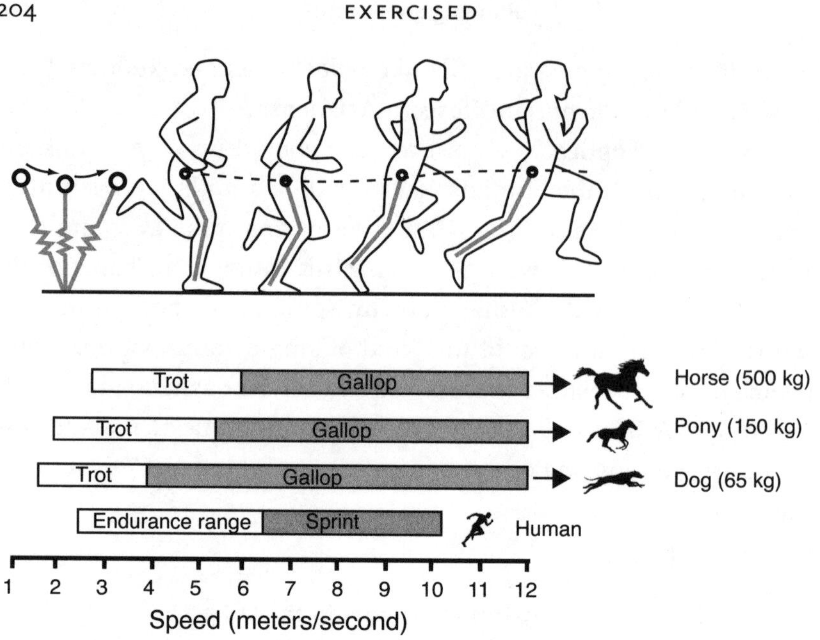

FIGURE 23 Running mechanics and speed. *Top:* springlike nature of running in which the body's center of mass falls during the first half of stance, storing up elastic energy in the leg's tendons and muscles. Then, during the second half of stance, these structures recoil, helping push the body back into the air. *Bottom:* endurance and sprint range of humans compared with the trot (endurance) and gallop ranges for dogs (greyhounds), ponies, and horses. Note that humans can run long distances above the trot speed of dogs, ponies, and sometimes even horses. (Modified from Bramble, D. M., and Lieberman, D. E. [2004], Endurance running and the evolution of *Homo, Nature* 432:345–52)

moves in sync with the opposite-side leg (your left arm and right leg swing forward together), trotting horses jump using the forelimb on one side and the hind limb on the other. The horse, however, can do something we bipeds cannot: gallop. When quadrupeds gallop, they alternate landing with their forelimbs and hind limbs, using not just their legs but also their spines as springs.[5]

With this basic understanding of running, let's explore how and why ordinary humans can outrun horses. To begin with, the white bars in figure 23 compare the speeds up to which humans can run marathon-length distances with the trotting speeds of greyhounds, ponies, and full-sized Thoroughbred horses.[6] I make this comparison because quadrupeds such as horses can run long distances only at a trot. So while horses, dogs, zebras, and antelopes can gallop faster than any human can sprint (the gray bars), they cannot gallop for more than a few miles before having to slow down to a walk or a trot, especially when it is hot.[7] Even some middle-aged professors can run a marathon well above the speed at which greyhounds, ponies, and even full-sized Thoroughbreds can trot the same distance.

In addition to being able to run long distances relatively fast, humans are unusual in habitually running long distances in the first place. Have you ever seen a wild animal run several miles on its own for no apparent reason? With the exception of social carnivores like wolves, dogs, and hyenas, which run up to ten miles to hunt, few animals willingly run more than a hundred yards or so without being forced to.[8] Antelopes and other prey on the savanna sprint to escape lions and cheetahs that are chasing them, but these mad dashes never last longer than a few minutes. Animals like dogs and horses can run many miles, but only when we coerce them with whips and spurs. Disturbing experiments from the 1930s showed that dogs could be forced to trot up to sixty miles on treadmills, and special endurance-trained horses can trot a hundred miles with a rider on their backs.[9] By these standards, humans are impressive. Millions of run-of-the-mill people like my mother lace on shoes and jog five miles a day several days a week, and at least half a million Americans complete a marathon every year, usually training more than thirty miles a week for months.[10]

As for cost, my professor was wrong that humans are as inefficient as penguins. When you look at large samples of people and correct for differences in body size, humans run as efficiently per pound as horses, antelopes, and other species well adapted for running.[11]

What enables ordinary humans who descended from flat-footed apes to be so good at endurance running?

Most obvious are our long, springy legs. Despite ending in clunky feet, human legs are lengthy for an animal our size—a fact easily demonstrated by standing next to a chimpanzee, sheep, or greyhound, all of which weigh the same or less than a typical human but have much shorter legs.[12] Just as important, human legs have lengthy elastic tendons like the Achilles. Even human feet have springlike tissues that run below the arch. Although tendons are unnecessary for walking, they function as springs during running. Every time your legs and feet land on the ground during a run, these tendons stretch as your hips, knees, and ankles bend and your arch flattens. When the tendons recoil, the energy they store is returned to help catapult you back into the air. All animals adapted for running, from kangaroos to deer, have legs with long, springy tendons, but these tendons are short in our close relatives the African apes. That means humans independently evolved long tendons such as the Achilles to help us run. According to one estimate, the Achilles tendon and the spring in the arch of the human foot together return about half the mechanical energy of the body hitting the ground.[13]

The next time you are huffing and puffing, feeling sloth-like, try to remember what figure 24 shows. Because speed is a function of how fast you move your legs (stride rate) times how far you travel with every stride (stride length), it graphs speed against both stride length and rate for a good human runner versus a similar-sized greyhound and a horse that weighs six times more. When Dennis Bramble and I first plotted these data, we were astounded. For a given endurance speed, good human runners have lower stride rates and similar stride lengths to a full-sized horse. In contrast, dogs have much shorter strides and faster stride rates. In other words, in the endurance speed range humans jump as well as horses. If the horse speeds up, however, we are toast because sprinting humans cannot further extend their stride lengths and can speed up only by increasing stride rate, which is costly and inefficient. Within seconds, the horse will leave the human in the dust. But the horse will have to slow down eventually, often because of heat.

Beyond our high-performance legs, the most vital and unique adaptation that enables humans to go the extra mile is our ability to perspire profusely. Running generates copious body heat, warming us pleasantly on cold days but turning dangerous in high heat and humidity. If we can't dump this heat, we must stop running or suffer from heatstroke because body temperatures above 41°C cook

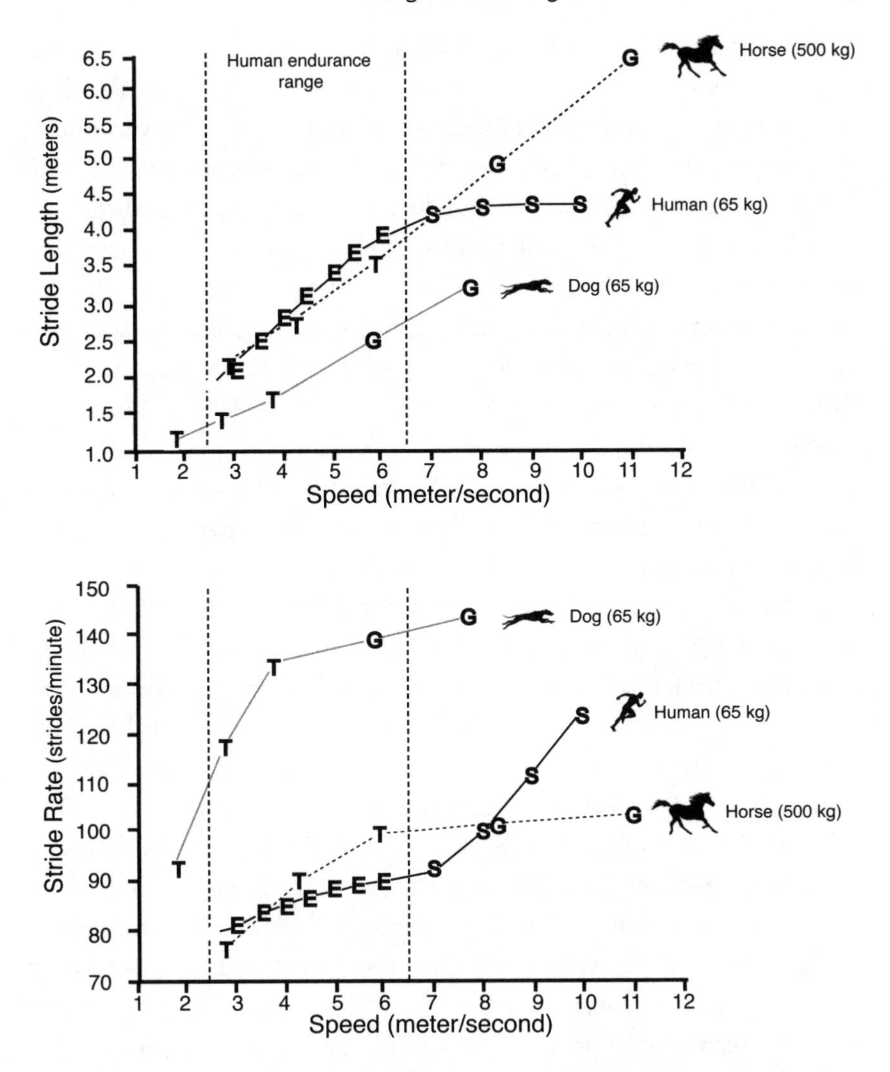

FIGURE 24 Speed versus stride length (*top*) and stride rate (*bottom*) for a good human runner, greyhounds, and horses. E, endurance range; S, sprint range; T, trot; G, gallop. Note that humans closely match horses despite being about seven times smaller. (Modified from Bramble, D. M., and Lieberman, D. E. [2004], Endurance running and the evolution of *Homo, Nature* 432:345–52)

cells in the brain and elsewhere. Like all mammals, we cool using the miracle of evaporation: when heat turns water into steam, the energy lost chills the skin underneath. Most animals take advantage of this natural refrigeration by panting—taking short, shallow breaths to evaporate saliva in their throats and on their tongues. As the water evaporates and cools the skin, blood in the veins just beneath is also cooled. This chilled blood then cools the rest of the body. Panting, however, suffers from two constraints. First, no matter how slobbery, tongues, mouths, and noses provide just a small surface area for cooling. Even more problematic, when dogs and other quadrupeds gallop, they lose the ability to pant because galloping is a seesaw gait that slams the guts like a piston into the diaphragm with every stride.[14] When a quadrupedal animal shifts from a trot to a gallop, it must stop panting and synchronize each breath with each stride. (You can test this by taking a dog for a run, but be sure not to make it gallop for too long on a sweltering day or you might overheat it.)

At some point, humans evolved a magnificent cooling system by taking advantage of special water-secreting glands that most animals have only on their paws. Monkeys and apes have small quantities of these so-called eccrine glands elsewhere on their bodies, but we alone have five to ten million sweat glands all over our skin, especially on our heads, limbs, and chests.[15] Sweating effectively turns the entire body into a giant, wet tongue. We also lost our fur, which helps air move along the skin's surface without any barrier, thus enabling us to rapidly dump prodigious quantities of heat. A few animals in the horse family as well as camels also sweat (lather), but these animals sweat less effectively and only with the oily glands that we too have in our armpits and groin. All told, humans are the sweating champions of the animal world. When running in the heat, humans can sweat one liter per hour (sometimes even more), enough to keep cool while racing a marathon in 90°F—something no other animal can do.[16]

Humans have even more adaptations. At rest, the heart pumps about four to six liters of blood each minute, but during running it must pump as much as five times more to supply hardworking muscles and cool the body. A typical runner's heart pumps twenty to twenty-four liters a minute, and an elite runner's can reach an impressive thirty-five liters a minute. My colleagues Rob Shave and Aaron Baggish and I showed that ordinary humans, like horses and other endurance-adapted animals, evolved voluminous and elastic heart chambers that differ markedly from the smaller, thicker, stiffer hearts of apes and that enable us to efficiently squeeze large volumes of blood with every beat.[17] We also have an elaborated blood supply to the brain to help cool this vital organ during exercise.[18] And the leg muscles of ordinary humans usually have 50 to 70 percent fatigue-resistant slow-twitch fibers, far more than chimpanzees, which range from 11 to 32 percent.[19] Humans who train for speed can increase the size of their fast-twitch fibers, but ordinary humans from every population are still slow-twitch dominated, and thus capable of more endurance than apes.[20]

Almost all the features I just reviewed that help humans run are what biologists term convergent, which means they evolved independently in humans and other animals adapted for running. However, because we are unsteady bipeds that evolved from tree-climbing apes, running humans are uniquely prone to falling. A slight shove or an unfortunately placed banana peel is more likely to topple a running human than any quadruped. Because a sprained ankle or broken wrist could be a death sentence in the Stone Age, it makes sense that humans evolved a suite of unique and crucial features for stabilization. My favorite of these adaptations is the enlarged gluteus maximus, the largest and plausibly most shapely muscle in the human body. If you walk a few steps with your hands on your buttocks, notice that this muscle is mostly dormant, but when you start to run, feel how it clenches up forcefully with every step. Experiments in my lab showed that this impressive muscle primarily acts to prevent the trunk from falling forward following each landing.[21] Other adaptations to keep us stable during running include the ability to rotate our trunks as we pump our arms opposite our legs,[22] and the previously mentioned nuchal ligament, which helps keep our heads from jiggling too much.[23]

Even if you dislike running, your body is loaded with features from head to toe that help you run long distances efficiently and effectively. Because many of these features don't help us walk or do anything else, they appear to have evolved as adaptations for running. It is no fluke that ordinary runners can compete with horses in marathons. But why?

Power Scavenging and Persistence Hunting

For decades, I have racked my brain to try to explain why our ancestors evolved so many adaptations to run long distances so exceptionally, and the only plausible explanation I can offer is to get meat.

In today's era of farming and supermarkets, you might be a vegetarian and, if not, I doubt you eat any dead animals you happen upon. Further, if you hunt, you probably do it chiefly for sport. But these are very modern attitudes. No hunter-gatherer spurns the chance to obtain free meat even if it isn't very fresh, and hunting is a highly valued way to acquire nutrient-rich food and achieve status. Yet until relatively recently, scavenging and hunting were difficult, dangerous, and almost impossible to do without running.

Scavenging probably came first. Imagine you are a hungry early hominin, maybe a *Homo habilis,* two or three million years ago. You are small, slow, and weak and lack any weapons more lethal than a stick or a rock. As you trek about the African landscape in search of food, it is highly unlikely you'll come across any dead animals worth scavenging because these are precious, evanescent resources. Almost as soon as an animal dies or its carcass is abandoned by lions or other predators, it becomes a locus of mayhem as vultures, hyenas, and other scavengers vie for whatever meat is left. Any interested, hungry scavenger had better be fast and willing to take on the snarling, vicious competition.

The likely solution for vulnerable hominins who had the ability to run moderate to long distances was a strategy called power scavenging that is still practiced by foragers such as the Hadza and the San.[24] A typical scenario begins when you see vultures circling far in the distance, a telltale sign of a carcass below. If it's the middle of the day, so hot, and you run, you have a good chance of getting there before the hyenas, which are less adapted to run in the heat. If you can then chase off vultures, you have a good chance of getting something to eat including the marrow-rich bones that lions cannot consume.

Hominins probably first started to eat meat by scavenging, yet by 2 million years ago there is clear archaeological evidence that they also hunted large animals like wildebeests and kudu.[25] This is easier said than done without serious weapons. The bow and arrow was invented less than 100,000 years ago, and putting stone points on spears was invented maybe 500,000 years ago.[26] Before these weapons, hominin hunters would have had to get close to their prey, sometimes by thrusting spears into them. Please don't try this. Meat is a nutritious food, but given the risk of being kicked or gored while getting up close and personal with an angry wildebeest, it's remarkable that more of our hunter-gatherer ancestors weren't vegetarian.

Early hominins might have hunted in several ways, but as David Carrier perceptively proposed in 1984, one strategy must have been persistence hunting. Although this ancient form of hunting is poorly known today, numerous anthropologists and explorers have described how people employed persistence hunting in different cultures and environments on every continent save Antarctica.[27] Some of the most detailed accounts come from Louis Liebenberg, a conservationist who has spent decades working with San hunter-gatherers of the Kalahari and has run with these persistence hunters.[28] My colleagues and I also interviewed elderly Tarahumara men in Mexico to record their experiences doing this kind of hunting when they were younger.[29] Other colleagues have described how hunters in the Amazon ran after peccaries. Despite the incredulity of many Westerners about this hunting method, these and other lines of evidence indicate that persistence hunting used to be widespread.

There are several ways to persistence-hunt. One is to take advantage of the distinctive human ability to not overheat while running. At the hottest time of the day, a group of hunters will chase an animal—the bigger the better because larger animals, like larger humans, overheat and tire faster. At first, the prey will inevitably gallop away faster than the hunters, who typically jog at a relaxed pace. Then while the poor animal pants to cool down, the hunters relentlessly track it, often while walking, so they can chase their prey again before it has cooled. This cat-and-mouse game of chasing, then tracking, is repeated again and again. Assuming the hunters resume the chase before the animal fully cools, its body temperature will gradually keep rising until, eventually, it reaches a state of heatstroke and collapses. A hunter can then walk right up to the animal and dispatch it safely without sophisticated weapons (sometimes using just a rock). According to Liebenberg's detailed records from more than a dozen hunts in the Kalahari, the average distance traveled was slightly longer than a half marathon, and the runners walked about half the time and ran at a ten-minute-per-mile pace, a moderate jogging speed. Liebenberg emphasizes that the most challenging aspect of these hunts is not the running but the ability to track using clues such as footprints, traces of blood, and knowledge of the animal's likely behavior.[30]

This method of chasing and tracking can also work in colder weather by driving animals to exhaustion or injury. The Tarahumara sometimes pursue deer in the winter over long distances, the Kalahari San chase large antelopes in sandy soils that tire out their prey, and Saami hunters in northern Scandinavia reportedly used cross-country skis to chase reindeer on powdery snow, which is especially fatiguing for the animals, eventually causing them to collapse.[31]

A related type of hunting that involves running, sometimes for long distances, is to drive animals into natural or artificial traps where the prey can be easily and safely killed. One common strategy, well documented among Native Americans, is for runners to pursue deer or other prey toward ravines, cliffs, or bogs or toward manufactured traps such as ditches, spikes, nets, or blinds that conceal waiting hunters. As with hunts in the heat, multiple hunters take part in these chases, working together strategically and using their knowledge of the environment and their prey.[32] The anthropologist Norman Tindale described how pairs of aboriginal hunters effectively chased kangaroos by "taking advantage of the animal's tendency to always run in the arc of a wide circle. One youth cuts across and takes up the running as the other becomes exhausted."[33]

Unsurprisingly, almost no one persistence hunts anymore unless they are obliged to like Alexander Selkirk, the inspiration for Robinson Crusoe, who ran down feral goats while marooned on a South American island.[34] Hunters today have guns, dogs, and other innovations, and wildlife is scarcer. The San have also been banned from hunting. However, if we were able to travel back in time a few thousand years, all over the world we would see people using hunting methods that involve running.

Happily, we now have safer, easier, and more reliable ways to get meat (or be vegetarians), but long-distance running had other benefits for our ancestors. People also ran to make war, honor gods, impress the opposite sex, and have fun. Among many Native American peoples, long-distance running is celebrated in footraces and sports like lacrosse, and some forms of running are a form of prayer like the Tarahumara men's footrace, the *rarájipari,* and the women's footrace, the *ariwete.*[35] These traditions remind us that running was never just for men. If you attend any major race today, women make up half the runners thanks to pioneers like Bobbi Gibb and Kathrine Switzer.[36] Although men do most of the hunting in hunter-gatherer societies, women also sometimes persistence-hunt, and they run races both sacred and secular.[37]

But our evolutionary history as runners raises a conundrum. If humans evolved to run, why do so many runners get injured?

Should I Run to the Doctor's Office?

When my midlife crisis hit, instead of leaving my wife and buying a sports car, I coped by training for my first marathon. Then came searing pain. As I rolled out of bed and took those first few stumbling steps to the bathroom, it felt as if some invisible fiend had burrowed into my feet and started to stab my soles with a scalpel. After a few minutes the twinges would subside, only to return later with a vengeance. Like many runners I had developed plantar fasciitis, inflammation of the thick band of connective tissue that runs like the string on a bow below the arch of the foot. According to advice I read on the internet, I needed to change my running shoes more frequently. As a shoe's elasticity deteriorates, the built-in arch support loses effectiveness, putting extra strain on the plantar fascia. So I rushed to my local running shoe store, bought new shoes, and the problem gradually cleared up. After that, I made sure to buy new shoes every three months despite the expense. Over the next year, however, I battled an irritated Achilles tendon and other mysterious pains that I worried might sideline me permanently.

Ironically, while I was obsessively buying new shoes, I was also beginning to study how people run without shoes. This research was kick-started at a public lecture on a dark and stormy night soon after Dennis Bramble and I had published our "Born to Run" paper in *Nature.* In the front row of the lecture hall was a bearded fellow wearing socks wrapped in duct tape. Following the talk he introduced himself as Jeffrey and asked an excellent question: "How come I don't like to wear shoes, even when I run?" My knee-jerk reactions were "You must be crazy to run barefoot!" and "Of course we must have evolved to run barefoot!" Because Jeffrey lived in Boston, I asked him if he wouldn't mind coming to the lab. A few days later he kindly showed up with his heavily callused feet, and we recorded data on how he ran. Unlike me and most of the runners I had measured, Jeffrey landed as light as a feather on the balls of his feet (a "forefoot strike") thus avoiding the impact peak and resulting shock wave normally caused by landing on the heel. It was a eureka moment. Because humans had been running for millions of years before shoes were invented, maybe we evolved to run this way to avoid the pain of landing on our heels without a cushioned shoe? If true, could this insight help prevent common running injuries?

Over the next few years my students and I started to study barefoot running. Jeffrey was part of a community of American runners, previously unbeknownst to me, who eschewed shoes. Barefoot Jeffrey introduced me to Barefoot Preston, Barefoot Ken Bob, and others who used the honorific "Barefoot," and soon we were studying barefooters from all over the United States, measuring how they ran, and working out the biomechanics of landing on the heel versus ball of the foot. Then we went to Kenya to study runners who had never worn shoes and published our research in another article in *Nature* that was featured on the cover with the title "Tread Softly: How We Ran in Comfort Before We Started Wearing Shoes."[38] Our paper presented and tested a model of how landing on the ball of the foot avoids the impact force caused by landing on the heel, and we showed that habitually barefoot people like Jeffrey usually (but not always) run this way. We speculated that humans evolved primarily to forefoot strike when running, and called for research to test if this running style, common among elite runners, might prevent injuries.

Many committed and would-be runners worry that running is intrinsically damaging. Apart from the dangers of falling and other traumatic mishaps, regularly pounding the pavement is widely thought to accumulate excess wear and tear, not unlike driving a car too many miles. The resulting damage is often termed an overuse injury. Studies claim such injuries afflict between 20 and 90 percent of runners in a given year, suggesting that millions of runners must be overdoing it.[39] The most prevalent site of these injuries is the knee; other common injuries include shin splints, tibial stress fractures, Achilles tendonitis, pulled calf muscles, plantar fasciitis, toe stress fractures, and lower back pain.[40] But most of all knees. I cannot count how many people (doctors included) have told me they blew out their knees by running too much. Is running really so injurious, and if we evolved to run long distances, why aren't our bodies better adapted?

One hypothesis out there is that running injuries are mismatch conditions like type 2 diabetes and myopia caused by our bodies being poorly adapted to the modern environments in which we now live. According to this way of thinking, running injuries would be less prevalent if we ran barefoot as we evolved to. While some running injuries may indeed be mismatches, a problem with this idealistic way of thinking is that almost everything, including physical activities, involves trade-offs and risks. Despite having evolved to get pregnant, eat, and walk, pregnant mothers often get back pain, people sometimes choke on their food, and people everywhere trip and sprain their ankles. Why would running be any different?

Another hypothesis is that our bodies are actually marvelously adapted to running and that its dangers have been exaggerated. If something like 80 percent of the world's millions of runners were dropping like flies from injuries, then doctor's offices would be overflowing with injured runners, and joggers would eventually be as rare as hen's teeth. Analyses of the combined evidence from hundreds of small studies show that running injury rates follow a U-shaped curve: the highest probabilities of injury are among novices radically increasing their mileage, competitive speedsters, and marathoners, but everyday runners in between these extremes are much less prone to problems.[41] For example, my colleagues and I found that three out of four of the middle- and long-distance runners on the Harvard cross-country team got significantly injured every year, but these athletes were running nearly two thousand miles per year at blistering speeds.[42] In contrast, only one in five novice Dutch joggers who run modest distances at sensible speeds experience any injury whatsoever, many of them trivial.[43]

We should do everything we can to prevent runners from getting hurt, but it also behooves us to puncture some pervasive myths about running injuries, the biggest of which is that wear and tear from hoofing too many miles will erode the cartilage in your knees and hips and give you osteoarthritis. Not so. Despite what many doctors and others assume, more than a dozen careful studies show that nonprofessional runners are no more likely to develop osteoarthritis than non-runners.[44] In fact, running and other forms of physical activity help promote healthy cartilage and may protect against the disease. A study from my lab showed that people's chances of getting knee osteoarthritis at a given age and weight have doubled over the last two generations as we have become *less,* not *more,* active.[45] Even so, injuries still do occur, most frequently in the knee. Leaving aside traumas like sprained ankles, is there anything we can do to prevent running injuries?

One obvious way to stay injury-free is to adapt one's body to the physical demands of running. Even a jogger like my mother who runs five miles five times a week takes two million steps a year, potentially leading to "repetitive stress injuries" (a better term than "overuse injuries") from innumerable repetitions of moderately forceful movements such as slamming into the ground or pushing overly hard off one's toes. Stresses from these actions can potentially cause minute damage that at first is imperceptible. But if I keep putting more loads on my shinbone than it can handle, tiny fractures will accumulate slowly and perniciously. Eventually, I will experience sore shins and then painful shin splints that, if ignored, can grow into a full-blown stress fracture. Similar damage can accrue in other bones as well as cartilages, tendons, ligaments, and muscles. However, if my tissues are sufficiently strong to handle these stresses without becoming damaged, I won't get injured. The problem is that connective tissues like bones, ligaments, and tendons adapt considerably more slowly than muscles and stamina. Novice runners, especially first-time marathoners, risk injury because they can increase their mileage or speed (or both) faster than their shins, toe bones, Achilles tendons, IT bands, and other vulnerable tissues can adapt. Many experts thus advocate increasing mileage only 10 percent a week.[46]

Another concern is muscle strength. We use muscles not just to push our bodies forward but also to control movements and reduce loads that can stress and injure tissues. Runners with plenty of stamina but weak core muscles and stabilizing muscles in their feet and legs are more at risk of injuries to the knee and elsewhere.[47] The muscles alongside the hip (termed hip abductors) that prevent the knees from collapsing dangerously inward during every step are a notorious weak link.[48]

But, clearly, the body can adapt. In 2015, I observed a demonstration of this principle when I followed eight amateur runners who ran 3,080 miles across the United States to raise money to combat childhood obesity. For six months, the runners, who ranged from their twenties to their seventies, ran about a marathon a day with only one day of rest per week. In addition to measuring their biomechanics, I asked these courageous souls to keep a daily log of their injuries. For the first few weeks they reported a typical list of afflictions from knee pain to blisters. After a month, however, their reported injuries slowed to a trickle as their bodies adapted. Of the fifty total injuries reported by all eight runners, three-quarters were in the first month and none were in the final month.

Beyond *how much* we run, another potential way to reduce our chances of injury is *how* we run. If repetitive stress injuries arise from innumerable recurrent, forceful movements, it stands to reason that some ways to run must be less stress inducing than others. Running lightly and gently, however, is easier said than done, and in my experience many runners are unaware of their form. We just lace on a pair of comfortable shoes and go. Many coaches also pay little attention to running form. This individualistic approach is epitomized by the hypothesis that each of us develops a preferred and most efficient running style involving our stride rate, how much we lean, how our foot strikes the ground, and how much we flex our hips, knees, and ankles. As long as we stick to this form, we are less likely to get injured.[49]

An anthropological approach combined with what we know about running biomechanics suggests a different perspective.[50] When I ask runners from different cultures if there is a best way to run, they invariably tell me they consider running a learned skill. As the anthropologist Joseph Henrich has shown, humans in every culture master critical skills by imitating people who are good at them.[51] Just as it makes sense to hit tennis balls like Roger Federer, doesn't it make sense to run like Eliud Kipchoge or other great runners? Tarahumara runners tell me they learn to run properly by following champions of the ball-game races. Kenyan runners do the same, often honing their skills in groups, which I have sometimes joined outside the city of Eldoret. Soon after the sun rises, about ten to twenty runners meet near a local church. One person always takes the lead as we start jogging slowly away from town, and as we follow him, I think, "Okay, I can do this!" But gradually we speed up until, gasping for breath, I have to drop out as the other runners laugh and wish me luck. Apart from drawing motivation from each other, participants in these group

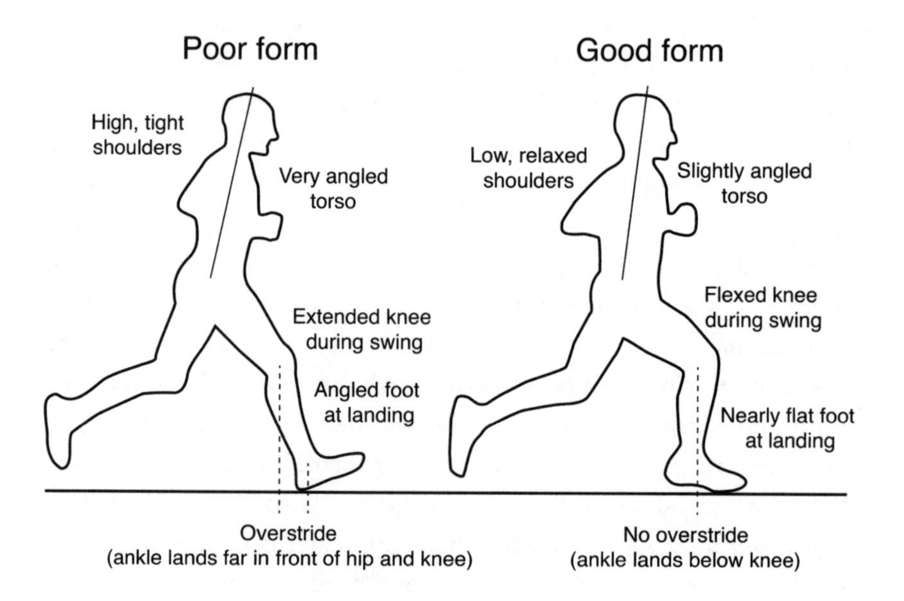

FIGURE 25 Good running form (*on right*) compared with common poor form (*left*).

runs learn running form. Watch ten Americans training and you'll generally see ten different running styles, but a group of Kenyans often looks more like a flock of birds with the leader not just setting the pace but also modeling how to run so that the runners appear to move in unison, adopting the same cadence, arm carriage, and graceful kick.

But what is that form? It is immensely challenging to study the relationship between injury and running form because repetitive stress injuries take months or years to accumulate, each body is different, and retrospective studies are complicated by not knowing if someone's form resulted from or caused his or her past injuries. Many studies measure only how running form relates to forces hypothesized to cause injury like how hard a runner hits the ground. With these caveats in mind, I think most experienced runners and coaches agree on four key, related elements illustrated in figure 25: (1) not overstriding, which means landing with your feet too far in front of your body; (2) taking about 170–180 steps a minute; (3) not leaning too much, especially at the waist; (4) landing with a nearly horizontal foot, thus avoiding a large, rapid impact force with the ground.[52]

1. Avoid overstriding. Get your knees up when you swing your legs forward so you land with a vertical shank and your foot below the knee, not too far in front of the hips. This prevents the legs from landing too stiffly and causing overly high breaking forces that slow you down.
2. Step rate usually increases with speed, but experienced endurance runners generally take 170–180 steps a minute regardless of speed. They thus speed up economically by jumping farther (running is jumping from one leg to another), and a high step rate prevents overstriding.
3. Lean forward slightly, but not too much at the waist. Too much upper-body lean requires you to spend more energy preventing your torso from toppling forward, and it encourages overstriding.
4. Land gently with your feet nearly horizontal. If you are barefoot and don't overstride, it is almost impossible not to land on the

ball of your foot before letting down your heel in what is called a forefoot or mid-foot strike. Forefoot and mid-foot strikes usually don't generate an impact peak on the ground—a rapid, large collisional force that is painful without shoes. Forefoot and mid-foot strikes also generate rotational forces (torques) that are lower in the knee but higher in the ankle, requiring strong calf muscles and Achilles tendons, which can lead to problems for people trying to transition to this way of running. If you change how your foot lands, do so gradually and build up strength.

I am repeatedly struck by how the aspects of good form seen in figure 25 characterize what I and others have observed among habitually barefoot runners in different parts of the world.[53] And that brings up the final, most controversial way to prevent injury: what we run on, namely shoes and running surfaces. Although some born-to-runners claim that cushioned, modern shoes inevitably cause injury, this is hyperbole. To be sure, shoes may lead to weak feet and encourage you to land hard on your heel, but millions of shod runners do just fine this way. Further, contrary to some claims, taking off your shoes doesn't necessarily make you run well, and plenty of people run beautifully in shoes. However, bare feet provide plentiful sensory feedback that is dampened in a shoe.[54] If you run far at a moderate speed on a hard surface without shoes, you simply have to run lightly and gently, which people usually accomplish by adopting a barefoot running style that includes a high step rate, not overstriding, and landing on the ball of the foot. I don't think it coincidental that this style, which most people probably used for millions of years until the modern running shoe was invented in the 1970s, may be a better way to run. Keep in mind that our running ancestors never ran on hard, flat roads in which every step was like the last, and they almost certainly ran less frequently and more slowly than committed runophiles.

They also never trained. As we have seen again and again, exercise is a thoroughly modern phenomenon, and no one in the Stone Age ever practiced running for months or years to prepare to stand on one painted line and then run as fast as possible for 13.1 or 26.2 miles to another. Yet to many people, the notion that you could run a marathon without practicing seems preposterous. How do nonexercising farmers like the Tarahumara and hunter-gatherers who do persistence hunts manage to "train" for distance running? One key factor must be hours and hours of long-distance walking and other forms of work that help build strength and endurance. But strange as it may seem, I think another important form of training was dancing.

Won't You Join the Dance?

In 1950, Laurence Marshall, having made a fortune at Raytheon Company, retired and decided to spend some time with his family by taking them on an adventure. After consulting with Harvard anthropologists, he traveled with his wife, Lorna, and their children, Elizabeth and John, to the remote Kalahari Desert on the other side of the world. Over the next eight years, the Marshall family returned many times to the Kalahari—once spending eighteen continuous months—observing and documenting the life of the San people, who were then still hunter-gatherers.[55] In addition to writing several books about the San, the Marshalls brought home detailed notes, objects, and photographs. John, a talented cinematographer, also captured thousands of hours of film.

Many of John Marshall's remarkable films show the San walking endless miles as they forage and hunt, and a few films document persistence hunts that involved running. But one of them (see figure 26) records an altogether different but no less important kind of endurance physical activity often ignored by exercise scientists: dancing.

To set the stage, imagine a world with no doctors, organized religion, television, radio, books, or any of the other institutions and inventions we depend on to minister to our physical and spiritual needs as well as entertain and educate us. Yet in every nonindustrial society ever studied, including the San, dancing helps people do these things and more. According to the Marshalls and other observers, dances were not just enjoyable social gatherings that united everyone in the group but also important, frequent, and physically intense rituals that helped ward off evil and heal the sick.[56]

San medicine dances occurred about once a week, typically beginning after dusk when everyone hangs out by the fire. As both men and women sing joyously and clap to ancient wordless songs, a handful of men start dancing in a winding, twisting line around the group, stomping out the song's beat, often adding extra light steps. Men do most of the dancing, but women also dance a turn or two when the mood is upon them. As the night draws on and the fervor of the hypnotic dance steadily increases, more men join in, and by dawn some begin to enter a trancelike state, which they call "half-death." Dancers in a trance emit unworldly sounds, and their movements become uncontrolled: their hands flutter, their heads shake, and they sometimes dash about or lie trembling on the ground. The San believe there is great power in this half-conscious state, which frees the medicine men's spirits to communicate between this and other worlds, to draw out manifest sicknesses and as yet unrevealed ills, and to protect people from unseen but lurking dangers.

FIGURE 26 San dance, Namibia. (Photo gift of Laurence K. Marshall and Lorna J. Marshall © President and Fellows of Harvard College, Peabody Museum of Archaeology and Ethnology, PM2001.29.14990)

The San don't dance to get fit, but dancing all night once a week requires and develops phenomenal endurance. Further, their dancing traditions are the rule, not the exception. I know of no nonindustrial culture in which men and women didn't dance for hours on a regular basis. The Hadza, for example, sometimes dance joyfully after dinner until the wee hours, doing line dances that involve some of the most sexually suggestive moves I have ever seen. On dark moonless nights the Hadza also perform the sacred *epeme* dance to heal social rifts and bring good hunting luck.[57] The Tarahumara have three or four different kinds of dances that often last between twelve and twenty-four hours and in some communities happen as many as thirty times per year.[58] As the Norwegian explorer Carl Lumholtz remarked of the Tarahumara in 1905, "Dance with these people is a very serious and ceremonious matter, a kind of worship and incantation rather than amusement."[59] Even the infamously repressed English used to dance much more than they now do. In Jane Austen's time, balls could go on all through the night. In *Sense and Sensibility,* Mr. Willoughby danced "from eight o'clock till four, without once sitting down."[60]

Dancing isn't running, but it's usually more fun and such a universal, valued form of human physical activity that we should consider it another gait akin to running. Indeed, while dancers sometimes use their legs like stilts as in a walk, most often they jump like runners from one foot to the other. And like long-distance running, dancing can go on for hours, requiring stamina, skill, and strength.

One rarely considered parallel between running and dancing is how both can induce altered states. Long periods of vigorous exercise stimulate mood-enhancing chemicals in the brain including opioids, endorphins, and, best of all, endocannabinoids (like the active compound in marijuana). The result is a runner's or dancer's high. I've never danced all night, but sometimes on a long, hard run I feel euphoric and relaxed, and my perception of sights, sounds, and smells becomes heightened. Blue things become bluer, and I hear every singing bird, honking car, and footstep with astonishing clarity. I hypothesize this intensified state of awareness evolved to help running hunters track animals. Ultra-runners report that after many miles they sometimes enter a trance state like San medicine dancers, and in one account Louis Liebenberg describes how he felt himself transform into a bull kudu on a persistence hunt with San hunters in the Kalahari.[61]

═══

For millions of years, humans regularly jumped from one leg to the other for hours on end as they ran or danced. Although these vigorous physical activities were never done as exercise to get fit and stay healthy, they nonetheless built up endurance capabilities for exactly the sort of running people once had to do: long, broken up, and not that fast.

Running and dancing are also lifelong pursuits. Elderly couples light up the dance floor all over the world, and I've seen Tarahumara runners in their eighties ticking along for mile after mile (remember Ernesto), not unlike the Boston legend Johnny Kelley, who ran the Boston Marathon every year from age twenty-one until eighty-four. A few years ago, near the finish line of the New York City Marathon, I saw a ninety-year-old man trotting along with the biggest grin I have ever seen, enjoying enthusiastic cheers from thousands of spectators lining the course. As I passed him, also applauding his effort, I remember thinking how I'd like to be like him should I ever be lucky enough to reach that age. Is his ability to run a marathon at ninety a fluke or a consequence of all those miles he ran over the course of his long life?

TEN

Endurance and Aging: The Active Grandparent and Costly Repair Hypotheses

MYTH #10 It's Normal to Be Less Active as We Age

"You are old, Father William," the young man said,
"And your hair has become very white;
And yet you incessantly stand on your head—
Do you think, at your age, it is right?"

—Lewis Carroll, *Alice's Adventures in Wonderland*

Everyone wants to live long, but no one wants to get old. So for centuries people have sought ways to slow aging and defer death. Desperation makes for lucrative business opportunities. Not long ago, quacks would have tried to lure you to consume tobacco, mercury, or ground-up dog's testicles to postpone your eternal rest; today's peddlers of immortality hawk human growth hormone, melatonin, testosterone, megadoses of vitamins, or alkaline food.[1] For millennia, however, the most sensible advice has always included exercise. Just about everyone knows what countless studies confirm: regular physical activity slows the aging process and helps prolong life. I doubt anyone was astounded when Hippocrates wrote twenty-five hundred years ago, "Eating alone will not make a man well; he must also take exercise."[2] Endurance promotes endurance.

But if we never evolved to exercise, why is it so beneficial? And how do we explain commonplace exceptions to the nearly universal advice that exercise can help us live longer? Consider, for example, the different fates of two men named Donald born at the end of World War II whose exercise habits couldn't have been more different.

Donald Trump needs little introduction. Born in 1946 to wealthy parents, he was sent to a military academy where presumably he had to participate in sports. Although a teetotaler and nonsmoker, Trump famously enjoyed eating abundant junk food and large steaks, drinking Diet Coke, getting little sleep, and avoiding any form of exercise apart from golf. According to biographers, "Trump believed the human body was like a battery, with a finite amount of energy, which exercise only depleted. So he didn't work out."[3] Trump became overweight in middle age and was prescribed medications to lower his cholesterol and blood pressure. Yet the medical evidence provided to the public in 2018 purported he was in good health with normal blood pressure (116/70) and satisfactory cholesterol levels.[4] Whatever your opinions of Trump, decades of avoiding vigorous exercise did not keep him from becoming the forty-fifth president of the United States at the age of seventy.

Donald Ritchie was born two years earlier than Trump on the other side of the Atlantic in Scotland. A competitive runner from childhood, Ritchie gradually worked his way from 440-yard races as a teenager to ultramarathons as an adult. A man who found marathons unchallenging, he set numerous world records including running 100 miles in 1977 in eleven hours thirty minutes and fifty-one seconds, an astonishing pace of just under seven minutes per mile. He once ran 844 miles from the northern tip of Scotland to the southwesternmost corner of England in just over ten days, averaging more than three marathons a day despite a nasty chest cold. By his own reckoning, Ritchie ran more than 208,000 miles over the course of his life.[5] Yet he developed diabetes at age fifty-one—a disease not associated with healthy, fit athletes. Ritchie's disease was a rare case of adult-onset type 1 diabetes in which his immune system destroyed the cells in his pancreas that made insulin. He kept running anyway. In one mind-numbing race when he was fifty-six, he ran 136 miles in twenty-four hours without stopping. In the end, however, Ritchie did have to stop because his high blood sugar levels triggered a series of cardiovascular problems including a blocked carotid artery, irregular heartbeats, high blood pressure, and a series of ministrokes. Ritchie died in 2018 at the age of seventy-three.

How do we square the contrasting medical fates of these two Donalds? I would be mad to argue that exercise doesn't slow aging and increase the chances of a longer life, but is exercise oversold as an anti-aging elixir? Was Donald Trump just lucky and Donald Ritchie unfortunate? Or perhaps Donald Ritchie would have died younger than seventy-three if he hadn't been so active, and maybe Donald Trump would be physically and mentally healthier in his seventies had he exercised?

Perhaps Trump's exercise is his work. Standing in front of rooms of people talking and gesticulating involves physical activity, although not vigorous, and Trump spent more time playing golf than any U.S. president in history (albeit using golf carts to get around the links).[6] Indeed, Trump remained active long past the age many people retire. And if there is any time to take it easy, shouldn't it be retirement? By the time you are sixty-five, don't you deserve to put your feet up, head to the golf course, play bridge, fish, go on a cruise, or whatever you find relaxing?

Not so according to the experts. They urge us to ignore aberrations like the two Donalds and pay attention to the mountain of evidence that the Fountain of Youth runs with sweat. That sweat, moreover, needs to keep flowing as we age.

One of the most venerable long-term studies on how exercise affects aging is the Cooper Center Longitudinal Study in Dallas, started in 1970 by the man who coined the term "aerobics," Dr. Kenneth Cooper. One of its analyses tracked more than ten thousand men and three thousand women older than thirty-five to test if physically fit exercisers lived longer and healthier lives. They generally did. After adjusting for age (because an older person is more likely to pass away in a given year than a younger person), Cooper found that the most physically fit men and women had mortality rates about one-third to one-fourth of those who were least fit.[7] Further, a subsample of those who were initially out of shape but started to exercise and increased their fitness halved their age-adjusted mortality rate compared to those who remained inactive and unfit.[8] Because there is more to health than not being dead, Cooper Center researchers also tracked over the decades more than eighteen thousand healthy middle-aged individuals to see who got chronic health conditions such as diabetes and Alzheimer's. Among both women and men, those who were more fit were about half as likely to suffer from chronic disease and, if they got sick, did so at an older age.[9] These and other such studies lend credence to the saying "Men do not quit playing because they grow old; they grow old because they quit playing."

Thinking back to our comparison of the two Donalds, I understand why many people are heedless or skeptical of the sorts of statistics I just cited. Everyone knows athletes who died tragically young and sedentary people who survived to old age. Further, as this book has relentlessly argued, the Trumps of this world who avoid needless physical activity are simply doing what we evolved to do, especially as we age. Finally, even if "exercise is medicine," *how* and *why* does physical activity affect how bodies age? As we have seen in previous chapters, such questions behoove us to look beyond studies of just Westerners and use evolutionary anthropological perspectives. They also require us to grapple with the age-old problem of why we get old in the first place. As it happens, humans age uniquely.

Old Age over the Ages

So many of my memories of my grandparents involve them feeding me and my brother. Top marks go to my mother's mother, whose specialty was breakfast. On weekends this first meal was a multicourse extravaganza that usually began with half a grapefruit, then hot cereal, then bagels with cream cheese and smoked salmon. Although less of a cook, my father's mother never appeared without her signature sugar-free oatmeal cookies. My grandfathers got in the game, too.

My mother's father would drive around Brooklyn on Sunday mornings stopping at one deli for the best smoked salmon, another place for whitefish, and yet another for the perfect bagels. My father's father always showed up with a giant salami and a tin of Dutch cocoa.

In hindsight, my grandparents were doing in their own Brooklyn way what human grandparents—alone among species—have been doing for millions of years: feeding their grandchildren. This unique behavior is strongly linked to our species' exceptional longevity in which we typically live beyond the age at which we cease to reproduce. Similarly long post-reproductive life spans are rare in the animal world. Chimpanzees, for example, seldom survive past the age of fifty, soon after females go through menopause and after the age when males sire offspring.[10] Kicking the bucket shortly after one has stopped producing and raising offspring makes sense from an evolutionary perspective. At this stage, organisms enter what the biologist Peter Medawar termed the "shadow of natural selection."[11] Theoretically, once an individual falls into this dreaded shadow, it becomes biologically and evolutionarily obsolete because natural selection should no longer act to combat natural processes of aging.

Thankfully, elderly humans are anything but biologically obsolete. To understand how our extraordinary reproductive strategy rescued us, at least partially, from the coldhearted shadow of selection, consider that ape females raise just one dependent offspring at a time without much help. Chimpanzee mothers, for example, cannot give birth to babies faster than once every five to six years because they forage only enough food every day to sustain their caloric needs plus those of one hungry youngster. Not until her juvenile is old enough to be fully weaned and forage for itself can she muster enough calories to become fertile again. Human hunter-gatherers, in contrast, typically wean their offspring after three years and become pregnant again long before their little ones are able to feed or fend for themselves, let alone stay out of danger. A typical hunter-gatherer mother, for example, might have a six-month-old infant, a four-year-old child, and an eight-year-old juvenile. Because she is usually capable of gathering only about two thousand calories a day, she cannot get enough food to provide for her own substantial caloric needs, which exceed two thousand calories, as well as the needs of her several offspring, none of whom are old enough to forage on their own.[12] She needs help.

Among those who lend her a hand are middle-aged and elderly folks. Anthropologists have shown that grandmothers, grandfathers, aunts, uncles, and other older individuals in foraging populations from Australia to South America remain active throughout life, gathering and hunting more calories every day than they consume, which they provide to younger generations.[13] This surplus food helps provide adequate calories to children, grandchildren, nieces, and nephews and reduces how much work mothers have to do. Elderly hunter-gatherers also help younger generations by contributing knowledge, wisdom, and skills for about two to three decades beyond childbearing years. Contrary to the widespread assumption that hunter-gatherers die young, foragers who survive the precarious first few years of infancy are most likely to live to be sixty-eight to seventy-eight years old.[14] That's not far off from the life expectancy in the United States, which is currently between seventy-six and eighty-one.

The evidence that hunter-gatherers stay physically active for several decades after they stop having children is fundamental for understanding the nature of human aging. Most especially, our unique system of intergenerational cooperation, especially food sharing, postpones Medawar's grim shadow. Instead of becoming obsolete, middle-aged and elderly hunter-gatherers bolster their reproductive success by provisioning children and grandchildren, doing child care, processing food, passing on expertise, and otherwise helping younger generations. Once this novel cooperative strategy—the essence of the hunting and gathering way of life—started to emerge during the Stone Age, natural selection had the chance to select for longevity. According to this theory, hardworking and helpful grandparents who looked out for others and who were blessed with genes that favored a long life had more children and grandchildren, thus passing on those genes.[15] Over time, humans were evidently selected to live longer to be generous, useful grandparents.[16] One version of this idea is known as the grandmother hypothesis in recognition of the evidence that grandmothers play especially important roles.[17]

In order to elucidate the links between exercise and aging, I propose a corollary to the grandmother hypothesis, which I call the active grandparent hypothesis. According to this idea, human longevity was not only selected for but also made possible by having to work moderately during old age to help as many children, grandchildren, and other younger relatives as possible to survive and thrive. That is, while there might have been selection for genes (as yet unidentified) that help humans live past the age of fifty, there was also selection for genes that repair and maintain our bodies when we are physically active. As a result, many of the mechanisms that slow aging and extend life are turned on by physical activity, especially as we get older. Human health and longevity are thus extended both by and for physical activity.

Another way of stating the active grandparent hypothesis is that human longevity did not evolve to enable elderly humans to retire to Florida, sit by the pool, and ride around in golf carts. Instead, old age in the Stone Age meant plenty of walking, digging, carrying, and other forms of physical activity. In turn, natural selection favored older individuals whose bodies stimulated repair and maintenance mechanisms in response to the stresses caused by these activities. And because middle-aged and elderly humans never had the opportunity to retire and kick up their heels, there was never strong selection to turn on these mechanisms to the same degree without the stresses caused by physical activity.

Once again let's travel to Hadzaland for a glimpse of what kinds and amounts of physical activity grandparents did in the Stone Age. A typical workday for a Hadza grandmother begins soon after dawn, tending to the fire and helping feed and care for her youngsters. A few hours later she and other women in camp head out into the bush. They bring with them infants under two, whom they carry in slings on their backs, and they are accompanied by children older than six or seven and sometimes an armed man or a couple of teenage boys to provide protection. Finding a good place to dig sometimes involves an hour-long trek. Once they find the vines that signal the presence of underground roots and tubers that are the staple of the Hadza diet, the women settle down to excavate. The main equipment is a digging stick, a thin piece of hardwood about the size of a cane whose end has been sharpened and hardened in a fire. Digging is arduous work because many tubers hide several feet deep under rocks that must be pried out, but the women chat as they work until midafternoon. Usually, everyone takes a break at midday for lunch. As with most Hadza cuisine, tubers are simply thrown on a fire and roasted for a few minutes and consumed there and then. After lunch comes yet more digging, and eventually the group heads home, carrying with them in slings whatever tubers were not yet consumed.

All Hadza women dig, but grandmothers dig more than mothers in part because they don't have to nurse or spend as much time taking care of little ones. According to measurements by Kristen Hawkes and colleagues, a typical Hadza mother forages about four hours a day, but grandmothers forage on average five to six hours a day.[18] On some days they dig less and spend more time collecting berries, but overall they work longer hours than mothers do. And just as grandmothers spend about seven hours every day foraging and preparing food, grandfathers continue to hunt and to collect honey and baobab fruits, traveling just as far on most days as younger men do. According to the anthropologist Frank Marlowe, "Old men are the most likely to fall out of tall baobab trees to their deaths, since they continue to try to collect honey into old age."[19]

How many elderly Americans dig several hours a day, let alone climb trees and hunt animals on foot? We can, however, compare how much Americans and Hadza walk. A study of thousands found that the average twenty-first-century woman in the United States aged eighteen to forty walks 5,756 steps a day (about two to three miles), but this number declines precipitously with age, and by the time they are in their seventies, American women take roughly half as many steps. While Americans are half as active in their seventies as in their forties, Hadza women walk twice as much per day as Americans, with only modest declines as they age.[20] In addition, heart rate monitors showed that elderly Hadza women actually spent *more* of their day engaged in moderate to vigorous activity than younger women who were still having children.[21] Imagine if elderly American women had to walk five miles a day to shop for their children and grandchildren, and instead of pulling items off the shelves, they had to dig for several hours in hard, rocky soil for boxes of cereal, frozen peas, and Fruit Roll-Ups.

Not surprisingly, hard work keeps elderly hunter-gatherers fit. One of the most reliable measures of age-related fitness is walking speed—a measure that correlates strongly with life expectancy.[22] The average American woman under fifty walks about three feet per second (0.92 meter per second) but slows down considerably to two feet per second (0.67 meter per second) by her sixties.[23] Thanks to an active lifestyle without retirement, there is no significant age-related decline in walking speed among Hadza women, whose average pace remains a brisk 3.6 feet per second (1.1 meters per second) well into their seventies.[24] Having struggled to keep up with elderly Hadza grandmas, I can attest they maintain a steady clip even when it is blisteringly hot. Older Hadza men also walk briskly.

Although elderly hunter-gatherers remain active as they age, they nonetheless age. Researchers have quantified the effects of aging on strength and fitness in hunter-gatherers from Africa and South America by measuring handgrip strength, maximum numbers of push-ups and pull-ups, and fifty-meter-dash times. These mini-Olympics show that men peak athletically in their early twenties, after which their strength and speed decline about 20 to 30 percent by their mid-sixties. Women are less strong and fast, but suffer smaller declines with age. Among the Aché, a foraging tribe in the Amazon, peak aerobic capacity (VO_2 max) in women stays impressively high throughout adulthood with no evidence of a decline among older individuals; VO_2 max falls with age among men, but even sixty-five-year-old Aché grandfathers have aerobic capacities well above the average for forty-five-year-old American men.[25] Overall, hunter-gatherers attain higher levels of strength and fitness than typical postindustrial Westerners and lose these capacities at a slower rate, remaining reasonably vigorous into old age. Debilitating muscle loss is not a problem among foragers.

The active grandparent hypothesis raises a classic chicken-or-egg question. How much do humans live to old age so they can be active grandparents helping younger generations, or how much does their hard work cause them to live long lives in the first place? Is human longevity a *result* of physical activity or an *adaptation* to stay physically active? In addition, how did our hunter-gatherer ancestors deal with the inevitable selective shadow when they could no longer hunt and gather? Some countries today have nursing homes, pensions, and government-funded health care to take care of senior citizens. Although elderly hunter-gatherers are afforded great respect, those who can't walk long distances, dig tubers, collect honey, and schlep stuff home presumably become burdens when food is limited. It follows that if humans were selected to live long after we stopped having babies, we were probably not selected to live those years in a state of chronic disability. From a Darwinian perspective, the best strategy is to live long and actively and then die fast when you become inactive. An even better strategy, however, would be to avoid any deterioration with age in the first place.[26]

The Essence of Senescence

Sometimes when I look in the mirror, I don't recognize the gray-haired fellow with a receding hairline who stares back. Happily, I don't yet feel as old as I look. Aging is inexorable, but senescence, the deterioration of function associated with advancing years, correlates much less strongly with age. Instead, senescence is also influenced strongly by environmental factors like diet, physical activity, or radiation, and thus can be slowed, sometimes prevented, and even partly reversed. The distinction between aging and senescence may seem obvious, but the two processes are frequently confused. Many conditions occur more commonly with advancing age, but only some are actually caused *by* age. Menopause, for example, is a normal consequence of aging that happens when a woman's ovaries run out of eggs. In contrast, type 2 diabetes occurs among some older people for reasons not intrinsic to the aging process itself but instead from factors like obesity and physical inactivity whose damaging effects accumulate with age.

Stated differently, some aspects of senescence are neither inevitable nor universal.[27] As we age, not all of us will get high blood pressure, dementia, or incontinence. In addition, some species seem immune to senescence. Bowhead whales, giant tortoises, lobsters, tortoises, and some clams can live and reproduce after hundreds of years. (The world record is a clam named Ming who researchers dredged up and, in the cruelest of ironies, then killed to determine it was 507 years old.)[28] How and why do certain animals and humans, including those who exercise, tend to senesce more slowly?

At a mechanistic level, we senesce from a multitude of nasty processes that damage cells, tissues, and organs. One worrying source of wear and tear arises from the chemical reactions that keep us alive. The oxygen we breathe generates energy in cells but leaves behind unstable oxygen molecules with free, unpaired electrons. These *reactive oxygen species* (charmingly also called free radicals) steal electrons indiscriminately from other molecules, thereby "oxidizing" them. That theft sets off a slow chain reaction by creating other unstable, electron-hungry molecules obliged to steal electrons from yet more molecules. Oxidation burns things gradually and steadily. Just as oxidation causes metal to rust and apple flesh to brown, it damages cells throughout the body by zapping DNA, scarring the walls of arteries, inactivating enzymes, and mangling proteins. Paradoxically, the more oxygen we use, the more we generate reactive oxygen species, so theoretically vigorous physical activities that consume lots of oxygen should accelerate senescence.

A related driver of senescence is *mitochondrial dysfunction*. Mitochondria are the tiny power plants in cells that burn fuel with oxygen to generate energy (ATP). Cells in energy-hungry organs like muscles, the liver, and the brain can have thousands of mitochondria. Because mitochondria have their own DNA, they also play a role in regulating cell function, and they produce proteins that help protect against diseases like diabetes and cancer.[29] Mitochondria, however, burn oxygen, creating reactive oxygen species that, unchecked, cause self-inflicted damage. When mitochondria cease to function properly or dwindle in number, they cause senescence and illness.[30]

Another self-sabotaging reaction that results from being alive and using energy is browning, technically *glycation*. Browning occurs when sugar and protein react with the help of heat. Glycation gives cooked foods like baked bread and roasted meat their dark, aromatic, tasty exteriors, but what's good for cookies is bad for kidneys. These reactions can damage tissues and produce compounds (*advanced glycation end products*) that stiffen blood vessels, wrinkle skin, harden the lenses in our eyes, clog up kidneys, and more. These and other kinds of damage then trigger inflammation.

As we have learned, the immune system stimulates inflammation to defend us from pathogens as well as self-inflicted damage caused by physical activity. In short bursts, inflammation is lifesaving, but low levels of inflammation that last for months or years are pernicious because they slowly attack our bodies. Over time, the destructive effects of chronic, simmering inflammation accumulate in cells and tissues from head to toe including neurons in the brain, cartilage in joints, the walls of arteries, and insulin receptors in muscle and fat cells.

If oxidation, mitochondrial dysfunction, mutations, glycation, and inflammation were not enough, plenty of other processes also contribute to senescence by damaging and degrading cells. Over time, tiny molecules glue themselves to the DNA in cells. These so-called epigenetic (on top of the genome) modifications can affect which genes are expressed in particular cells.[31] Because environmental factors like diet, stress, and exercise partly influence epigenetic modifications, the older we are, the more of them we accumulate.[32] Most epigenetic modifications are harmless, but the more you have for a given age, the higher your risk of dying.[33] Other forms of senescence include cells losing the ability to recycle damaged proteins,[34] inadequately sensing and acquiring nutrients,[35] and (less likely) being unable to divide because the little caps (telomeres) that protect the ends of chromosomes from unraveling have become too short.[36]

If this list of aging mechanisms alarms you, it should. Altogether, these mechanisms slowly wreak havoc. Plaque builds up in blood vessels, causing them to stiffen and clog. Receptors on cells become clogged. Muscles lose their mojo. Crud builds up around neurons and other key cells. Brain cells die. Membranes tear. Bones dwindle and crack. Tendons and ligaments fray. Our immune systems become less able to fight infections. Unless we repair these and many other forms of damage, our bodies become more vulnerable to breaking down like cars driven too many miles.

But there is hope. Aging and senescence are not inextricably linked, because to varying degrees most of these destructive processes can be prevented or slowed and the damage they cause can be mended. Oxidation, for example, is halted by antioxidants, compounds that bind with reactive oxygen species, thus rendering them harmless. Some antioxidants like vitamins C and E come from food, but our bodies synthesize many other antioxidants in abundance. Similarly, mitochondria can be regenerated, and some products of glycation can be repaired by enzymes that scavenge or break down these compounds.[37] Inflammation can be turned off by anti-inflammatory proteins produced by white blood cells and muscles; telomeres can be lengthened; DNA can be repaired; and cells can be induced to restore or repair dozens of functions. Indeed, almost every cause of aging in almost every tissue (with a few glaring exceptions like hardening of the eyes' lenses) can be countered, fixed, or prevented by one mechanism or another.

The body's multitude of anti-aging mechanisms raises a conundrum: Why don't more humans—who we've already seen were selected to live longer than most animals—employ them earlier and more often to slow senescence and keep useful grandparents healthier for even longer?

Evolutionary biologists have been pondering this question for generations, and to cut to the chase, the best explanation by far is that natural selection becomes weaker as we age.[38] On account of diseases, predators, harsh weather, and other cruelties of nature (what biologists euphemistically term "extrinsic mortality"), older individuals inevitably become less common. As a result, natural selection acts less intensely on genes in elderly people that prolong life and promote repair. So even if middle-aged and old people help out younger generations, thus postponing Medawar's selective shadow, the older we get, the less natural selection cares about fighting the accumulation of wear and tear that comes with age.[39] Like it or not, the shadow eventually comes. But thankfully, its arrival and severity can be slowed and reduced by physical activity.

The Costly Repair Hypothesis

In the mid-1960s, a team of physiologists in Dallas decided to compare the effects of sedentariness with those of exercise on health by paying five healthy twenty-year-olds first to spend three weeks in bed and then to undergo an intensive eight-week exercise program. The bed rest was ruinous. When they were finally allowed to arise from their beds, the volunteers' bodies resembled forty-year-olds' by many metrics: they were fatter, had higher blood pressure, higher cholesterol levels, less muscle mass, and lower fitness.[40] The eight ensuing weeks of exercise, however, not only reversed the deterioration but in some cases led to net improvements. To the lead researcher, Bengt Saltin, the take-home message was simple: "Humans were meant to move." Time marches on, however, and to evaluate how aging affects the effects of inactivity, researchers had the bright idea of restudying the same five volunteers thirty years later.

Three decades of typical American lifestyles had not been kind to the original volunteers: they had each gained about fifty pounds, had higher blood pressure and weaker hearts, and were less fit and healthy in numerous ways. But they agreed to be studied once more as they tried to undo the consequences of thirty sedentary years with a six-month program of walking, cycling, and jogging. Fortunately, this second late-in-life exercise intervention helped the volunteers lose about ten pounds and, most astoundingly, largely reversed their decline in cardiovascular fitness. After six months of moderate exercise, the average volunteer's blood pressure, resting heart rate, and cardiac output returned to his twenty-year-old level.[41] Many other studies confirm the anti-aging benefits of exercise.[42] But few of them explain *why*.

The most common explanation for why exercise slows and sometimes turns back the gradual slide toward poor health is that physical activity prevents or ameliorates bad things that accelerate senescence. Top of the list is fat. Exercise staves off and sometimes reverses the accumulation of excess fat, especially belly fat, a chief cause of inflammation and other problems. Exercise also lowers bloodstream levels of sugar, fat, and unhealthy cholesterol that slowly contribute to hardening of the arteries, damage proteins, and otherwise gum up the works. And as trials like the Dallas Bed Rest and Training Study show, exercise also improves cardiovascular function, lowers levels of stress hormones, revs up metabolisms, strengthens bones, and more. Yet these and other salubrious effects of exercise explain only *how* but not *why* physical activity combats senescence. To understand *why* physical activity activates dozens of processes that maintain function and repair some of the damage that accumulates with age, we need to explore what I term the costly repair hypothesis.

To introduce this idea, let's follow (with her permission and the help of figure 27) what happened to my wife when she included a hard gym workout as part of a typical Saturday. Figure 27 plots time on the *x*-axis against the calories she expended on the *y*-axis. I've broken down her total energy expenditure (TEE) into the calories she spent on her resting metabolism (RMR) versus being active (her active energy expenditure, AEE). As you can see, for the first few hours that day, my wife was either sedentary or did light physical activities. Then, at 10:00 a.m., she went to the gym and did forty-five minutes of vigorous cardio before a demanding forty-five-minute workout with weights. Unsurprisingly, my wife's AEE shot up during her ninety minutes of exercising; afterward, she was not only tired but also slightly sore. Crucially, however, once she stopped exercising, her RMR did not immediately return to its previous level. Instead, her RMR remained slightly elevated for hours in a state technically known as excess post-exercise oxygen consumption (EPOC) but informally called an afterburn.

That afterburn may help explain how and why physical activity can help slow senescence. Importantly, my wife's exercise session was not only calorically costly but also physiologically stressful. As she struggled to complete her cardio and weight workouts, her body's "fight and flight" system released cortisol, epinephrine, and other stress-related hormones to speed up her heart and mobilize her energy reserves. As her muscles rapidly consumed calories, they pumped out waste compounds that compromised her cells' functions, and her mitochondria leaked an abundance of harmful reactive oxygen species that damaged DNA and other molecules throughout the body. To add injury

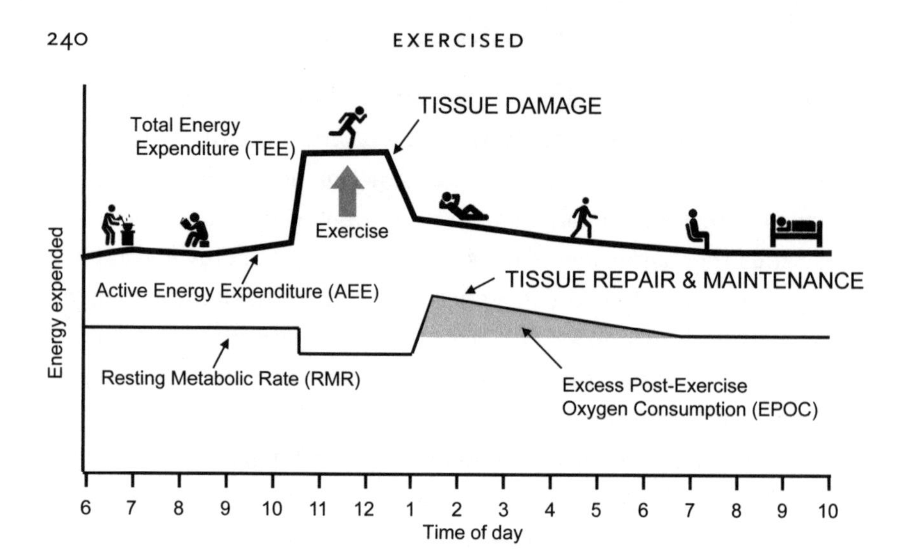

FIGURE 27 Costly repair hypothesis. Representation of total energy expenditure (TEE), resting metabolic rate (RMR), and active energy expenditure (AEE) over the course of a day showing how energy use changes before, during, and after a bout of exercise. AEE is low before exercise, goes up during exercise, and then falls again. However, RMR can remain elevated for several hours after exercise as the body recovers, replenishes energy stores, and repairs damage.

to insult, her hardworking muscles also developed microtears as she struggled with heavy weights. All in all, beyond causing discomfort, my wife's strenuous workout generated some short-term damage.

If exercise is so destructive, why is it healthy? One explanation is that once she stopped exercising, my wife's body reacted by repairing whatever harm she caused and, crucially, also repairing some of the damage that she had accumulated beforehand when she wasn't exercising. As a result, she restored many tissues to their previous state. Among these repair and maintenance responses, her "rest and digest" system slowed her heart rate, lowered her cortisol levels, and shuttled unused energy back into her muscle and fat cells, replenishing her energy reserves. To deal with the tissue damage caused by her workout, she mounted an initial inflammatory response followed by a later anti-inflammatory response. She also produced copious, powerful antioxidants to mop up the reactive oxygen species unleashed by her mitochondria. And she turned on a host of other processes to rid her cells of waste products and repair DNA mutations, damaged proteins, and epigenetic modifications, as well as mend cracks in her bones, replace and add mitochondria, and more.[43] Although these maintenance and repair processes are not nearly as costly as her workout, they nonetheless require calories, thus elevating her resting metabolic rate very slightly for some time. Studies show that people's afterburns can last from two hours to two days depending on the intensity and duration of the physical activity.[44]

While exercise restores most structures (what biologists term homeostasis), in some cases it may make things even better than before (this is termed allostasis). For example, demanding physical activities can increase the strength of bones and muscles, increase cells' abilities to take up glucose from the blood, and both augment and replace mitochondria in muscles. In addition, repair mechanisms sometimes overshoot the damage induced by exercise, leading to a net benefit. It's like scrubbing the kitchen floor so well after a spill that the whole floor ends up being cleaner. Among other effects, while physical activity initially stimulates inflammation, especially via muscles, it subsequently causes muscles to produce an even stronger, more lasting, and more widespread anti-inflammatory response whose long-term effect is less inflammation not just in the affected muscle but elsewhere.[45] As a result, physically active people tend to have lower baseline levels of inflammation. In addition, exercise causes the body to produce more antioxidants than necessary, decreasing overall levels of oxidative stress.[46] Exercise also causes cells to clean out damaged proteins, lengthen telomeres, repair DNA, and more. All in all, the modest physiological stresses caused by exercise trigger a reparative response yielding a general benefit, a phenomenon sometimes known as hormesis.[47]

If you are entrepreneurial, hate exercise, or both, these beneficial responses might have ignited a lightbulb. Instead of going through the bother and discomfort of exercising, why not find some easier, preferably consumable way to turn on the same maintenance and repair mechanisms?[48] Why not just take a pill? Without breaking a sweat, I can buy vitamins C and E and beta-carotene to boost my antioxidant levels, and purchase capsules loaded with turmeric, omega-3 fatty acids, and polyphenols that fight inflammation. These and other elixirs are sometimes marketed with the blessing of doctors and scientists. Linus Pauling, who won two Nobel Prizes, wrote a book modestly titled *How to Live Longer and Feel Better,* which claimed you could extend your life by twenty to thirty years with massive doses of vitamin C.[49]

Pauling was a brilliant chemist, but his advocacy of vitamin C was quackery. Dozens of studies have found that taking antioxidant pills is no substitute for physical activity to fight senescence. A comprehensive review published in 2007 examined sixty-eight clinical trials that compared the effects of commonly prescribed antioxidants like vitamin C with placebos on more than 230,000 people. Three or four studies reported a modest benefit, but the rest found that antioxidants provided no benefit or even increased the risk of dying.[50]

To add insult to injury, additional studies suggest antioxidants may sometimes do more harm than good when combined with exercise. This head-turning conclusion followed from a groundbreaking 2009 experiment by Michael Ristow. Ristow's team asked forty healthy young males with varied fitness levels to undergo four weeks of supervised exercise. Half the participants were given large doses of vitamins C and E; the other half received a placebo. Muscle biopsies taken before and after their exercise bouts showed that, as expected, physical activity induced plenty of oxidative stress, but those who took antioxidants incurred *more* oxidative damage because their bodies produced much lower levels of their own antioxidants.[51] The antioxidant pills apparently suppressed the body's normal antistress response probably because oxidative damage from exercise itself is needed to trigger the body's health-promoting antioxidative defense mechanisms. Along the same lines, it is possible that eating lots of carbohydrates during exercise diminishes the body's anti-inflammatory response.[52]

An alternative to coddling one's body with products that mimic the effects of exercise is to try non-physically active forms of suffering. This kind of "no pain, no gain" philosophy has inspired a dizzying array of self-inflicted hardships thought to ward off aging (an added benefit is their aura of virtue). Hoping to live longer, people take cold showers, restrict their caloric intake, endure long periods without eating, shun carbohydrates, burn their digestive tracts with spicy food, and more.[53] Some of these strategies are downright questionable, and, with the exception of intermittent fasting, none is yet supported by solid evidence as a way to extend human longevity.[54]

Why is regular physical activity the best way to delay senescence and extend life?

Recall that according to the costly repair hypothesis, organisms with restricted energy supplies (just about everyone until recently) must allocate limited calories toward either reproducing, moving, or taking care of their bodies, but natural selection ultimately cares only about reproduction. Consequently, our bodies evolved to spend as little energy as possible on costly maintenance and repair tasks. So while physical activities trigger cycles of damage and restoration, selection favors individuals who allocate enough but not too much energy to producing antioxidants, ramping up the immune system, enlarging and repairing muscles, mending bones, and so on. The challenge is to maintain and repair any damage from physical activity just enough and in the right place and the right time.

Evolution's stingy solution to this problem is to *match capacity in response to demand*. In this case, the demand is the stress caused by physical activity, especially reactive oxygen species and other damaging processes that stiffen arteries, mutate genes, and gunk up cells. The capacity is the ability to maintain, often through repair, a stable internal environment so we can adequately and effectively perform those functions needed for survival and reproduction. And, crucially, the maintenance and repair mechanisms activated by physical activity don't cease to function as we age. Although some become less responsive, they keep on ticking, allowing physically active post-reproductive individuals to slow or delay senescence.

Unfortunately, this marvelous system has one big flaw. Apparently, we never evolved to activate these maintenance and repair responses as effectively in the absence of regular physical activity. As we have seen, almost no one in the Stone Age, least of all grandparents, managed to avoid hours of walking, running, digging, climbing, and other manual labors. Hunter-gatherers of all ages would have stimulated their body's natural reparative mechanisms nearly every day in response to the demands posed by their way of life. So just as our species never evolved to diet or cope with jet lag, we never evolved to counter many aging processes to the same degree without physical activity. Absence of regular physical activity thus becomes a mismatch condition as we age by allowing us to senesce faster.

Unless of course you are Donald Trump or one of the millions of others who live into old age without walking ten thousand steps a day, let alone running marathons and pumping iron in the gym. Do growing legions of evidently healthy elderly exercise avoiders suggest that the long-term benefits of physical activity may be exaggerated?

Extending or Compressing Morbidity

My high school required us to take a class in health education that replaced a semester of physical education. Instead of climbing ropes or playing basketball, we were sent to a dingy classroom underneath the school gym where our portly, florid teacher often paced up and down the aisles, thumbs tucked behind his suspenders, and lectured loudly about health. I don't remember a single fact or counsel he imparted apart from his declaring that he could tell from our earlobes if we smoked pot and that 90 percent of all smokers who get lung cancer die. When a smart aleck friend (not me, really) asked if it wasn't the case that *all* people who get lung cancer eventually die, he roared back that, no, only 90 percent died. Then, one shocking day, we showed up to class and were informed by a substitute that our health teacher had died of a heart attack—an unintended, posthumous lesson.

We all die of something, and if you are following my argument in this chapter skeptically—as you should—that includes physically active people who eat sensibly and do everything else they are supposed to. In fact, despite being told to exercise, more inactive people like Donald Trump are living longer and in better health today than ever before.

To evaluate this conundrum, let's look closely at the probabilistic relationship between death (mortality) and illness (morbidity). Too often, statistics on aging focus on life span without also considering health span (the length of time spent in good health without morbidity). A useful way to think about both life span and health span is to graph functional capacity (a measure of health) on the *y*-axis versus time on the x-axis, as shown in figure 28. Someone who is generally healthy is at nearly 100 percent functional capacity most of the time, despite occasional, temporary illnesses. Then, at some point, age-related senescence commences and functional capacity declines because of serious illness, eventually leading to death.

For thousands of generations, the health span and life span of a typical hunter-gatherer who did not die in infancy probably looked something like the top graph in figure 28. The graph is based on medical surveys of hunter-gatherers, who mostly die from respiratory and infectious diseases, violence, and accidents and have a relatively low incidence of long-term chronic noninfectious conditions.[55] These and other data indicate that about two-thirds of older hunter-gatherers remain at high functional capacity with limited morbidity until just before death, which most often occurs in the seventh decade. Accordingly, their health span and life span are very similar.

Advances in public health and medical science have changed health span and life span in ways both good and bad, illustrated by the bottom graph in figure 28. The bad news is that despite impressive advances in preventing and treating infectious diseases, many people today get sick from chronic noninfectious diseases that involve many years of morbidity prior to death. In medical jargon, this longer period of illness prior to death is termed the *extension of morbidity*. Among westernized populations, many people become sick for a long time before they die from heart disease, type 2 diabetes, Alzheimer's, and chronic respiratory disease; many also suffer from osteoarthritis, osteoporosis, and a growing list of autoimmune diseases.[56] At least one in five Americans over the age of sixty-five is in fair or poor health. Despite this high morbidity, we nonetheless live much longer than our farmer ancestors, and a little longer than hunter-gatherers. The average American in 2018 lives to be seventy-eight years old, almost twice as long as one a hundred years ago.[57]

This shift, in which more of us live longer but die from chronic rather than infectious diseases, thus extending morbidity, is known as the epidemiological transition and widely hailed as medical progress. By not dying rapidly from smallpox in our youth, aren't we fortunate to die slowly from heart disease at an older age? This thinking is

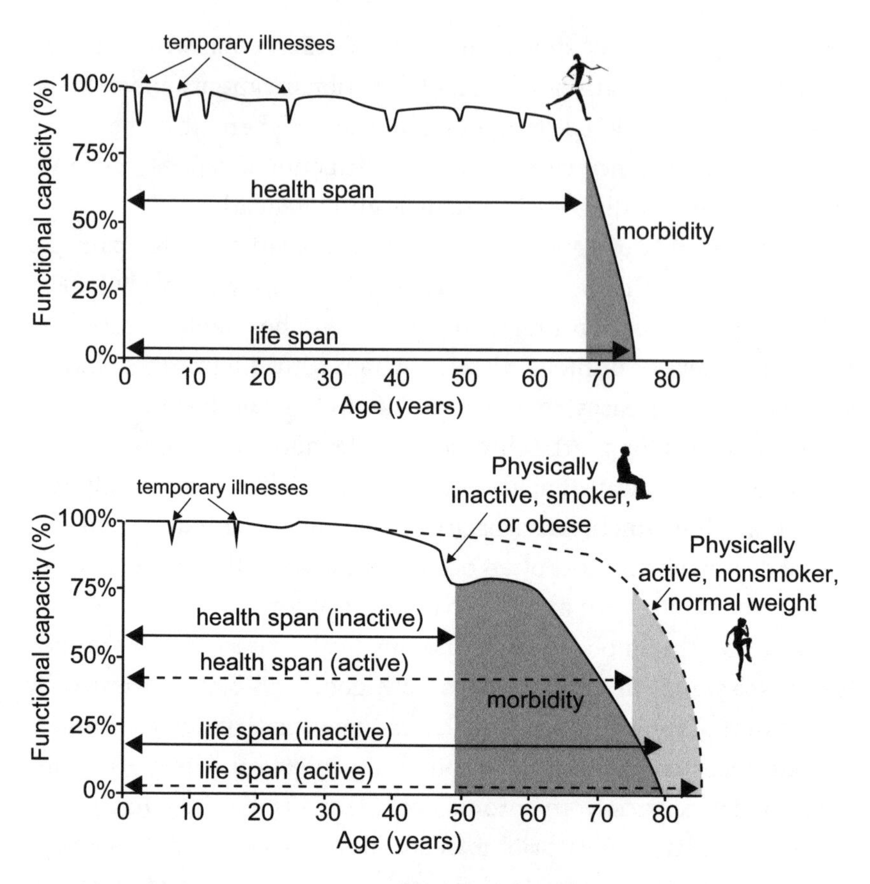

FIGURE 28 How physical activity affects health span (morbidity) more than life span (mortality). Typical health span and life span in hunter-gatherers (*top*) compared with industrialized people (*bottom*) who are physically active or inactive.

mistaken. My last book, *The Story of the Human Body,* made the case that many of the diseases that kill us slowly today are *mismatch diseases* caused by our bodies being imperfectly or inadequately adapted to modern environmental conditions like smoking, obesity, and physical inactivity.[58] Although these diseases are commonly classified as diseases of aging because they tend to arise when we are middle-aged, they are not *caused* by age, nor should they be considered inevitable consequences of aging. Plenty of people live to old age without getting these diseases, which rarely if ever afflict elderly hunter-gatherers and many aged people who live in subsistence societies.

If many so-called diseases of aging are preventable, it follows that a slow demise at the end of life is not inevitable. In a celebrated study, the Stanford medical professor James Fries showed that preventive medicine could help people stay healthier for longer through a *compression of morbidity*. Fries initially based his argument on a massive study that measured life span, disability, and three risk factors for disease (high body weight, smoking, and lack of exercise) among more than twenty-three hundred alumni from the University of Pennsylvania. Predictably, the alums with two or more risk factors died 3.6 to 3.9 years earlier than those with one or no risk factors; perhaps more impressively, they lengthened their period of disability prior to death by 5.8 to 8.3 years.[59] Simply put, an unhealthy lifestyle affects morbidity twice as much as mortality.

Fries's compression of morbidity model, illustrated by the bottom graph in figure 28, is a useful way to think about the effects of physical activity on aging. In a nutshell, persistent physical inactivity along with smoking and excess body fat are the biggest three factors that influence the likelihood and duration of the major illnesses that kill most people who live in industrial, westernized contexts.[60] Although two of three Americans' death certificates state they died of heart disease, cancer, or stroke, the deeper underlying causes of these illnesses were most likely smoking cigarettes, obesity, and physical inactivity.

Because people who are inactive are often overweight and sometimes smoke, it can be difficult to isolate the effects of just physical activity on morbidity and mortality. One effort to do this is the Stanford Runners Study conducted by, once again, James Fries. In 1984, he and his students began studying more than five hundred members of amateur running clubs along with more than four hundred healthy but physically inactive controls. Back then, the subjects were over the age of fifty and healthy: few smoked, none drank heavily, and none were obese. Then, for the next twenty-one years, Fries and his colleagues patiently kept track of each subject's physical activity habits, administered a yearly disability questionnaire that measures functions like the ability to walk, dress, and do routine activities, and recorded the year and cause of every death that occurred.

Fries and colleagues had to wait two decades for the results, but I have summarized them for you instantaneously in figure 29. They are worth a careful look. The top graph plots the runners' and non-runners' probability of not dying in a given year against time; the graph below plots disability against time. As you can see, the healthy non-runners died at increasingly faster rates than the runners and by the study's end were about three times more likely to pass away in a given year. In terms of cause of death, the non-runners were more than twice as likely to die of heart disease, about twice as likely to die

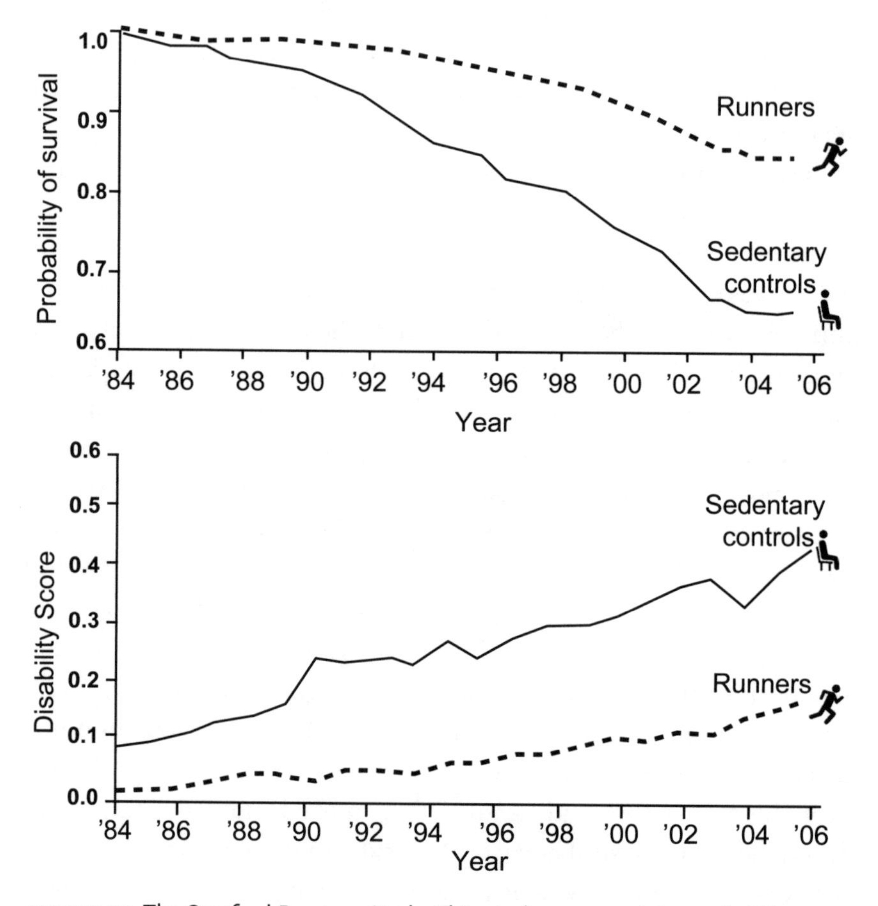

FIGURE 29 The Stanford Runners Study. This study measured the probability of surviving in a given year (*top*) and disability (*bottom*) over two decades in a group of amateur runners over fifty in 1984 compared with a group of healthy but sedentary controls. After more than twenty years the runners had 20 percent higher survival rates and 50 percent less disability. (Modified with permission from Chakravarty, E. F., et al. [2008], Reduced disability and mortality among aging runners: a 21-year longitudinal study. *Archives of Internal Medicine* 168:1638–46)

of cancers, and more than three times as likely to die of neurological diseases. In addition, they were more than ten times as likely to die of infections like pneumonia. Just as important, the disability scores plotted on the bottom show that the non-runners lost functional capacity at double the rate of the runners. By the end of the study their disability scores were more than twice as high as the runners', indicating that the runners' bodies were approximately fifteen years younger by this measure. In sum, running caused a compression of morbidity, thus also extending lives.

As Hippocrates would have predicted, scores of other studies on the effects of physical activity on morbidity and mortality yield similar results.[61] That doesn't mean, however, that physical activity is a surefire Fountain of Youth, and remember it doesn't delay mortality by preventing aging per se. Instead, physical activity triggers a suite of mechanisms that increase the chances of staying healthy with age by retarding senescence and preventing many chronic diseases that contribute over time to mortality. This logic raises three vitally important insights that help explain the Donald Trumps of the world who don't die young in spite of being sedentary and overweight.

First, and most fundamentally, the mortality and morbidity statistics I have been citing are *probabilities*. Eating sensibly and exercising don't guarantee long life and good health; they just decrease the risk of getting sick. By the same token, smokers have a higher risk of getting lung cancer, and individuals who are unfit or obese are more likely to get heart disease or become diabetic, but plenty don't.

Second, advances in medical care are shifting the relationship between morbidity and mortality.[62] Conditions like diabetes, heart disease, and some cancers are no longer imminent death sentences but instead can be treated or held at bay for years with drugs that maintain blood sugar levels, decrease harmful cholesterol levels, lower blood pressure, and combat mutant cells. In Donald Trump's case, for example, his reportedly normal blood pressure and cholesterol levels likely reflect the medications he takes to lower these risk factors.[63]

Finally, many complex environmental and genetic factors contribute to the probability of getting a disease, making it difficult to unravel causation. Twin studies have found that only about 20 percent of the variation in longevity up to the age of eighty can be explained by genes. If you make it to that age, however, your genes play a greater role in determining whether you will be a centenarian.[64] But that doesn't mean genes don't play an important role in disease. Genetic variations influence plenty of chronic diseases such as coronary artery disease, heart arrhythmias, type 2 diabetes, inflammatory bowel disease, and Alzheimer's.[65] In these and other cases, genes help load the gun, but environment pulls the trigger. In addition, the genes underlying most of these diseases tend to be rare and of tiny effect. I might have inherited hundreds of genes that increase my likelihood of getting heart disease, but each one's contribution to my getting sick is barely measurable.

Medical science has made impressive strides to keep us alive and kicking as we age, but practically speaking, the best advice for staying healthy as we age hasn't changed in centuries: don't smoke, avoid obesity, eat and drink sensibly, and of course stay physically active. Accordingly, we'll look more closely in later chapters at how much and what kind of exercise stimulate the repair and maintenance mechanisms that, with luck, may help us become active grandparents. Maybe even great-grandparents.

My wife's grandmother so dearly wanted to be a great-grandmother that she offered us a thousand dollars to have a child before we were married. We politely declined the offer and instead simply got married in good time and eventually produced a great-granddaughter that brought her much joy before she died at the ripe age of ninety-three. Had my grandmother-in-law been a hunter-gatherer, perhaps she would have offered us tubers and berries, but her basic urge to have and provision grandchildren and great-grandchildren is testimony to a deep and uniquely human instinct that extends back tens of thousands of generations.

To return to a previous question, the proclivity of our species to live for decades after we stop having babies may be partly an adaptation to provision children and grandchildren (and great-grandchildren) but also a consequence of staying physically active to provision these younger generations.[66] Either way, the upshot is that exercise is not just for the young. We evolved to be physically active as we age, and in turn being active helps us age well. Further, the longer we stay active, the greater the benefit, and it is almost never too late to benefit from getting fit. People who decide to turn over a new leaf and get fit after the age of sixty significantly reduce their mortality rate compared with others who remain sedentary.[67]

For most of us, however, the problem is not recognizing the benefits of physical activity but overcoming natural disinclinations to exercise at any age and figuring out how much and what kind of exercise to do.

PART IV

Exercise in the Modern World

ELEVEN

To Move or Not to Move: How to Make Exercise Happen

MYTH #11 "Just Do It" Works

Everybody is striving for what is not worth the having!

—W. M. Thackeray, *Vanity Fair*

The playwright and singer Noël Coward quipped he'd try everything once in his life except incest and folk dancing. Although my list of taboos is longer, I strive to try new experiences. So at 7:00 one morning I joined a group of health-care providers for an Ironstrength Workout, led by Dr. Jordan Metzl, a tall, lean, tireless advocate for exercise who does Ironman triathlons in his spare time. We were all attending a sports medicine conference, and beyond being curious, I signed up because we were assured it would be incredibly fun. Also, I was damned if I was going to be the only person at a conference on how "Exercise Is Medicine" who didn't show up to exercise.

So there we were before breakfast in the hotel garden beneath palm trees, all wearing our matching conference T-shirts. The sound of waves crashing into the hotel beach was drowned out by a boom box playing loud electro workout music to pump us up: exuberant, high-octane tunes with pulsating rhythms that keep building to new crescendos. After dividing into teams, we spent the next forty-five minutes racing from one exercise to the next—planks, squats, sit-ups, sprints, and burpees (a combined squat, push-up, and vertical jump)—constantly high-fiving each other and shouting encouragements. At the end, everyone was exhausted, and we all congratulated each other for our efforts, agreeing vociferously how much fun it was.

I enjoyed myself, but was it fun? I did the exercises as best I could, but what I actually enjoyed was the camaraderie, the beautiful setting, the high-fiving, and even the music. Afterward, I also enjoyed the feeling of having exercised intensely. But frankly, the planks, squats, sit-ups, sprints, and burpees were hard. The routine brought to mind the running guru George Sheehan's observation that "exercise is done against one's wishes and maintained only because the alternative is worse."

My reasons for exercising that morning accord totally with this book's mantra that we never evolved to exercise—that is, do optional physical activity for the sake of health and fitness. I participated because I felt I had to and it was supposed to be fun. For generation after generation, our ancestors young and old woke up each morning thankful to be alive and with no choice but to spend several hours walking, digging, and doing other physical activities to survive to the next day. Sometimes they also played or danced for enjoyment and social reasons. Otherwise, they generally steered clear of nonessential physical activities that divert energy from the only thing evolution really cares about: reproduction. The resulting paradox is that our bodies never evolved to function optimally without lifelong physical activity but our minds never evolved to get us moving unless it is necessary, pleasurable, or otherwise rewarding. Plunk us down in a postindustrial world, and we struggle to replace physical activity with exercise—an optional and often disagreeable behavior. Despite being badgered to exercise by doctors, trainers, gym teachers, and others, we often avoid it.

According to a 2018 survey by the U.S. government, almost all Americans know that exercise promotes health and think they should exercise, yet 50 percent of adults and 73 percent of high school students report they don't meet minimal levels of physical activity, and 70 percent of adults report they never exercise in their leisure time.[1]

Can an evolutionary anthropological approach help us do better?

If we evolved to be physically active because it was either *necessary* or *fun,* then isn't the solution to make exercise *necessary* and *fun* like my Ironstrength Workout?

If only things were that simple. Because exercise is defined as voluntary physical activity, it is inherently unnecessary. And for many people, especially those who are unfit, exercise simply isn't fun. That said, our social institutions try to accomplish these two goals for most youngsters. Throughout the world, recess, physical education, or sports are mandatory in some primary and secondary schools, and for some students these respites from the classroom are times to have fun.[2] Adults, however, are different, and I know of only one place in the world—an unusual company in Stockholm, Sweden—that has attempted to make exercise utterly necessary and also fun for every adult employee. Curious and a little bit skeptical, I swung an invitation to the Björn Borg sportswear company to see for myself.

Sports Hour at Björn Borg

If you were casting about for an actor to play a heroic Viking who does insane stuff like ski to the North Pole, you couldn't do much better than Henrik Bunge, CEO of the Björn Borg sportswear company, named after the legendary tennis player.[3] Henrik (who in fact *did* set a record skiing to the North Pole) is a tall, lean man with a chiseled face, blond hair, piercing blue eyes, and a broad, muscular torso that bulges out of his sweater. And if you want to work at his company, you'd better be ready to participate in the compulsory "Sports Hour." There are no exceptions, not even board members or visitors like me during the few dark, cold days I spent in December 2018 at the company's headquarters in downtown Stockholm.

Sports Hour at Björn Borg is every Friday at precisely 11:00 a.m. Before this mandatory weekly ritual, the company is eerily quiet. From 9:00 to 10:00, employees sit silently for an hour of reflection to contemplate their goals and think about how they can become better. As "Quiet Hour" ends, folks get cups of coffee, converse in hushed voices, and return to work for about forty-five minutes when the calm mood evaporates. All of a sudden the entire company is animated. Everyone in the building—from the CEO to the mail-room employees—grabs a sports bag and marches to a gym just a few blocks away. As I follow the throng, a few employees clap me on the back and assure me I'm going to love Sports Hour.

Thankfully, Henrik had supplied me with some Björn Borg exercise clothing (including their "high-performance underwear") so neither I nor my private parts feel entirely out of place in the large, cheerful gym where everyone is wearing the company's high-tech shirts, shorts, and underwear emblazoned with its distinctive logo. A handful of employees look as fit and athletic as Henrik, but most are ordinary Swedes in their thirties, forties, and fifties. Several are overweight, and one woman is very pregnant. Everyone gossips while stretching until precisely 11:00, when our trainer, Johanna, shows up, turns on loud pulsing music, and tells us to pair up with a partner for a CrossFit-type workout.

What a workout. I team up with Lena Nordin, the company's HR director, who is as passionate about the mandatory exercise program as Henrik, and for the following hour we do a nonstop exhausting mix

FIGURE 30 Sports Hour at Björn Borg. Henrik Bunge, the CEO, is at the front left. (Photo © Linnéa Gunnarsson)

of cardio and weights that I have to admit is fun. Instead of doing ordinary planks (which I hate), Lena and I do them together as a kind of tug-of-war. Johanna next has us do other paired exercises with weights and giant balls, then games involving squats, lunges, and curls. For the finale, the entire company divides into two teams for a challenging burpee contest. And all the while, we endlessly high-five each other to throbbing electro workout music. Everyone participates, although a few (including the pregnant woman) work out less intensely. Then, at noon on the dot, the music stops and everyone rushes back to the locker room to shower and return to their desk jobs.

How would you feel if your boss required you to exercise? Henrik says he wants his employees to exercise because he wants them—and by extension his company—to be the best people they can be, and that requires exercise. It's why, like Quiet Hour and other unusual aspects of Björn Borg culture, Sports Hour is compulsory. To gauge the program's success, every employee's fitness is evaluated twice a year (the information is provided to management on a group, not individual, basis). Exercise, moreover, is part of a general fitness culture at Björn Borg. Instead of a boozy Christmas party, the entire company goes sledding and then drinks hot chocolate. Every summer they do a six-mile "fun run" through the streets of Stockholm.

To be honest, I had qualms about Henrik's approach. Philosophically, I classify myself as an admirer of "libertarian paternalism," the idea that companies, governments, and other institutions should help us act in our own best self-interests while respecting our freedom of choice.[4] Libertarian paternalists favor nudges over coercion. Instead of forcing people to exercise, libertarian paternalists provide incentives. Rather than relying on us to "opt in" to be organ donors or tip waiters, libertarian paternalists ask us to "opt out" of these programs or have credit cards automatically remind us to tip when we pay the bill. In lieu of banning tobacco, libertarian paternalists slap dire warnings on cigarette packages and advocate heavy taxes. Why should exercise be any different from smoking? Just as we have a right to smoke despite its unhealthy effects, don't we also have the right not to exercise?

Informed by this way of thinking, I spent several days at Björn Borg interviewing any employee willing to share her or his thoughts.

How would you feel if Henrik banned cigarettes or made you eat only vegetarian food? What if he made you exercise twice or three times a week? Does Sports Hour ever make you feel ashamed about your body or your level of fitness? Do you feel coerced? What about employees with disabilities? Do you know anyone who quit because they didn't want to be forced to exercise, let alone shower in the gym in front of their naked hyper-muscular boss?

What I heard was both expected and surprising. When Henrik became Björn Borg's "Head Coach" in 2014, some staff members were upset. One longtime employee told me his initial reaction to Henrik was "Shit—get this guy out of my face!" About 20 percent quit. But just about everyone who remains now thinks the benefits outweigh the costs. To be sure, a few employees joined the company because they like to exercise, but many confessed that Sports Hour was often their only exercise all week. One woman told me she was injured in one activity. Regardless of their level of enthusiasm for exercising, all the people I met say that the mandatory Sports Hour helps them become healthier and that exercising with management encourages a sense of community, camaraderie, and shared purpose. As one Björn Borger put it, "We Swedes are shy, and usually don't open up without a drink in our hands, but this works even better. It's special."

As I flew home a few days later, sore as hell, I was of two minds about Henrik's exercise policy. On the one hand, he has managed to get his employees to exercise according to what I consider Paleolithic precepts: making it necessary and fun. Further, by any objective measure, his policy benefits both his employees and his company. On the other hand, Henrik's compulsory Sports Hour violates the widely held principle that it is wrong to coerce adults. Mandatory exercise may be good for you, but it is illiberal. In my opinion, an evolutionary anthropological perspective points to other effective but less authoritarian ways to help us exercise.

I Would Prefer Not To

Imagine you are the CEO of a company with numerous sedentary employees and your health-care costs are escalating partly because of their inactivity, but unlike Henrik you don't want to make exercise compulsory. Perhaps you are also a parent struggling to budge a surly, reluctant teenager to exercise. And maybe you have been striving unsuccessfully to get yourself off the couch more often. How do you succeed?

Everyone copes with the urge to postpone or avoid exercise, so environments that neither require nor facilitate physical activity inevitably promote inactivity.[5] If I have to choose between sitting comfortably in a chair or slogging through a sweaty workout, the chair is almost always more appealing. The slow, rational part of my brain knows I should exercise, but my instincts protest, "I would prefer not to," and another, enticing voice queries, "Why not exercise tomorrow?"[6] Perhaps I don't have the time or energy, and I have to go out of my way to get physical activity because I am stressed for time, my town lacks sidewalks, or the stairway in my building is dingy and inaccessible. To add to these impediments, maybe I inherited genes that predispose me to being physically inactive. Scientists have bred laboratory mice that are instinctively addicted or averse to running on treadmills, and studies of humans suggest that some of us inherited tendencies to be slightly less inclined to exercise.[7]

To find ways to overcome natural disinclinations to exercise, hundreds of experiments have tested an exhaustive list of interventions designed to entice non-exercisers to get moving. Some studies evaluate the effect of giving people information. This can involve lectures, websites, videos, and pamphlets about how and why to exercise, or providing devices like Fitbits so subjects know how much activity they are getting. Other experiments try to influence people's behaviors. These studies include having doctors personally prescribe specific doses of exercise, providing free gym memberships, paying people to exercise, fining them for not exercising, boosting their confidence, or pestering them with phone calls, texts, and emails. Finally, some studies try to encourage people to exercise by altering their environments. Examples include funneling people toward stairs instead of elevators and building sidewalks and bicycle paths. You name it, someone's tried it.[8]

The good news is that some of these interventions can and do make a difference. A typical example is a 2003 study that enrolled about nine hundred very sedentary New Zealanders between the ages of forty and seventy-nine. Half of them received normal medical care, but the other half were personally prescribed exercise by doctors, followed up by three phone calls over three months plus quarterly mailings from exercise specialists. After a year, the individuals prescribed exercise averaged thirty-four minutes of more physical activity per week than the standard care controls.[9]

The bad news is that big successes are the exception rather than the rule. While the extra thirty-four weekly minutes achieved by prescribing those New Zealanders exercise is progress, all that extra effort amounted to only five more minutes of physical activity per day. Comprehensive reviews that have examined hundreds of high-quality studies find that many interventions fail, and those that succeed tend to have only similarly modest effects.[10] Further, interventions that work in one study often don't work in others. If you plow through the studies used to inform the U.S. Department of Health and Human Services' thorough 2018 review of just about every kind of exercise intervention ever tried, the phrase you keep seeing over and over again is "small but positive effect."[11] Please don't think I am arguing that we should abandon these efforts. Quite the contrary: even small changes can improve people's health, and sometimes interventions cause veritable U-turns. One of my friends was diagnosed with type 2 diabetes in his forties and worried he might not live to be a grandparent. His doctor prescribed a strict exercise regime that he heeded so diligently he now runs half marathons and no longer needs medication. But for every success story like his, there are many more failures. There is no surefire way to persuade or coax non-exercisers to exercise substantially.

But didn't we already know that? If there were an effective, dependable way to transform sedentary people into regular exercisers, it would spread like wildfire. Why aren't any of these interventions more likely to succeed than our generally ill-fated New Year's resolutions?

One reason is the complexity and variety of human nature. Even among westernized, industrialized populations, people are dazzlingly diverse in terms of psychology, culture, and biology. Why would a strategy that works on a college student in Los Angeles succeed for an elderly woman in London or a time-stressed parent in the suburbs of Tokyo? Do we really expect the same action plan to work for people who are overweight or thin, shy or outgoing, insecure or confident, men or women, college graduates or less educated, rich or poor, urban or rural, stubborn or docile? Indeed, studies that try to figure out who does and doesn't regularly exercise find few factors common to exercisers apart from some really obvious ones: having a prior history of exercising, being healthy and not overweight, having confidence in the ability to exercise, being more educated, and both liking and wanting to exercise.[12] That list of attributes is about as illuminating as figuring out that people who go to art museums tend to be people who already like art.

In my opinion, if we want to promote exercise effectively, we need to grapple with the problem that engaging in voluntary physical activity for the sake of health and fitness is a bizarre, modern, and optional behavior. Like it or not, little voices in our brains help us avoid physical activity when it is neither *necessary* nor *fun*. So let's reconsider both of these qualities from an evolutionary anthropological perspective.

First, *necessity*. Everyone, including the billion or so humans who regularly don't get enough exercise, knows that more exercise would be good for them. Many of these non-exercisers feel frustrated or bad about themselves, and annoying exercists who nag and brag about exercise rarely improve matters by reminding them to jog, take long walks, go to the gym, and take the stairs. Part of the problem is the distinction between "should" and "need." I know I *should* exercise to increase the probability I will be healthier, happier, and live longer with less disability, but there are numerous, legitimate reasons I don't *need* to exercise. In fact, it is patently obvious one can lead a reasonably healthy life without exercise. As the Donald Trumps of the world attest, the 50 percent of Americans who get little to no exercise aren't doomed to keeling over prematurely. To be sure, insufficient exercise increases their chances of getting heart disease, diabetes, and other illnesses, but most of these diseases tend not to develop until middle age, and then they are often treatable to some degree. Even though more than 50 percent of Americans rarely if ever exercise, the country's average life expectancy is about eighty years.

Not only is exercise inherently unnecessary, the modern mechanized world has eliminated other formerly necessary forms of non-exercise physical activity. I can easily spend my days without ever having to elevate my heart rate or break a sweat. I can drive to work, take an elevator to my office floor, spend the day in a chair, and then drive home. I regularly accomplish formerly laborious chores like getting water and food, making dinner, and washing clothes with little effort, often by just pressing a button or turning a tap. I can even buy a robot to vacuum my floors. These laborsaving devices are sometimes scorned as decadent and corrupting, but they are popular because we actually like them.

In addition to being unnecessary, exercise takes precious *time,* keeping us from other, higher-priority activities. I'm lucky to have a short commute and a flexible work schedule, so I can almost always find time to fit in a run or nip home to walk the dog. Many of my friends, however, commute long distances, their inherently sedentary office jobs are fixed in terms of hours, and they have other time-consuming obligations including child care and elder care. Paradoxically, for the first time in history, wealthier people get more physical activity than the working poor.[13] When free time is scarce, optional activities like exercise are relegated to weekends, and by then a week's worth of accumulated fatigue can make it hard to muster the energy to exercise. When people are asked what keeps them from exercising, they almost always list time as a main barrier.

Which brings up *fun*. Lack of time can be stressful, but even the busiest people I know manage to find time to do things they enjoy or find rewarding like watch TV, surf the web, or gossip. I suspect millions of non-exercisers would succeed in making exercise a greater priority if they found it more enjoyable, but for them exercise is often emotionally unrewarding and physically unpleasant. These negative reactions are probably ancient adaptations. Like most organisms, we have been selected to enjoy and desire sex, eating, and other behaviors that benefit our reproductive success and to dislike behaviors like fasting that don't help us have more babies. If our Stone Age ancestors found unnecessary physical activities like optional five-mile jogs unpleasant, they would have avoided squandering limited energy that could have been allocated toward reproduction.

That may be a "just-so story," but few would disagree that non-exercisers are not entirely irrational because exercise is a modern behavior that is by definition unnecessary and often unrewarding physically and emotionally. For many, it is also inconvenient and inaccessible. If we can't make exercise necessary and fun, perhaps we can make it *more* necessary and *more* fun.

How Do We Make Exercise More Fun?

The least fun exercise experience I ever had was the 2018 Boston Marathon. I know that sounds boastful and preposterous (how can a marathon be fun?), but please bear with me because the conditions that day were so dreadful they illustrate an important point. Boston weather at the end of April is sometimes nice, sometimes chilly, sometimes warm, or sometimes rainy, but the nor'easter that battered Boston that day was unusually brutal. By 10:00 a.m., when the race began, it had been pouring steadily for hours, the temperature was a few degrees above freezing, and there was a relentless, fierce headwind that gusted up to thirty-five miles per hour. Normally, no sum of money would entice me to run in such miserable weather, and as I obsessively checked the forecast before the race, I considered staying home despite having trained for months. On race day, however, I smeared my entire body with Vaseline, dressed in several waterproof layers, tucked a shower cap under two hats, donned supposedly waterproof gloves, encased my shoes in plastic bags to avoid getting them wet before the start, and like a lemming boarded the bus in downtown Boston to travel far out to the little town of Hopkinton, where the race begins.

The scene in Hopkinton made me think of one of those movies that depicts soldiers in the trenches before a World War I battle. The high school's sports field, where runners wait before starting, had been churned into mud by twenty-five thousand cold, wet, miserable marathoners. I am always jittery before a race—a mix of apprehension, anxiety, and excitement—but this time I was worried. How was I going to make it home without getting hypothermia? Yet when my appointed time came, I stood hunched in the pelting rain and biting wind behind the starting line, glumly ate my good luck blueberry muffin (a ritual), and waited for the starting gun so I could begin trudging along with thousands of other anxious, miserable runners.

The next 26.2 miles were horrid. At times, the headwind and rain were so fierce it was hard to take a step forward, my waterlogged shoes made each step sound like an elephant's, and every inch of my body felt raw. Within a few miles, I decided that the sole reason to keep on running despite the drenching rain, puddles, and unabating wind was that not stopping was the fastest way to get home and avoid getting even colder. My primary urge on crossing the finish line was to crawl into bed as fast as possible to warm up, which is exactly what I did.

Over the next few days as I recovered physically and mentally, I thought about why I and twenty-five thousand other lunatics ran through that storm. If my goal was simply to run 26.2 miles, I could have waited until the next day and enjoyed nearly perfect weather. The only explanation I can give is that I ran for social reasons. Like a soldier in battle, I wasn't alone but instead part of a collective doing something difficult together. The Boston Marathon has been a revered tradition since 1897 and has become even more meaningful since terrorists attacked the race in 2013. I felt I was running not just for myself but also for others, including the hundreds of thousands of spectators who braved the storm to cheer us on. Finally, shameful as it may be to admit, I ran because I didn't want to face the social disapprobation that comes from being a coward or a quitter. Peer pressure is a powerful motivator.

And therein lies an important lesson about why we exercise. Because exercise by definition isn't necessary, we mostly do it for emotional or physical rewards, and on that horrid April day in 2018, the only rewards were emotional—all stemming from the event's social nature. For the last few million years humans rarely engaged in hours of moderate to vigorous exertion alone. When hunter-gatherer women forage, they usually go in groups, gossiping and otherwise enjoying each other's company as they walk to find food, dig tubers, pick berries, and more. Men often travel in parties of two or more when they hunt or collect honey.[14] Farmers work in teams when they plow, plant, weed, and harvest. So when friends or CrossFitters work out together in the gym, teams play a friendly game of soccer, or several people chat for mile after mile as they walk or run, they are continuing a long tradition of social physical activity.

I think there is a deeper evolutionary explanation for why almost every book, website, article, and podcast on how to encourage exercise advises doing it in a group. Humans are intensely social creatures, and more than any other species we cooperate with unrelated strangers. We used to hunt and gather together, and we still share food, shelter, and other resources, we help raise one another's children, we fight together, we play together. As a result, we have been selected to enjoy doing activities in groups, to assist one another, and to care what others think of us.[15] Physical activities like exercise are no exception. When we struggle with fatigue or lack of skill, we encourage and help one another. When we succeed, we praise each other. And when we think of quitting, being in a group can deter us. My hardest workouts have always been in groups, and I have often shown up for a run or a workout only because I had previously arranged to meet a friend. Of course, exercise is also sometimes enjoyable without socializing. A solitary walk or run can be meditative, and working out while listening to podcasts or watching TV in the gym (a distinctly modern phenomenon) can be diverting. But for most people exercising with others is more emotionally rewarding. For this reason, sports, games, dancing, and other types of play are among the most popular social activities, and regular exercisers often belong to clubs, teams, and gyms. To entice customers, the gym down the street has a big sign, "Never Work Out Alone!" Some of the most popular, effective ways to exercise are group workout experiences like CrossFit, Zumba, and Orangetheory.

Exercise can also make us feel good, which helps make it enjoyable. After a good workout I feel simultaneously alert, euphoric, tranquil, and free from pain—not unlike taking an opioid. Actually, natural selection did adopt this drug-pushing strategy by having our brains manufacture an impressive cocktail of mood-altering pharmaceuticals in response to physical activity.[16] The four most important of these endogenous drugs are dopamine, serotonin, endorphins, and endocannabinoids, but in a classic evolutionary design flaw these primarily reward people who are already physically active.

Dopamine. This molecule is the linchpin of the brain's reward system. It tells a region deep in the brain "do that again." Evolution thus geared our brains to produce dopamine in response to behaviors that increase our reproductive success including having sex, eating delicious food, and—surprise—doing physical activity. But there are three shortcomings in this reward system for today's non-exercisers. First, dopamine levels go up only while we exercise. So they don't get us off the couch. Worse, dopamine receptors in the brain are less active in people who haven't been exercising than in fit people who are regularly active.[17] And to add insult to injury, people who are obese have fewer active dopamine receptors.[18] Consequently, non-exercisers and obese individuals must struggle harder and for longer (sometimes months) to get their receptors normally active, at which point they can cause what is sometimes considered "exercise addiction." If you exercise regularly, you know the feeling when you have to endure several days without exercise: you get twitchy, irritable, and crave physical activity to satisfy your hungry dopamine receptors. In extreme cases exercise addiction can be a serious dependency, but the term is usually applied to a normal, harmless, and generally beneficial reward system.[19]

Serotonin. This still mysterious neurotransmitter helps us feel pleasure and control impulses, but it also affects memory, sleep, and other functions. Our brains produce serotonin when we engage in beneficial behaviors like having physical contact with loved ones, taking care of infants, spending time outdoors in natural light, and, yes, exercising.[20] Elevated levels of serotonin induce a feeling of well-being (the drug ecstasy exaggerates this feeling by boosting serotonin levels sky-high), and we become better at controlling nonadaptive impulses. Low serotonin is thus associated with anxiety, depression, and impulsivity. Although some people with depression take pharmaceuticals to maintain normal serotonin function, exercise has been shown to be often as effective as any prescription.[21] However, as with dopamine, non-exercisers are at risk of having lower serotonin activity, making them more vulnerable to being depressed and unable to overcome the impulse to avoid exercise, which in turn keeps serotonin levels low.

Endorphins. Endorphins are natural opioids that help us tolerate the discomfort of exertion.[22] The body's own opioids are less strong than heroin, codeine, and morphine, but they too blunt pain and produce feelings of euphoria. Opioids allow us to go for a long hike or run without noticing our muscles are sore and our feet have blisters. They may also contribute to exercise addiction. But, once again, there is a catch. Although their effects can last for hours, endorphins aren't produced until after twenty or more minutes of intense, vigorous activity, making them more rewarding for people who are already fit enough to work out that hard.[23]

Endocannabinoids. For years, endorphins were thought to cause the infamous runner's high, but it is now evident that endocannabinoids—the body's natural version of marijuana's active ingredient—play a much greater role in this phenomenon.[24] Despite causing a truly pleasurable high, this system has little relevance for most exercisers because it usually takes several hours of vigorous physical activity before the brain releases these mood- and sensory-enhancing drugs. Further, not everyone has the genes that make a runner's high possible.[25] I suspect the runner's high evolved primarily to increase sensory awareness to help hunters track animals during persistence hunting.

While these and other chemicals released by exercise help us exercise, their drawback is they mostly function through virtuous cycles. When we do something like walk or run six miles, we produce dopamine, serotonin, and other chemicals that make us feel good and more likely to do it again. When we are sedentary, however, a vicious cycle ensues. As we become more out of shape, our brains become less able to reward us for exercising. It's a classic mismatch: because few of our ancestors were physically inactive and unfit, the brain's hedonic response to exercise never evolved to work well in persistently sedentary individuals.

So what should we as a society and you and I as individuals do? How can we make exercise more fun and rewarding especially if we are out of shape?

First and foremost, let's stop pretending exercise is necessarily fun, especially for habitual non-exercisers. If that describes you, start by choosing types of exercise you either enjoy the most or dislike the least.[26] Just as important, figure out how to distract your mind while you exercise with other things you find fun. At the very least, such diversions will help make the exercise less disagreeable. Commonly recommended, sensible methods to make exercise more fun (or less unfun) include:

- Be social: exercise with friends, a group, or a good, qualified trainer.[27]
- Entertain yourself: listen to music, podcasts, or books, or watch a movie.
- Exercise outside in a beautiful environment.
- Dance or play sports and games.
- Because variety is enjoyable, experiment and mix things up.
- Choose realistic goals based on time, not performance, so you don't set yourself up for disappointment.
- Reward yourself for exercising.

Second, if you are struggling to exercise, it is useful to remember how and why exercising takes time to become enjoyable or less unpleasant. Because we never evolved to be inactive and out of shape, the adaptations that make physical activity feel rewarding and become a habit develop only after the several months of effort it takes to improve fitness. Slowly and gradually, exercise switches from being a negative feedback loop in which discomfort and lack of reward inhibit us from exercising again to being a positive feedback loop in which exercise becomes satisfying.

So, yes, exercise can become more rewarding and fun. But let's not deceive ourselves or others. No matter what we do to make exercise more enjoyable, the prospect of exercising usually seems less desirable and less comfortable than staying put. Every time I plan to exercise, I first struggle to prevail over instincts to not exercise. Afterward, I never regret it, but to overcome my inertia, I usually have to figure out how to make it seem necessary.

How Do We Make Exercise Seem Necessary?

For years, a friend of mine tried to exercise regularly but could never make it stick. She tried New Year's resolutions, bought gym memberships, and made exercise schedules. After each new wave of effort and enthusiasm, she settled back into her sedentary life. Frustrated, she decided to try a completely different approach: sticks instead of carrots. Here's how it worked. She sent the website StickK.com a thousand dollars, pledged to walk four miles a day, and designated her husband as her official referee. For every week she didn't make her goal, as verified by her husband, the website would send twenty-five dollars to the National Rifle Association (NRA), the controversial organization that opposes gun control in the United States. According to her, "There were many days when I didn't want to go for a walk, but I was damned if the NRA would get a penny. I had no choice." It worked: for a year she never missed her goal and now is a dedicated walker. The chief difference between my friend and the employees of Björn Borg is that she found a way to coerce herself, whereas Henrik Bunge coerces his workers.

How do you feel about coercion? If you are like me, you are probably generally opposed to coercing others. Forcing people to exercise doesn't respect their rights to make decisions about their lives. It violates the Golden Rule. Just as I have the right to not take vitamins, avoid vegetables, or not floss my teeth, I have the right to not exercise.

And yet there are several noncontroversial exceptions to the principle of not forcing people to exercise. One exception includes people like first responders and soldiers whose occupations require certain levels of fitness. Soldiers, for example, obviously must exercise to be strong and fit enough to fight. When they enlist, soldiers know drill masters will scream at them to do compulsory push-ups, sit-ups, and pull-ups and run laps in boot camp. Failure to exercise leads to punishment. The other major exception is children, whom we often force to exercise because it's good for them. Because experts agree that children should get at least one hour of moderate to vigorous exercise a day, almost every country in the world has mandatory physical education in school.[28] So what makes it acceptable to force children to exercise for their benefit but not adults like me who aren't soldiers or firefighters?

One justification is that children, unlike adults, are incapable of making decisions in their own self-interest. It is universally acceptable to coerce children to do all sorts of things that are good for them like eat healthy food, go to bed, attend school, sit in car seats, and get vaccinated and to prohibit them from smoking and drinking alcohol. At some point, we allow adults to make such decisions for themselves, but with some exceptions. While adult Americans have the right to not exercise and smoke all they want, they are prohibited from snorting cocaine and still must wear seat belts.

From a purely utilitarian perspective, how is requiring exercise different from mandating seat belt use? According to the National Transportation Safety Board, seat belts prevent approximately 10,000 deaths a year in the United States.[29] According to the Centers for Disease Control, inadequate physical activity causes about 300,000 deaths a year in the United States—thirty times more.[30] Worldwide, physical inactivity causes about 5.3 million deaths per year—about as many as caused by smoking.[31] But psychologically, these deaths are very different. You have only to drive by a serious car accident or see images on TV of dead, mangled bodies in cars to realize you are better off being forced to wear that belt, but deaths from congestive heart failure or type 2 diabetes usually happen quietly and out of sight in hospitals. In addition, a twenty-year-old's life cut short by a car accident is widely considered more tragic than a seventy-year-old's death from colon cancer or a heart attack. We've also become habituated to being forced to wear seat belts. When seat-belt laws were first enacted, my father-in-law refused to wear them because they were an assault on his liberty, but my daughter's generation thinks the requirement is totally normal.

Despite its utilitarian benefits, I oppose mandating universal exercise because adults have the right to make unhealthy decisions. The dilemma is that most people who struggle unsuccessfully to exercise also *want* to exercise.[32] It's not anyone's fault that we inherited tendencies (some of us more than others[33]) to avoid unnecessary physical activity but were born to a world in which physical activity is no longer required and increasingly hard to do thanks to commuting, desk jobs, elevators, shopping carts, streets without sidewalks, buildings without easily accessible stairs, and so on. Instead of being shamed and blamed for being inactive, we deserve compassionate help to make exercise more necessary. The most acceptable way to do that is to find ways of coercing ourselves through agreed-upon *nudges* and *shoves.*

Nudges influence our behaviors without force, without limiting our choices, and without shifting our economic incentives.[34] Typical nudges involve changing default options (like opting out of being an organ donor instead of opting in) or small changes to the environment (like placing healthier foods prominently at the front of the salad bar). Predictably, many would-be exercisers are advised to try various nudges to make the act of choosing exercise more of a default, simpler, and less of a hassle. Examples include

- Put out your exercise clothes the night before you exercise so you wear them first thing in the morning and are ready to go (alternatively, sleep in your exercise clothes).
- Schedule exercise so it becomes a default.
- Use a friend or an app to remind you to exercise.
- Make the stairs more convenient than taking the elevator or escalator.

Shoves are more drastic forms of self-coercion, along the lines of my friend's walking to avoid sending money to the NRA. They are unobjectionable because you do them to yourself voluntarily, but they are more forceful than nudges. Examples of shoves include

- Schedule exercise with a friend or a group beforehand. You then become socially obligated to show up.
- Exercise in a group such as a CrossFit class. If you waver, the group will keep you going.
- Sign a commitment contract with an organization like StickK .com that sends money to an organization you dislike if you don't exercise (a stick) or to one you like if you do (a carrot).

- Sign up (and pay) for a race or some other event that requires you to train.
- Post your exercise online so others see what you are (or are not) doing.
- Designate a friend, a relative, or someone you admire or fear as a referee to check up on your progress.

Note that all of these methods share one essential quality: they involve social commitment. Whether you plan to exercise with a friend, a yoga class, a team, a platoon of CrossFitters, or fellow runners in a 5K race or report your exercise accomplishments (or lack thereof) online, you are pledging to others that you will be physically active. In return you get both carrots in the form of encouragement and support and sticks in the form of shame or disapprobation. For evidence that social commitment works, you have only to look at our most popular and durable social institutions that help us behave as we aspire: marriage, religion, and education. To varying degrees, all involve a public display of commitment to that institution and its principles in return for some benefit along with social support and censure. While marriage and religion are not good models, I think we should treat exercise more like education.

For children, we already do. Just as we compel children to attend school, we require them to exercise (although rarely enough). As with school, we try to make exercise fun by making it social. So why not do the same for adults by treating exercise like college? Going to college is essentially a highly social commitment contract for adults that includes carrots and sticks. Students in my university pay a fortune to have professors like me compel them to read, study, and work under penalty of getting bad grades or failing. My students compete for and agree to these conditions because they know they would not learn as much without the school's nudges, shoves, and requirements. In return, they enjoy a social experience that is usually fun, involves support from fellow students and staff, and encourages them to participate in something larger than themselves. Can this kind of commitment contract model help promote exercise, especially among youth?

Focusing on Youth

Young people need to move. Because our thrifty physiologies evolved to build capacity in response to demand, sufficient physical activity during the first decades of life is indispensable for developing a healthy body. At least an hour a day of moderate to vigorous physical activity reduces children's risk of obesity and helps them grow healthy muscles, bones, hearts, blood vessels, digestive systems, and even brains. Children who get more physical activity learn more and are smarter, happier, and less prone to depression and other mood disorders.

But we are failing our children miserably. In the United States, less than one in four children get at least an hour of physical activity a day.[35] Girls exercise even less than boys, and older children are more sedentary than younger children.[36] According to the World Health Organization, the picture worldwide is generally worse with more than 81 percent of children not getting an hour of daily physical activity.[37] Many factors are to blame. Children today spend more time glued to screens both large and small, they walk less often to school, in some neighborhoods parks and streets are dangerous, and growing numbers of schools have let physical education slide to paltry levels. While most school districts require some physical education, only a shockingly tiny fraction provide enough. Just 11 percent of elementary school districts in the United States have regular classroom physical activity breaks during the school day; among high schools, the percentage plunges to 2 percent.[38] And, true to my own experience, students typically spend more than half their time in classroom physical activities inactively, sitting on a bench or waiting in line to bat or dribble a ball.[39] To make matters worse, competitive sports in many schools are exclusionary, leading to what Bradley Cardinal terms an "inverted" system in which the further students advance, the likelier they are to be sidelined or eliminated.[40] All in all, we face a dire epidemic of physical inactivity among youth.

My own state of Massachusetts is no exception. Massachusetts General Law 71.3 states, "Physical education shall be taught as a required subject in all grades for all students." But in 1996, the state board of education repealed minimum hours of physical education in order to increase time spent on standardized test preparation. According to a local newspaper, Massachusetts students average just eighteen to twenty-two minutes of physical education per day during the school week.[41]

This is a failure of priorities and policies. Beyond widespread ignorance about the long-term consequences of childhood inactivity, parents and educators appear to be more worried about test scores, discipline, and safety—all of which would be *increased, not decreased,* by having appropriate amounts of physical activity.[42] It's as if we have forgotten just how deeply the body and the mind are interconnected.

Universities sadly illustrate our collective amnesia and disregard of the ancient wisdom that healthy bodies foster healthy minds. Educators have always known that students benefit from being physically active, and nearly all four-year colleges and universities in America used to require moderate levels of physical education.[43] Those requirements have mostly been dropped, and the few that remain have been watered down. My own university, Harvard, first mandated physical education in 1920 but totally abandoned the requirement in 1970. Now, as at most universities, only about one-quarter of our students get baseline levels of regular exercise, despite sky-high rates of mental health problems like depression and anxiety. So far, my efforts to restore some kind of physical activity program for our undergraduates have failed. The most common criticisms are that mandating exercise is coercive, that our job is to educate minds, not bodies, that students don't have enough time, and that exercise recommendations can be traumatic and discriminatory to students with disabilities or those who are overweight or feel uncomfortable about their bodies.

Some of these concerns are real, but all can be addressed. The least persuasive concern is coercion. As we already saw, universities thrive on a commitment contract model. When students enroll, they willingly commit themselves to being forced by professors and deans to fulfill a lengthy set of requirements. If they don't like those requirements, they can apply to other schools without them. I also disagree that it is not within our purview to ask students to be physically active. Our primary mission is to educate, and physical activity helps students thrive intellectually, socially, and personally. Exercise helps young adults stay mentally healthy and fosters good habits that appear to last. According to one study, 85 percent of students who exercised regularly in college continued to exercise later in life, but 81 percent of those who were physically inactive in college remained sedentary as older adults.[44] To avoid backfiring, however, physical education needs to be mandatory as well as positive. Studies show that making physical activity optional paradoxically reinforces inactivity by primarily attracting students who are already active and motivated, and that negative physical education experiences (like always being picked last and mostly warming the bench) lessen a student's likelihood of later exercising as an adult.[45]

As for time, I am sympathetic but unconvinced. Students lead busy lives, but with the exception of occasional crunch times when papers are due and exams must be taken, the inability to find thirty minutes five times a week to exercise is mostly a function of exercise being a lower priority than other extracurricular activities (including many hours spent on social media).[46] In fact, skipping exercise is sometimes counterproductive for those who are pressed for time. Randomized controlled studies on college students reinforce what most of us instinctively know: even short bouts of moderate-intensity exercise improve memory and concentration.[47]

Finally, I agree wholeheartedly that we need to be sensitive to students with disabilities and those who feel insecure, unfit, or uncomfortable about exercise. Different students have different needs, and it is wrong and counterproductive to engage in body shaming or fitness shaming. Unfit students, however, aren't best served by not being helped to exercise, because the benefits are substantial for everyone, especially those who are the least fit. The challenge is to support and assist everyone at every level nonjudgmentally and in ways and degrees they find acceptable and rewarding.

In short, we all need nudges. So, assuming we have figured out how to make exercise happen, the next problem is what type and how much to do.

TWELVE

How Much and What Type?

MYTH #12 There Is an Optimal Dose and Type of Exercise

All things are poison and nothing is without poison; only the dose permits something not to be poisonous.

—Paracelsus (1493–1541)

Imagine a great king in a far-off country who loves his only daughter more than anything else in the world. The princess is kind and compassionate, and she excels in philosophy, mathematics, history, and languages. But, like him, she is sedentary and unfit. Advisers, trainers, nurses, and governesses all fail to entice the feeble king and his daughter to exercise. The king knows this is a problem, so when it comes time for her to marry, he invites dozens of princes from lands both near and far to a special one-day competition to win her hand. Instead of having the suitors joust, fence, or wrestle, he sits them down in the castle's great hall for a written examination. At the stroke of 9:15 a.m., each prince opens his blue book and has three hours to answer the same question: "What is the best way to exercise?"

The kingdom is never so quiet. As the mighty princes wield their pens, nary a dog barks, horse whinnies, nor door creaks as every living thing for miles around the castle holds its breath. And then, at 12:15 p.m., pens go down, and the exams are collected, scanned, and posted on the web for all and sundry to read and post comments while the king's judges meet behind closed doors to pick the winner.

What a diverse, brilliant set of answers. One of the most popular is a twelve-step program, "Be as Strong as a Lioness," that alternates many repetitions of moderate weights with fewer repetitions of heavier weights. Another clever prince writes "Walk, Run, and Live Forever," a ten-step plan that begins with long walks and then adds short runs that gradually increase to ten miles. Other crowd favorites are "Seven Minutes or Your Life," which promises "optimal health" from just seven minutes a day of high-intensity interval training, and "Live Longer than a Caveman," which replicates a paleo fitness regime with barefoot walking, tree climbing, and rock lifting. Yet more plans advocate stretching, swimming, biking, jogging, dancing, boxing, yoga, and even pogo sticking. Some of the prescriptions consider genetic variation, others have different plans for men and women, many are designed to maximize weight loss, and one is cleverly tailored to integrate with a woman's monthly cycle. While the judges ponder, journalists, bloggers, celebrities, enthusiasts, and trolls fiercely argue the merits of every entry. With each day it seems there is a new consensus favorite.

Then, finally, after a week of waiting and debating, the day of decision arrives. At noon, just two sentences are posted on the royal website: "After much deliberation, the judges have determined there is no best way to exercise. Come back next year for a better question."

After the previous eleven chapters, I hope you agree the fairy-tale judges' decision was as Solomonic as the question was ill-considered. Despite claims to the contrary, how can there be a best or optimal amount and type of exercise? What does "best" even mean? Best in terms of how many years of life it adds? Best in terms of time efficiency? Best for preventing heart disease? Losing weight? Avoiding injury? Averting Alzheimer's? Even if there were a way to choose a best plan for one of these goals, would the same plan be best for everyone regardless of age, sex, weight, fitness level, and history of injury?

Although there can be no optimal exercise prescription, physical activity nonetheless stimulates growth, maintenance, and repair mechanisms that build capacities and slow aging. We have thus medicalized exercise. Accordingly, even though exercise is an odd kind of medicine, we prescribe certain dosages and types. But how much and what type? Regardless of whether we exercise for fun or fitness, some amounts and kinds of exercise are obviously better or worse for health depending on our goals and circumstances. We do weights for muscles, cardio for hearts, and bungee jumping to terrify parents. The next and final chapter will consider how and why different exercise prescriptions affect the most common diseases likely to kill or disable us. But first, we should consider the general problem of what and how much to do.

By any measure, the relationship between exercise dose and health is confusing. I know Nobel Prize–winning scientists who are bewildered by the cacophony of conflicting recommendations out there. Many experts advise us simply to do whatever we enjoy because any exercise is better than none. Others claim we get the most bang for our buck with short intervals of high-intensity training. In terms of cardio, there are passionate advocates of running, walking ten thousand steps a day, swimming, or low-impact machines like ellipticals. For weights, some prescribe using our own body weight, others recommend free weights, yet others push weight machines. But of all the prescriptions, by far the most commonly and widely promoted—advocated by almost every major health organization in the world—is that we do at least 150 minutes of moderate-intensity or 75 minutes of vigorous-intensity aerobic exercise per week, supplemented by two sessions of weights.[1]

150 Minutes per Week?

The exercise guru Jack LaLanne (who lived to ninety-six) liked to say, "People don't die of old age, they die of inactivity."[2] That's hyperbole, but ever since the dawn of civilization and probably before, it has been obvious that physical activity promotes health. Still, until the pioneering studies of Dr. Ralph S. Paffenbarger Jr., known affectionately as

Paff, no one had ever shown a medicine-like relationship—a dose-response curve—between how much you exercised and how long you lived. Born in 1922, Paffenbarger began his career working on polio vaccinations for the U.S. government but switched his focus to studying chronic disease while teaching at Harvard and then Stanford Medical Schools. Paffenbarger published many brilliant studies (including one contending that regular chocolate consumption adds nearly a year to your life[3]), but his big breakthrough came when he cleverly realized he could take advantage of the way universities never lose touch with their alumni in order to pester them for money. Beginning in 1962, Paffenbarger persuaded Harvard and the University of Pennsylvania to solicit fifty thousand alumni for information about their physical activity habits and health. And then he waited patiently for decades as the alumni aged and many of them died. In the end, he was able to use data from more than seventeen thousand individuals.

Figure 31 (the left side) reproduces the key results from Paffenbarger's landmark 1986 paper in *The New England Journal of Medicine.*[4] The *x*-axis is dose, expressed as the average number of calories spent on physical activity per week; the *y*-axis is the rate at which the alumni died. The numbers written above each point are *relative risks* (the probability of dying compared with the sedentary individuals within each age-group). As expected, the oldest alumni died at more than ten times the rate of the youngest alumni, but notice the different slopes of the dose-response relationships for the age categories: the older the age-group, the steeper the slope. Middle-aged alumni who exercised more than two thousand calories per week had a 21 percent lower risk of dying than their sedentary classmates, and those who were over seventy and exercised the same amount had *half* the risk of dying in a given year as their inactive classmates. Yes, half. This study when published was the first unequivocal evidence for a powerful dose-response relationship between exercise and mortality. Exercise is no panacea, but the more you exercise, the longer you are likely to live, and the effects of physical activity on longevity become vastly greater as we age.

As the years went by, Paffenbarger and colleagues continued to add to these results. By 1993 they had enough data to show that the

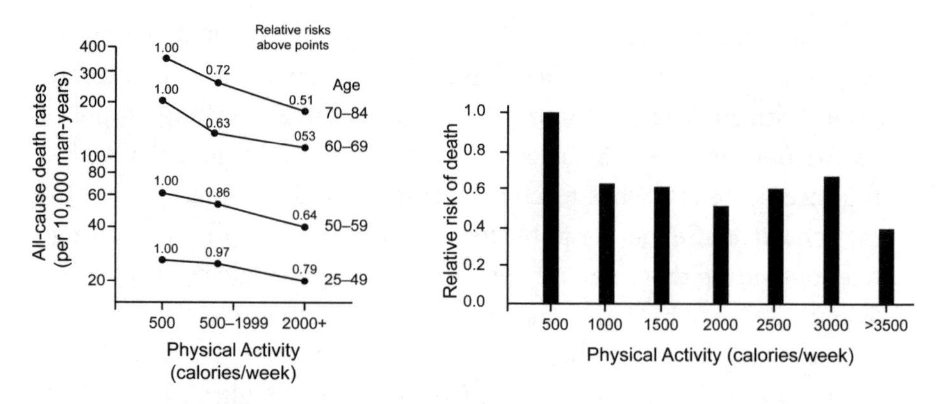

FIGURE 31 Dose-response effects of exercise on risk of death among Harvard alumni. Mortality rates among Harvard alumni grouped by age and physical activity (*left*) and by physical activity levels for combined age-groups (*right*). (Modified from Paffenbarger, R. S., Jr. [1986], Physical activity, all-cause mortality, and longevity of college alumni, *New England Journal of Medicine* 314:605–13; Paffenbarger, R. S., Jr., et al. [1993], The association of changes in physical-activity level and other lifestyle characteristics with mortality among men, *New England Journal of Medicine* 328:538–45)

relationship between longevity and physical activity is not a straight line, as shown in the right side of figure 31.[5] This graph plots the same group of alumni's relative risk of dying against activity level for all age-groups combined. Note that even modest levels of physical activity (a thousand calories a week) lower the rate of death by nearly 40 percent and that twice as much is even better. However, as the dose increases, the benefit diminishes. Another finding not shown in the graph is that alumni who reported they exercised moderately or vigorously did better than their classmates who exercised only lightly. Finally, alumni who took up exercise later in life had similarly lower rates of mortality compared to those who had been active all along. It's never too late to start.

Since Paffenbarger's pioneering investigations, scores more studies have examined associations between health and exercise dose in westernized countries like the United States. Many of these studies, like Paffenbarger's, look at rates of death or disease in large samples of individuals with different levels of exercise. Others are randomized control experiments that measure the effects of varying prescribed doses of exercise on factors that predict health outcome like blood pressure, cholesterol, or the ability to digest sugar. By the 1990s, so many studies had accumulated that three major health organizations decided to convene expert panels to review the evidence and make recommendations. In 1995 and 1996, all three panels published essentially the same advice: to reduce the overall risk of chronic disease, adults should engage in at least 30 minutes of moderate-intensity exercise at least five times a week.[6] They also concluded that children should engage in 60 minutes of physical activity a day. Since then, these prescriptions—150 minutes per week for adults and 60 minutes a day for kids—have been revisited, confirmed, and only slightly modified many times.

Let's look at the most recent update by the U.S. Department of Health and Human Services (HHS) in 2018. Among the many results of this thoughtful, comprehensive report is a revised analysis of Paffenbarger's famous dose-response relationship between exercise and mortality, shown in figure 32 (left side).[7] This graph incorporates data from more than one million adults! Similar to Paffenbarger's study, the *x*-axis plots exercise as the cumulative dose of aerobic activity in minutes per week; the *y*-axis plots the relative risk of dying at a given age corrected for factors like sex, smoking, alcohol consumption, and socioeconomic status. As you can see, the biggest reduction in mortality, about a 30 percent drop, is between sedentary individuals and those who exercise sixty minutes a week. However, the risk of death continues to fall with higher doses of exercise. People who report three and six weekly hours of exercise lower their risk of death by about another 10 percent and 15 percent, respectively. Because the analysis also examined exercise dose in terms of intensity, the study further concluded that half an hour of vigorous exercise and an hour of moderate exercise confer the same benefit.[8]

In the end, the 2018 HHS panel concluded that some physical activity is better than none, that more physical activity provides additional health benefits, and that for "substantial health benefits" adults should do *at least* 150 minutes per week of moderate-intensity or 75 minutes per week of vigorous-intensity aerobic physical activity, or an equivalent combination of the two. (Moderate-intensity aerobic activity is defined as between 50 and 70 percent of your maximum

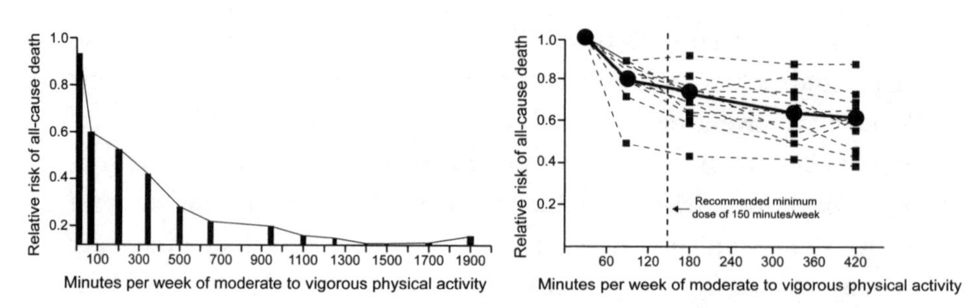

FIGURE 32 Large-scale studies of the dose-response relationship between weekly physical activity and the relative risk of mortality. *Left:* combined results from many studies totaling more than one million people. The benefits of even a little exercise are substantial, but the benefits eventually level off. The small rise at nineteen hundred minutes per week (an extreme level—more than thirty hours) is not statistically significant. *Right:* variation among twelve different studies around the median dose-response relationship (*thick line*). (Modified from Physical Activity Guidelines Advisory Committee [2018], *Physical Activity Guidelines Advisory Committee Scientific Report* [Washington, D.C.: U.S. Department of Health and Human Services]; Wasfy, M. M., and Baggish, A. L. [2016], Exercise dosage in clinical practice, *Circulation* 133:2297–313)

heart rate; vigorous-intensity aerobic activity is 70 to 85 percent of your maximum heart rate.) They also reaffirmed the long-standing recommendation that children need an hour of exercise a day. Finally, they recommended everyone also do some weights twice a week.

So Paffenbarger was right. But let's look more closely and skeptically at the recommended minimum for adults. Remember that the dose-response curve on the left side of figure 32 comes from many studies, a dozen of which are plotted separately on the right side of figure 32 by my colleagues Meagan Wasfy and Aaron Baggish. As before, the *x*-axis plots the median number of minutes per week of moderate to vigorous physical activity, and the *y*-axis plots the relative risk of death compared with those who exercised less than one hour per week.[9] The thick line represents the most common (median) value for all twelve studies.

I'd like to draw your attention to several noteworthy insights evident from this figure. The first is the variation among the studies. Physical activity had half the effect on the risk of death in some populations than others, probably because of factors like age and exercise type. Second, despite this variation, the dose-response relationship between physical activity and mortality follows a common pattern. In every study, the largest benefit came from just ninety weekly minutes of exercise, yielding an average 20 percent reduction in the risk of dying. After that, the risk of death drops with increasing doses but less steeply. If we assume the studies' median to be a reasonable guide, to attain another 20 percent reduction in risk beyond the benefits of ninety weekly minutes, we'd have to exercise another five and a half hours for a total of seven hours per week.

In the final analysis, exercising a minimum of 150 minutes per week is as good a prescription as any and has the advantage of being a clear, attainable dose. But there is no optimal, most beneficial dose of exercise. People who exercise the least have the most to gain from just modest added effort, more is better, and the benefits of additional exercise gradually tail off. So can you exercise too much?

Can You Exercise Too Much?

One day in early February 2015, my in-box filled with a mix of angry, puzzled, and "gotcha" emails citing a widely reported paper hot off the press in the prestigious *Journal of the American College of Cardiology*.[10] Beginning in 2001, Danish researchers compared more than a thousand self-identified runners from Copenhagen with about four thousand age-matched sedentary Danes. When the researchers tabulated deaths among these groups over the succeeding twelve years, they found that joggers who ran slowly over moderate distances had 30 percent lower death rates than sedentary individuals, but the serious runners who ran the most and the fastest died at the same rate as the non-exercisers. Headlines around the world blared, "Fast Running Is as Deadly as Sitting on the Couch," "Good News for Couch Potatoes," and "Slow Runners Come Out Ahead."

So far we have examined how various doses—including the nearly ubiquitous recommendation of 150 minutes per week—can overcome the detrimental effects of too little physical activity, but can you get too much of a good thing? From an evolutionary perspective, it is within reason to expect a U-shaped relationship between exercise dose and mortality. Because hunter-gatherers are generally only moderately physically active without being either couch potatoes or ultramarathoners, we are probably adapted for moderate rather than extreme exercise. While few of us think it sensible to run across the United States or swim the Atlantic, plenty of people were comforted to read the Copenhagen City Heart Study's conclusion that staying on the couch is just as healthy as running marathons.

As the philosopher George Santayana once quipped, "Skepticism, like chastity, should not be relinquished too readily." When it comes to health news, a dose of incredulity is especially necessary because science and journalism are no less susceptible to humanity's flaws than other endeavors. Unfortunately for those who wanted to hear they were better off not exercising than running, the Copenhagen City Heart Study offered more truthiness than truth. Although the researchers sampled more than a thousand runners, only eighty (7 percent) engaged in strenuous exercise, and of that tiny sample only two died during the study. In addition, the researchers never looked at cause of death, making no distinction between traffic accidents and heart attacks. You don't need a degree in statistics to realize the study's conclusions were meaningless and misleading.

Fortunately, better studies have been published. And contrary to predictions of a U-shaped curve between exercise dose and mortality, there is little solid evidence that extreme levels of exercise are either harmful or additionally healthy. A number of studies have found that elite athletes, especially those who do endurance sports, live longer and require less medical care than nonathletes.[11] In case you are worried that athletes might have better genes than the rest of us, thus protecting them from the rigors of extreme exercise, a study that followed nearly 22,000 ordinary nonathletes for fifteen years found that the highest dose exercisers did not have higher or lower rates of death including by heart disease than those who exercised moderately.[12] An even larger analysis of more than 600,000 individuals found that extremists who exercised more than ten times the standard recommended dose of 150 minutes per week did not have significantly higher rates of death than those who exercised between five and ten times the standard dose.[13] In summarizing these data, Meagan Wasfy and Aaron Baggish conclude that "these findings reinforce the notion that light to moderate doses of exercise have a substantial positive impact on health but that continued dose escalation appears neither incrementally better nor worse."[14]

To be honest, extreme exercise might cause harm, but so few people attain such levels of physical activity that the effects of over-exercising are hard to study rigorously. But if by rare chance you or a loved one runs ultramarathons or competes in the Tour de France, you probably still have worries.

One long-standing and very legitimate concern is the potential effect of too much exercise on the immune system. In 1918 as the Spanish Flu was sweeping across the world, causing millions of deaths, Dr. William Cowles suggested that fatigue from "violent exercise" contributed to higher rates of pneumonia based on his experiences treating the staff and students at the Groton School just outside Boston.[15] This concern gained new traction in the 1980s following studies which found that marathoners and ultramarathoners had higher rates of self-reported respiratory tract infections following their grueling races than fit individuals who exercised more moderately.[16] Additional studies found lower levels of disease-fighting white blood cells in the bloodstream and saliva immediately following intense bouts of vigorous exercise.[17] These and other data led to the hypothesis that the energetic demands of extreme exercise create a temporary "open window" for infection.

The open window hypothesis is commonsensical but just how much exercise is too much needs further study. When researchers repeated studies of respiratory tract infections among marathoners and ultramarathoners using medically based rather than self-reported diagnoses, they found no elevated incidence of infection following acute bouts of exercise.[18] In addition, new sophisticated experiments have tracked how immune cells move throughout the body after prolonged bouts of exercise instead of measuring their abundance in just the bloodstream. According to these studies, long and hard workouts do lower bloodstream levels of key immune cells that fight infections but also redeploy some of these cells to the mucus-lined surfaces of the lungs and other vulnerable tissues, thus potentially providing heightened surveillance and protection.[19] As we will see in chapter 13, there is evidence that regular, moderate physical activity can help protect against some contagious diseases, but we need more clinical data on the extent to which high doses of exercise suppress the immune system's ability to ward off infections and under what conditions.[20] That said, there is no question that anyone fighting a serious infection should avoid overexertion. An experiment that gave mice a deadly form of influenza and then forced them to exercise before their symptoms developed found that low levels of moderate exercise (twenty to thirty minutes of daily running) doubled their rate of survival compared to sedentary mice, but extremely high levels of exercise (two and a half hours of running a day) caused them to die at even higher rates.[21] Every physician I know recommends rest when combating a full-blown infection, especially one below the neck.

Another big concern is heart damage. Every once in a while, someone dies tragically from a heart attack in a marathon or some other athletic event, prompting scary articles about the dangers of overexercising. You can also read that some extreme endurance athletes have abnormally enlarged hearts or show signs of damage such as calcified coronary arteries and too much fibrous tissue.[22] Everything, including exercise, involves trade-offs, so it would be surprising if extreme exercise didn't carry some risks for the cardiovascular system. Apart from increased levels of musculoskeletal injuries, the most well-documented hazard of very high exercise doses appears to be an increased likelihood of developing atrial fibrillation, an abnormally rapid heart rhythm.[23] However, many other supposedly worrisome risk factors reported in athletes appear to be misinterpretations of evidence by doctors who compare the hearts of athletes with those of "normal" sedentary individuals with no diagnosis of disease. As we have repeatedly seen, being sedentary is by no means normal from an evolutionary perspective, and such individuals are more likely to develop chronic illnesses and die at a younger age than more active people. The medical habit of erroneously considering sedentary individuals "normal" controls has led to some diagnostic blunders such as mistaking normal repair mechanisms for signs of disease. A prime example is coronary calcification.

When he turned sixty-five, Ambrose "Amby" Burfoot decided to get a thorough heart checkup. By any standard, Burfoot counts as an extreme exerciser. Prior to walking into the doctor's office on that day in 2011, he had clocked more than 110,000 miles of running and had raced more than seventy-five marathons (winning the Boston Marathon in 1968), not to mention countless shorter races. As the much-admired editor of *Runner's World* who writes frequently about the science and health implications of running, he also knows more than most people on the planet about the benefits and risks of running. But he wasn't prepared for the bad news he received from his doctor. On the scan of his heart were many bright shiny white spots in the coronary arteries that supply blood to the heart. These calcified plaques can cause a heart attack if they block an artery. Because plaques contain calcium, which shows up nicely in a CT scan, doctors routinely score plaques by their calcium content: a coronary artery calcium (CAC) score. CACs above 100 are generally considered cause for concern. Burfoot's CAC score was a staggering 946, which according to other studies put him at more risk than 90 percent of men his age.[24]

Burfoot left the doctor's office terrified by his CAC score. "Driving to my *Runner's World* office 10 minutes later, I felt lightheaded, dizzy. My palms left a damp smear on the steering wheel." Burfoot, however, was otherwise totally healthy with excellent cholesterol levels and no other evidence of heart disease, and it turns out he is hardly unusual for an extreme athlete and probably shouldn't be worried. For some time, doctors have noted that many competitive runners have CAC scores greater than 100 and assumed these patients were at elevated risk for heart disease.[25] But these risk estimates are based on nonathletes and do not take into consideration the size and density of the plaques, the size of the coronary arteries around them, or the likelihood that the plaques will grow, detach, or do anything else that could cause a heart attack. An alternative, evolutionary perspective suggests that plaque calcification is one of the body's many normal defense mechanisms, not unlike a fever or nausea. And when researchers look more carefully, they find that the dense coronary calcifications commonly found in athletes like Burfoot tend to differ from the softer, less stable plaques that are indeed a risk factor for heart attacks. Instead, they appear to be protective adaptations—kind of like Band-Aids—to repair the walls of arteries from high stresses caused by hard exercise.[26] One massive analysis of almost twenty-two thousand middle-aged and elderly men found that the most physically active individuals had the highest CAC scores but the lowest risk of heart disease.[27]

Burfoot's CAC score scare is a characteristic example of how fears about high doses of exercise tend to be based on poorly understood risk factors rather than actual deaths associated with those risk factors. Another example is the so-called athlete's heart. Endurance athletes like Burfoot tend to have enlarged, more muscular chambers of the heart that allow each contraction to pump more blood. One consequence is a low resting pulse (forty to sixty beats a minute). Because these big, strong hearts at first glance resemble the dilated hearts of individuals suffering from congestive heart failure, worries persist that too much exercise causes pathological expansion of the heart. Big was thought to be bad. But the superficial similarities in heart size between athletes and those who suffer from heart failure have different causes and consequences. Apart from potential arrhythmias (especially atrial fibrillation), there is no evidence that a big, strong heart poses any health risks.[28]

Stay tuned for more on these and other worries about the effects of too much exercise on the heart and other organs, but even if new concerns emerge, overexercising will never be a major public health problem. That said, high levels of exercise still expose an underlying paradox. As we have repeatedly seen, regular exercisers, including those who engage in extremes, are less likely to die young than non-exercisers, but very physically stressful activities like shoveling snow after a blizzard or running a marathon do increase the risk of sudden death.[29] These deaths, however, mostly occur because of an underlying congenital condition or acquired disease, and without exercise some of these individuals might have died even younger.[30] You might be more likely to die while running than watching a marathon, but training for the marathon likely adds years to your life.

So can you exercise too much? Perhaps at extreme levels, and most certainly if you are sick with a serious infection or injured and need to recover. You also increase your risk of musculoskeletal injuries if you haven't adapted your bones, muscles, and other tissues to handle the stresses of repeated high forces of Olympic-level weight lifting, playing five sets of tennis a day, running marathons, or overdoing some other sport that obsesses you. In other respects, the negative effects of too much exercise appear to be ridiculously less than the negative effects of too little. As my wife points out, the biggest risk of exercising too much is ruining your marriage, to which I would add that the biggest risk of exercising too little is not being around long enough to enjoy your marriage.

Mix It Up?

Apart from choosing how much to exercise, many of us wonder what type and how intensely to do it. This is a very modern problem. Although people, mostly the privileged, have been exercising for the sake of health since at least the time of Socrates, few exercisers until recently planned what mixture of cardio and weights to do every week. Instead, they figured out ways to have fun moving, usually outside, often with workouts as varied as the activities they did for a living. One of the Founding Fathers of the United States, Benjamin Franklin, loved to swim, walk, leap, lift and swing weights, and take "air baths" that apparently involved exposing his naked body to cold air.[31] Teddy Roosevelt, twenty-sixth president of the United States, famously boxed, rode horses, lifted weights, hiked, and swam in icy rivers. Twenty years later, Herbert Hoover, concerned that he wouldn't be able to walk or ride while serving as president, had the bright idea of requiring his staff to play a game with him from precisely 7:00 to 7:30 every morning, rain or shine. The game, dubbed Hoover-ball, was a cross between tennis and volleyball that involved lobbing and then catching a six-pound medicine ball over an eight-foot-high net on the White House lawn, earning his staff the nickname "Medicine Ball Cabinet" and helping Hoover lose twenty-two pounds while in office.[32]

But after World War II, exercise gradually began to be medicalized. As evidence accumulated, doctors and medical scientists increasingly viewed being sedentary as a pathological condition, and exercise became a form of treatment. Because of medicalization, everyday people began to choose or were prescribed specific volumes, intensities, and types of exercise—cardio versus weights—primarily on the basis of medical evidence. Exercise became medicine. And no one influenced this shift more than Dr. Kenneth Cooper, who helped make moderate-intensity aerobic physical activity the bedrock of most exercise regimes.

Moderate-Intensity Aerobic Exercise

Cooper was an all-star track and basketball athlete at the University of Oklahoma who stopped exercising in medical school and then rapidly became so obese and unfit that he suffered from heart problems in his late twenties. Terrified, he changed his diet and started running. A year later and forty pounds thinner, he ran his first marathon, coming in last with a time so slow his wife had to persuade the race officials to hang around at the finish to record his time of 6:24. As he got back in shape, Cooper became interested in the problem of how to measure fitness and the effects of exercise. Fortunately, he had the perfect job for this. As director of the U.S. Air Force's Aerospace Medical Laboratory in San Antonio, Cooper was charged with training astronauts to overcome the muscle-depleting, bone-wasting effects of being in gravity-free space. As he forced astronauts to walk, run, cycle, and swim, he developed a point system that ultimately turned into a twelve-minute test that measured cardiorespiratory fitness. Cooper published his test and the science behind it in a 1968 book titled *Aerobics* (a term he coined) that became an international best seller and was a major impetus behind the 1970s fitness boom.[33] To this day, when people think of exercise, they usually have in mind sustained moderate-intensity aerobic exercise. Because that's a mouthful of jargon, let's use the term "aerobic exercise."

Aerobic exercise is sustained physical activity fueled by burning oxygen. The key metrics are heart rate and oxygen use. By convention, aerobic exercise elevates your pulse to between 50 and 70 percent of maximum (most people's maximum heart rate is between 150 and 200 beats per minute depending on fitness and age).[34] Another way to measure exercise intensity is the percentage of the maximum rate of oxygen use, VO_2 max. Regardless of how we measure it, aerobic exercise causes breathing that is fast and deep enough to make singing impossible but not hard enough to prevent conversing in normal sentences. Typical aerobic exercises include fast walking, jogging, cycling, or (ever since Jack LaLanne and Jane Fonda) working out at home in front of the TV. If you are fit, you can also sustain for lengthy periods of time more vigorous aerobic exercise, which is conventionally defined at 70 to 85 percent of maximum heart rate. During vigorous aerobic activities like fast running (but not sprinting), one can usually speak a few words, but full sentences are impossible.

Thousands of studies since 1968 have firmly established the many diverse benefits of aerobic exercise. We will consider their effects on diseases later, but to summarize quickly, the most obvious benefits are cardiovascular, hence the term "cardio." Because the fundamental challenge of aerobic activity is to deliver more oxygen at a faster rate to muscles and other organs, this demand stimulates the chambers of the heart to grow stronger, more capacious, and more elastic. These adaptations in turn increase the heart's cardiac output, the product of heart rate and the volume of blood pumped per contraction. In the blood, aerobic exercise augments the red blood cell count but also increases the volume of plasma, reducing viscosity so the heart can pump blood more easily. Sustained increased cardiac output also stimulates the expansion of the many small arteries and capillaries where oxygen exchange occurs in muscles everywhere including the heart's muscle itself. And aerobic exercise raises so-called good cholesterol (HDL) and lowers so-called bad cholesterol (LDL) and circulating fats (triglycerides). Altogether, these many effects help keep hearts strong; arteries clear, supple, and unclogged; and resting blood pressure low.

Aerobic exercise additionally stimulates the growth and upkeep of just about every other system in the body. Within muscles, it increases the number of mitochondria, promotes the growth of muscle fibers, and increases their ability to store carbohydrates and burn fat. In terms of metabolism, it burns harmful organ fat, improves the body's ability to use sugar, lowers levels of inflammation, and beneficially adjusts the levels of many hormones including estrogen, testosterone, cortisol, and growth hormone. Weight-bearing aerobic activities (alas, not swimming) stimulate bones to grow larger and denser when we are young and to repair themselves as we age, and they strengthen other connective tissues. In moderation, aerobic exercise stimulates the immune system, providing enhanced ability to ward off some infectious diseases. And last but not least, aerobic exercise increases blood flow to the brain and elevates the production of molecules that stimulate brain cell growth, maintenance, and function. A good cardio workout really does improve cognition and mood.

High-Intensity Aerobic Exercise

Most of the physical activity we do is sustained, low- to moderate-intensity aerobic. However, not all cardio is fully aerobic. Even if we never exercise, we sometimes exert maximum, gasp-inducing effort for short periods like running up several flights of stairs. Or sprinting after a giraffe. In the extraordinary 1957 documentary *The Hunters,* John Marshall followed and filmed a group of desperately hungry San hunters in the Kalahari who were having no luck hunting until they encountered a herd of giraffes. In a riveting scene, one of the hunters dashes full speed, barefoot, for about a minute through the grass after the giraffes in order to get a decent shot with a poison-tipped arrow. His shot succeeds, but he and his companions must then track the wounded and poisoned giraffe for more than thirty miles—mostly walking but sometimes jogging slowly—as the suffering giraffe tries to flee. That initial sprint illustrates unforgettably how occasional short bursts of high intensity were vitally important complements to more usual low- to moderate-intensity aerobic activities.[35]

Short bursts of intense cardio elevate heart rate and oxygen consumption close to their upper limit, usually above 85 or 90 percent of maximum rate. Athletes have long known that repeated surges of this intensity, termed high-intensity interval training, are an effective way to improve performance. HIIT usually involves short bouts, anywhere from ten to sixty seconds, of maximum effort that leaves one breathless (but not dangerously so) interspersed with periods of rest. HIIT workouts became especially popular among runners and other endurance athletes after the great Finnish middle- and long-distance runner Paavo Nurmi (the "Flying Finn") trained for and won nine Olympic gold medals in the 1920s by doing short four-hundred-meter runs over and over as fast as he possibly could.[36]

In recent years, HIIT has also gone mainstream as exercise scientists started to study, appreciate, and laud its many potential health benefits for ordinary people. One influential researcher behind this shift was the Canadian physiologist Martin Gibala, whose lab started to compare college students asked to do six sessions of HIIT (six repetitions of thirty seconds of maximal effort followed by a short rest) over two weeks with those who did more conventional long-term aerobic training. Astonishingly, HIIT had as much or more effect on the students' cardiovascular fitness and metabolic functions such as the ability to use blood sugar and burn fat.[37] Since then, hundreds of studies have investigated and confirmed the effects of HIIT in men and women regardless of age, fitness, obesity, and health. Because HIIT stresses the cardiovascular system more acutely than moderate-intensity aerobic exercise, it can yield rapid, dramatic benefits. Done properly, HIIT can substantially elevate aerobic and anaerobic fitness, bring down blood pressure, lower harmful cholesterol levels, burn fat, improve muscle function, and stimulate the production of growth factors that help protect the brain (more on this in chapter 13).[38]

If you regularly do the same thirty-minute leisurely jog or bike ride several times a week, consider adding a little HIIT to your weekly routine. (But please consult a doctor if you are thinking of trying this.) By some measures, a few minutes of HIIT provides as much benefit as, if not more benefit than, thirty minutes of conventional aerobic exercise, and it has the virtue of improving rather than just maintaining fitness. HIIT is also mercifully short-term, hence less tedious than hours of trudging. A quick HIIT session of sprinting, stair running, or whatever else you can manage is especially useful if you have little time to exercise.

But does this mean you should do only HIIT? I wouldn't. HIIT done properly requires one to push really, really hard and is seriously uncomfortable as well as potentially inadvisable for individuals who are unfit or have health issues like joint pain or impaired cardiovascular function. In addition, it is unwise to do HIIT more than a few times per week, it doesn't burn as many calories, and it may increase susceptibility to injury. Most of all, it doesn't deliver all the diverse benefits of regular aerobic activity. Have you ever heard of anyone who got fit doing just a few minutes a week of intense exercise? All in all, HIIT is a faster way to improve fitness and a key complement to moderate aerobic exercise, but not the only way to get and stay in shape. Furthermore, as a form of intense cardio, many (but not all) HIIT workouts involve little to no weights.[39]

Resistance Exercise

Some exercises involve using muscles against an opposing, heavy weight that resists their efforts to contract. It bears repeating that when working against substantial loads, muscles can shorten (concentric contractions), but they are more stressed and grow larger and stronger in response to forceful contractions in which they stay the same length (isometric contractions) or stretch (eccentric contractions). Humans have always had to do some demanding resistance-generating activities that involve all three kinds of contractions. Remember those San hunters I described above? After they finally killed that giraffe, they had to butcher it, but giraffes are heavy. The film shows how strenuously they labor to cut the enormous animal into pieces, remove its thick skin, and then carry hundreds of pounds of meat. Other common resistance activities in the Stone Age included digging and climbing.

Few of us today need to butcher huge animals, let alone carry heavy things, dig, or do much else that involves resistance. To replace these activities, we must do exercises like push-ups and pull-ups against our body weight, or we lift special-purpose weights. In the eighteenth century it was fashionable to lift church bells that were silenced (made "dumb") by having their clappers removed, hence the term "dumbbells." Today's gyms are stocked with an assortment of dumbbells, free weights, and contraptions that can be adjusted to place a constant level of resistance on muscles throughout their entire range of motion.

However you do them, resistance activities are critical for maintaining muscle mass, especially fast-twitch fibers that generate strength and power. Resistance exercise can also help prevent bone loss, augment muscles' ability to use sugar, enhance some metabolic functions, and improve cholesterol levels. As a result, every major medical health organization recommends we supplement cardio with weights, especially as we age. A consensus suggestion is two sessions per week of muscle-strengthening exercises involving all major muscle groups (legs, hips, back, core, shoulders, and arms). Space these sessions several days apart to permit recovery, and they needn't involve large weights but should include eight to twelve repetitions of each exercise tiring enough to make you want to stop; two or three sets of exercises are more effective than just one.[40]

Putting it all together, I know gym rats who avoid cardio like the plague, and devotees of aerobics who wouldn't touch a barbell if you paid them. Yet everyone benefits from mixing it up because weights, moderate-intensity aerobic exercise, and HIIT have different, complementary effects on the body. Given that each of us is an "experiment of one" with different backgrounds, goals, and predilections that change with age, there can be no optimal mixture of exercise type any more than there can be an optimal amount.[41] So despite the strange, modern nature of exercise, an evolutionary perspective makes the same commonsensical recommendations for physical activity that people have followed for centuries, albeit using different terminologies: *exercise several hours a week, mostly cardio but also some weights, and keep it up as you age.* If you want a concrete prescription, the tried-and-true 2018 HHS recommendations appear to be a reasonable minimum.

Ultimately and understandably, exercise will always be medicalized to some extent. Exercise can foster vigor, vitality, and fun, but the majority of us also exercise because we worry about our weight and about conditions like heart disease, cancer, and Alzheimer's. So to conclude, let's consider how and why different amounts and types of exercise affect the diseases most likely to kill or disable us.

THIRTEEN

Exercise and Disease

> I take my only exercise acting as a pallbearer at the funerals of my friends who exercise regularly.
>
> —Mark Twain

Exercise may be medicinal, but it is no elixir. Perhaps no one epitomizes this truth more infamously than James "Jim" Fixx. Born in New York in 1932, Fixx became a magazine editor and authored several books of tricky puzzles designed to challenge "superintelligent" minds. As a young man, he smoked two packs of cigarettes a day, feasted on a junk-food diet of burgers, fries, and milk shakes, and burgeoned to 220 pounds. Aware that his equally unhealthy father survived a first heart attack at the age of thirty-five but died of a second at forty-three, Fixx decided to turn over a new leaf when he turned thirty-five. He gave up cigarettes, improved his diet, and started jogging. Three years later, he ran his first five-mile race, finishing last. But he kept training, became a marathoner, and eventually lost sixty pounds. A passionate believer in the power of running to improve health and increase longevity, Fixx published *The Complete Book of Running* in 1977. The book was a best seller, helped ignite the running boom, and made Fixx famous. Seven years later he died of a massive heart attack while running alone on the roads of Vermont at the age of fifty-two.[1]

Over the years, I've heard my share of snarky comments about Fixx's death being evidence that running is risky for your health. Rather, given Fixx's previous history of obesity, smoking, and a possible congenital heart defect, his heart attack is a sad reminder that you can't outrun a bad diet and the lasting damage caused by decades of smoking two packs of cigarettes a day. In all probability, he might have died younger had he not started running.[2]

Fixx's death notwithstanding, only the most incorrigible skeptic doesn't believe that exercise promotes health. It behooves us, however, to remember that exercise is a truly odd sort of medicine. It is largely medicinal because the absence of physical activity is unhealthy. Further, exercise not only is an abnormal behavior from an evolutionary perspective but also never evolved to be therapeutic. Instead, we evolved to spend energy—much more than our ape cousins do—on physical activity primarily out of necessity and for other social reasons and otherwise sensibly reserve scarce calories for the chief thing natural selection cares about: reproductive success. To use energy frugally, many of the genes that maintain our bodies thus depend on the stresses caused by being active. When we are young, physical activity prompts us to develop capacities like strong bones and improved memory; as we grow older, physical activity triggers many key maintenance and repair mechanisms that help us stay vigorous into middle and old age. And so, for countless generations, our ancestors rested as much as possible but also spent many hours a day walking, carrying, and digging, and occasionally they also ran, climbed, threw, danced, and fought. Their lives were challenging, and plenty of them died young, but physical activity helped many of those who survived childhood to become active, productive grandparents.

Then, in a blink of the eye, we invented the modern postindustrial world. Suddenly some of us can take it easy 24/7 in ways unimaginable to earlier generations. Instead of walking, carrying, digging, running, and throwing, we sit for most of the day in ergonomically designed chairs, stare at screens, and press buttons. The only catch is we still inherited our active ancestors' thrifty genes that rely on physical activity to grow, maintain, and repair our bodies. Incessant sitting combined with modern diets and other novelties thus contributes to evolutionary mismatches, defined as *conditions that are more common and severe today than in the past because our bodies are poorly adapted to novel environmental conditions.*[3] The twenty-first-century world, of course, is not without extraordinary benefits. Today, nearly seven billion of us live longer and healthier lives than most of our Stone Age forebears ever did, many of us enjoying comforts beyond the imaginations of pharaohs and emperors of yore. But just as we never evolved to cope with jet lag or guzzle gallons of soda, we never evolved to be persistently physically inactive. As we age, daily hours of minimal physical activity—typically in chairs—render us more vulnerable to a litany of chronic illnesses and disabilities that used to be rare or unknown such as heart disease, hypertension, many cancers, osteoporosis, osteoarthritis, and Alzheimer's. It is commonly assumed that these conditions are the inevitable by-products of more of us living to be older. But this is not entirely true. Exercise may not be an elixir, but by stimulating growth, maintenance, and repair, it can reduce our susceptibility to many of these mismatches. In this sense, exercise is medicinal. And unlike other medicines, exercise is free, has no side effects, and is sometimes fun. So to stay healthy and fit, many of us exercise.

But how much and in what way will exercise help ward off disease? Over the last twelve chapters we have considered how exercise affects aging, metabolism, weight, muscle function, knee injuries, and other issues related to health. But we haven't focused on how and why exercise affects the diseases most likely to kill or disable us. Thus to conclude on a practical but somewhat alarming note, let's once again use the lens of evolutionary anthropology to review briefly how and why different doses and types of physical activity affect vulnerabilities to major health conditions, physical and mental, hypothesized to be mismatches. For each major condition, let's ask three questions: *Is the condition more common today than in the past because of less physical activity? How does physical activity help prevent or treat the condition? What kind and dose of exercise are best?*

A few caveats. First, this chapter is a sort of compendium where you can look up a particular condition to read how it is affected by physical activity and how exercise can help prevent or treat the problem. In addition, while this explores some of the major ways that exercise promotes health, it offers no prescriptions for how or how much to exercise. If you are planning to start an exercise program, please consult a physician, especially if you have a medical condition or are unfit, and consider hiring an experienced professional to help. Finally, this cannot be a comprehensive review, because exercise affects hundreds of conditions. I've briefly zeroed in on the few most concerning, widely recognized mismatches unambiguously affected by physical activity. That means our first stop needs to be the world's biggest and fastest-growing risk factor for chronic disease: obesity.

Obesity

In 2013, the American Medical Association ignited controversy by classifying obesity as a disease. Physicians define obesity using the body mass index (BMI): weight (in kilograms) divided by height (in meters) squared. BMIs are not always the best way to measure body composition, but by convention a BMI between 18.5 and 25.0 is considered normal, 25.0–30.0 is overweight, and above 30.0 is obese.[4] Classifying obesity as a disease was intended to send a clear warning about its manifold health risks, to change the way the medical industry pays for treatment, and to destigmatize the condition. Despite these laudable goals, the classification remains contentious. While obesity is a risk factor for many diseases, not all obese people suffer from ill health. Further, it is staggering to categorize one-third of Americans as diseased, and labeling obesity as a disease potentially suggests it is a fixed, unalterable state. We must never blame or stigmatize anyone for being obese, but we also need to find compassionate ways to help each other prevent excess weight gain and lose weight. Should these include exercise?

What Is the Hypothesized Mismatch?

If mismatches are caused by harmful interactions between genes and environments in which environments rather than genes recently changed, it's hard to find a bigger example than obesity. Although some of us carry genes that make us more likely to become obese, the role of environment is uncontested. Obesity is almost unknown among foraging populations and was much less common a few generations ago, but nearly two billion people are now overweight or obese.

While obesity is patently a mismatch, the relationship between exercise and obesity is debated. It's worth remembering that energy balance links obesity and physical activity. When we are in *positive energy balance* from consuming more calories than we expend, we convert surplus calories into fat that we store in fat cells. When we are in *negative energy balance* from spending more calories than we consume, we burn some of this fat. This calories-in-calories-out equation, however, is regulated by hormones, which in turn are strongly affected by diet and by other factors including psychosocial stress, the microbes in our gut, and, of course, physical activity.

While the uncontested chief culprit for obesity is diet, especially processed foods that are low in fiber and brimming with sugar, the efficacy of exercise for affecting weight gain or loss is controversial. Many experts and others contend that exercise plays little role in weight loss. The most common arguments against using exercise to control weight are that calories from diet dwarf those spent on physical activity and that exercise increases hunger and fatigue, thus supposedly causing us to compensate by eating more and becoming couch potatoes after we exercise. A two-mile walk burns about 100 more calories than sitting, but that refreshing Coca-Cola afterward contains 140 calories. However, studies show that people who exercise more don't necessarily compensate by eating more and they usually don't become less active for the rest of the day.[5] It is untrue you can't lose weight by exercising. Instead, weight loss from exercise is much slower and more gradual than weight loss from dieting. Over the course of a year, walking an extra two miles a day can potentially lead to five pounds of weight loss. In addition, exercise definitely helps prevent weight regain following a diet, and likely plays a major role in preventing weight gain in the first place.[6]

Regardless of how people become obese, the harmful effects of obesity are not in question. Aside from overloading joints and interfering with breathing, excess fat cells overproduce hormones that alter metabolism, and when swollen, they become invaded by white blood cells that ignite chronic low-grade inflammation, damaging tissues throughout the body. Big deposits of enlarged fat cells in and around organs (so-called visceral, abdominal, or organ fat) are especially hazardous because they react sensitively to hormones and connect more directly to the bloodstream. All in all, obesity, especially too much organ fat, is an important risk factor for cardiovascular disease including heart attacks and strokes, type 2 diabetes, some cancers, osteoarthritis, asthma, kidney disease, and Alzheimer's and plays a key role in the development and progression of many other conditions.

How Does Physical Activity Help?

Beyond debates over how much physical activity helps people lose weight and prevent weight gain, hefty quantities of ink have been spilled over how much exercise protects against the adverse effects of being overweight or obese. Is it okay to be "fat but fit"?

Several observations underlie this disagreement. On the one hand, dozens of studies show that overweight people who exercise are healthier and live longer than overweight individuals who do not.[7] On the other hand, not everyone who is overweight or obese gets sick or dies prematurely.[8] According to a few studies, individuals who become a little plump in old age (but not obese or extremely overweight) tend to live slightly longer, perhaps because they have more energy reserves to help them survive serious illnesses like pneumonia.[9] On the surface, none of these observations should incite controversy. How many people are surprised that exercise is healthy regardless of one's weight, that extra pounds don't necessarily condemn you to an early grave, and that many healthy elderly (think Queen Elizabeth II) gain a few pounds?

As far as I can tell, "fat but fit" became a controversy by sometimes being spun as proof that obesity is not a health concern for those who exercise. That is untrue. While overweight people who exercise and are physically fit lessen their risk of chronic disease, if you must choose between being fit and fat or unfit and lean, the evidence overwhelmingly indicates you should gamble on being unfit and lean.[10]

One of the largest efforts to tease apart the independent effects of physical inactivity and weight is the Nurses' Health Study, a prodigious undertaking begun in 1976 that has tracked the habits, health, and deaths of more than 100,000 nurses who volunteered to share their life and death experiences with Harvard researchers. Among the many lessons learned is that nurses of the same weight who are physically active have mortality rates (deaths per year) about 50 percent lower than those who are inactive, while nurses who are similarly active but obese have 90 percent higher mortality rates than those who are lean.[11] If so, obesity has nearly twice the effect on death rates as physical inactivity. Even better is to avoid both risk factors: nurses who are lean and fit have 2.4 times lower mortality rates than those who are obese and unfit.

All in all, being active doesn't cancel out the higher risk of death associated with obesity, but being active is still beneficial if one is obese. This is an important message because so many people struggle to lose weight but can still manage to exercise. In doing so, they lessen or counteract many harmful consequences of obesity such as chronic inflammation.

How Much and What Kind of Exercise Are Best?

This one is easy: cardio is better than weights for obesity. As we will see later, weights help counteract some of the metabolic consequences of obesity, but cardio is better for preventing and reversing excess weight. One randomized control study that compared the effects of cardio and weights on overweight and obese adults found that individuals prescribed just weights barely lost any body fat but those prescribed twelve miles a week of running lost substantial amounts of fat, especially harmful organ fat.[12] How much the intensity of cardio matters for weight loss, however, is up for debate. Although individuals vary widely in their responses, higher-intensity activities generally burn more calories than lower-intensity activities, but they are also harder to maintain for long and thus sometimes end up consuming less total energy.[13] What matters most is probably cumulative dose. A hundred and fifty minutes of walking a week is probably not enough to lose much weight.

Metabolic Syndrome and Type 2 Diabetes

Forgive me for asking, but have you ever sipped another person's urine? Disgusting as that may sound, if you were a doctor back in the old days, you'd be a pee connoisseur. As a matter of routine, you would collect your patients' "liquid gold" to examine its taste, color, smell, and consistency. Much of what doctors discerned from urine was nonsense, but an exception was its sweetness. The English physician Thomas Willis (1621–1675) coined the term "diabetes mellitus" (Latin for "honey sweetened"), what we now call diabetes, from urine that was "wonderfully sweet as if it were imbued with honey or sugar."[14]

For better or worse, doctors no longer drink your urine, but routinely have your blood analyzed by a lab, and they also measure your blood pressure, weight, height, and waist circumference. By convention, someone has *metabolic syndrome* if they have most of the following characteristics: high levels of blood sugar, high levels of cholesterol, high blood pressure, and a large waist.[15] These characteristics, which commonly occur as a package with a fatty liver and other forms of obesity, are telltale signs of troubled metabolism. Metabolic syndrome, in turn, often leads to type 2 diabetes.

What Is the Hypothesized Mismatch?

Metabolic syndrome and type 2 diabetes are unambiguous mismatch conditions. They are essentially unrecorded among hunter-gatherers, rare among subsistence farmers, and only recently have become epidemic.[16] An astounding 20 to 25 percent of the world's adults have metabolic syndrome, and that percentage is projected to double in coming decades.[17] Metabolic syndrome is a prime risk factor for many scary conditions including cardiovascular disease, strokes, and dementia, but the poster child is type 2 diabetes (also known as adult-onset diabetes). Type 2 diabetes (which differs from type 1 and gestational diabetes[18]) is now the fastest-growing disease in the world. Prevalence of the disease increased more than sevenfold between 1975 and 2005, and there will be more than 600 million type 2 diabetics by 2030.[19]

Although too much sugar in your urine or blood is a sign of type 2 diabetes, the disease's root cause is a problem termed insulin resistance. Imagine you just wolfed down a dozen cookies. As the sugar from the cookies floods your bloodstream, blood sugar levels rise. Because too much sugar is toxic to many cells, excess sugar stimulates your pancreas to release the hormone insulin, whose basic function is to cause the body to store energy. Among its many actions, insulin directs special molecules on the surfaces of fat and muscle cells to transport sugar from the bloodstream into those cells to be stored or burned. Type 2 diabetes arises when the effects of metabolic syndrome prevent insulin receptors on these cells from binding with insulin (a phenomenon termed resistance). A vicious cycle ensues. When insulin binding doesn't happen, the glucose transporters don't take up sugar from the bloodstream. Then, as blood sugar levels rise, the brain desperately commands the pancreas to produce yet more insulin, but with diminishing effect, causing blood sugar levels to stay dangerously high. Symptoms include frequent thirst and peeing, nausea, tingly skin, and swollen feet. Eventually, the overworked pancreas fails, requiring injections of insulin to avoid death.

As with other mismatch diseases, many genes increase the chances of getting type 2 diabetes, but the primary environmental trigger for the disease is too much positive energy balance from some combination of that pernicious quartet of modern, Western, industrial lifestyles: obesity, poor diet, stress, and physical inactivity. It bears repeating that too many swollen fat cells, especially in the liver and other organs, cause inflammation and high levels of triglycerides that provoke insulin resistance, which further worsens these problems. Bad diets promote obesity and deluge the bloodstream with sugar and fat. Stress elevates cortisol, which releases blood sugar, causes organ fat to accumulate, and facilitates inflammation. Last but not least, persistent sedentariness contributes to metabolic syndrome by elevating blood sugar and fat levels and failing to dampen inflammation.

How Does Physical Activity Help?

Having type 2 diabetes raises a person's risk of dying, in some cases to a small degree, in other cases substantially, but it is treatable using drugs, diet, and exercise. Although drugs help, they aren't always necessary. Diet and exercise can sometimes allow the body to heal itself.

In one dramatic test of this concept, ten overweight Australian aborigines with type 2 diabetes reversed their disease after just seven weeks of returning to an active hunting and gathering lifestyle.[20]

The mechanisms by which physical activity helps prevent and treat type 2 diabetes are well studied. Most basically, exercise (in conjunction with diet) can ameliorate every characteristic of metabolic syndrome including excess organ fat, high blood pressure, and high levels of blood sugar, fat, and cholesterol. In addition, exercise lowers inflammation and counteracts many of the damaging effects of stress. And most remarkably, exercise can reverse insulin resistance by restoring blocked insulin receptors and causing muscle cells to produce more of the transporter molecules that shuttle sugar out of the bloodstream.[21] The effect is akin to unclogging a drain and flushing out the pipes. Altogether, by simultaneously improving the delivery, transport, and use of blood sugar, exercise can resuscitate a once resistant muscle cell to suck up as much as fiftyfold more molecules of blood sugar. No drug is so potent.

How Much and What Kind of Exercise Are Best?

Because physical inactivity is never the only cause of metabolic syndrome and type 2 diabetes, exercise is rarely sufficient as a sole treatment. Yet moderate to high amounts of exercise are a powerful complement to dieting and medication. Physicians and patients alike have been disappointed by the lackluster results of clinical trials that prescribed modest doses such as 150 minutes a week or less of walking.[22] However, trials that prescribed more than 150 minutes a week of moderate-intensity exercise have been more successful.[23] In one compelling study, Danish researchers randomized patients with type 2 diabetes into two groups: both were given advice on how to eat a healthy diet, but one group also labored through five or six 30- to 60-minute-long sessions of aerobic exercise a week plus two or three weight sessions per week. After a year, half of those who exercised were able to eliminate their diabetes medications, and another 20 percent were able to reduce their medication levels. Further, the more they exercised, the more they recovered normal function. In contrast, just one-quarter of the dieters were able to reduce their medication, and 40 percent had to increase their medication levels despite receiving excellent, standard health care.[24] As we have repeatedly seen, some exercise is better than none, and more is better.

As for exercise type, because metabolic syndrome and type 2 diabetes are so strongly linked to persistent positive energy balance, cardio remains the bedrock of most treatment plans. If, however, you think plodding daily on a treadmill a form of torture, it's heartening to know that you can and probably should mix it up. HIIT cardio is especially efficient and effective for countering metabolic syndrome.[25] In addition, many studies find that weights are also effective for restoring muscle sensitivity to insulin, lowering blood pressure, and improving cholesterol levels.[26] As one clever study showed, a combination of exercise types appears to be the best prescription.[27]

Cardiovascular Disease

According to the odds, you and I are most likely to die from some form of cardiovascular disease. Fortunately, ever since the groundbreaking studies of Dr. Jeremy Morris, we have solid evidence of just how potently lifestyle can substantially decrease this risk. Born in 1910, Morris grew up in the slums of Glasgow, became a physician, and served in the Royal Army Medical Corps during World War II. When the war ended, he moved to London and became interested in why heart attacks were becoming more prevalent. According to Morris, almost nothing was known about this disease in 1946: "No literature—a wonderful situation! You could go to the Royal Society of Medicine library and read the literature before you had tea."[28] By collecting data from morgues and hospitals, Morris noticed that the drivers of London's famous double-decker buses suffered more heart attacks than the conductors who walked up and down the aisles and stairs collecting tickets. Curious, he initiated a large-scale study. In a pair of charmingly written papers published in 1953, he showed that the more sedentary drivers had twice the rate of heart attacks as the conductors.[29] Morris also showed that postal workers who sat all day in offices were twice as likely to have a heart attack as those who walked around London delivering letters. His results have since been confirmed and furthered to explain how and why not enough physical activity—along with bad diet, smoking, chronic stress, and other novel environmental conditions—is bad for our plumbing.

What Is the Hypothesized Mismatch?

The heart is essentially a muscular pump connected to an elaborate network of branching tubes. Although there are several kinds of cardiovascular disease, almost all arise from something going wrong in either the tubes or the pump. Most problems start with the tubes, primarily the arteries that carry blood from the heart to every nook and cranny of the body. Like the pipes in a building, arteries are vulnerable to getting clogged with unwanted deposits. This hardening of the arteries, termed atherosclerosis, starts with the buildup of plaque—a gloppy mixture of fat, cholesterol, and calcium—within the walls of arteries. Plaques, however, don't simply accumulate in arteries like crud settling in a pipe. Instead, they are dynamic, changing, growing, shifting, and sometimes breaking. They develop when white blood cells in arteries trigger inflammation by reacting to damage usually caused by a combination of high blood pressure and so-called bad cholesterol that irritates the walls of the artery. In an effort to repair the damage, white blood cells produce a foamy mixture that incorporates cholesterol and other stuff and then hardens. As plaque accumulates, arteries stiffen and narrow, sometimes preventing enough blood from flowing to the tissues and organs that need it and further driving up blood pressure. One potentially lethal scenario is when plaques block an artery completely or detach and obstruct a smaller artery elsewhere. When this happens, tissues are starved of blood (also called ischemia) and die. Plaques can also cause the artery wall to dilate, weaken, and bulge (an aneurysm) or to tear apart (a rupture), which can lead to massive bleeding (a hemorrhage).

Blocked and ruptured arteries create trouble anywhere in the body, but the most vulnerable locations are the narrow coronary arteries that supply the heart muscle itself. Heart attacks, caused by blocked coronary arteries, may damage the heart's muscle, leading to less effective pumping of blood or triggering an electrical disturbance that can stop the heart altogether. Other highly vulnerable arteries are in the brain, which cause strokes when blocked by blood clots or when they rupture and bleed. To this list of more susceptible locations we should also add the retinas, kidneys, stomach, and intestines. The most extreme consequence of coronary artery disease is a heart attack, which, if one survives, leaves behind a weakened heart unable to pump blood as effectively as before, leading to heart failure. Arrhythmias are additional common causes of problems and death, and the heart can also be damaged by infections, birth defects, drugs, and faulty wiring. But atherosclerosis is by far the leading culprit, and chronically high blood pressure, hypertension, is a close second.

Hypertension is a silent condition that relentlessly strains the heart, arteries, and various organs. At least 100,000 times a day, the heart forces about five liters of blood through thousands of miles of arteries that resist each squeeze, generating pressure. When we exercise, blood pressure rises temporarily, causing the heart's muscular chambers to adapt, mostly by becoming stronger, larger, and more elastic so it can pump more blood with each stroke.[30] Just as important, arteries also adapt to exercise to keep blood pressure low, primarily by expanding, multiplying, and staying elastic.[31] However, when blood pressure is chronically high, the heart defends itself by developing thicker muscular walls. These thicker walls stiffen and fill with scar tissue, and eventually the heart weakens. A vicious cycle then ensues. As the heart's ability to pump blood declines, it becomes harder to exercise and thus control high blood pressure. Blood pressure may rise as the heart progressively weakens until the failing heart cannot support or sustain a normal blood pressure. Death usually ensues.

Coronary artery disease is ancient and has even been diagnosed in mummies.[32] But research on nonindustrial populations provides powerful evidence that coronary artery disease and hypertension are largely evolutionary mismatches. Although many medical textbooks teach doctors that it's normal for blood pressure to rise with age, we have known since the 1970s this is not true among hunter-gatherer populations like the San and the Hadza.[33] The average blood pressure in a seventy-year-old San hunter-gatherer is 120/67, no different from a twenty-year-old. Lifelong low blood pressure also characterizes many subsistence farming populations. My colleagues Rob Shave and Aaron

Baggish and I measured more than a hundred Tarahumara farmers of every age and found no difference in blood pressure between teenagers and octogenarians.[34] By the same token, blood pressure can also stay normal into old age among industrialized people who eat sensibly and stay active.[35]

Low blood pressure combined with healthy levels of cholesterol prevent active nonindustrial populations from developing coronary artery disease. When Hillard Kaplan, Michael Gurven, and colleagues examined CT scans of the hearts of more than seven hundred middle-aged and elderly Tsimane forager-farmers from the Amazon, they found no trace of threatening plaques in the coronary arteries of even the oldest individuals.[36] Predictably, as these populations become industrialized and change their lifestyles, their incidence of coronary artery disease and hypertension skyrockets.[37] In the last 120 years, coronary artery disease has exploded more than two-and-a-half-fold to become a leading cause of death worldwide.[38] Since Jeremy Morris's pioneering study on London bus conductors first pointed the way, it has become indisputable that coronary artery disease is a largely preventable mismatch caused by a combination of formerly rare risk factors: high cholesterol, high blood pressure, and chronic inflammation.[39] These harbingers of disease, in turn, are affected by genes but are mostly caused by the same interrelated behavioral risk factors we keep encountering: smoking, obesity, bad diets, stress, and physical inactivity.

How Does Physical Activity Help?

In 2018, Dave McGillivray, the beloved race director of the Boston Marathon, underwent a successful triple bypass surgery to avert an imminent heart attack. Despite having run hundreds of marathons, often for charity, Dave is the first to admit he ate untold quantities of junk food for decades. As with Jim Fixx, his heart disease illustrates how physical activity doesn't shield you from a bad diet. That said, it is possible Dave might have died earlier had he not been so physically active.

To explore how physical activity helps but doesn't entirely prevent cardiovascular diseases, let's return to the trinity of intertwined factors that are the root causes of the problem: high cholesterol, high blood pressure, and inflammation.

Cholesterol. A cholesterol test usually measures the levels of three molecules in your blood. The first is low-density lipoprotein (LDL), often termed bad cholesterol. Your liver produces these balloon-like molecules to transport fats and cholesterol throughout your bloodstream, but some LDLs have a harmful tendency to burrow into the walls of arteries, especially when blood pressure is high. These intrusions cause an inflammatory reaction that generates plaques. The second type of cholesterol is high-density lipoprotein (HDL), sometimes called good cholesterol, because these molecules scavenge and return LDLs back to the liver. The third type are triglycerides, fat molecules that are floating freely in the bloodstream and a signpost for metabolic syndrome. To make a long story short, diets rich in sugar and saturated fats contribute to cardiovascular disease because they promote high levels of plaque-forming LDLs. Conversely, physical activity helps prevent cardiovascular disease by lowering triglycerides, raising HDL levels, and to a lesser degree lowering LDL.

Blood pressure. A blood pressure test gives you two readings: the higher (systolic) number is the pressure your heart's main chamber overcomes when it squeezes blood throughout your body; the lower (diastolic) number is the pressure your heart experiences as its main chamber fills with blood. By convention high blood pressure is a reading greater than 130/90 or 140/90. Blood pressures above these values are concerning because, unabated, they damage the walls of arteries, making them vulnerable to invasion by plaque-inducing LDLs. As we already saw, once plaques start to form, blood pressure can rise, potentially stimulating yet more plaques. Chronically high blood pressure also strains the heart, causing it to thicken abnormally and weaken. By forcing more blood to flow more rapidly through arteries, physical activity stimulates the generation of new arteries throughout the body and helps keep existing arteries supple, protecting against high blood pressure.

Inflammation. Plaques don't form out of the blue but instead occur when white blood cells in the bloodstream react to the inflammation caused by LDLs and high blood pressure. Chronic inflammation also increases one's likelihood of developing plaques from high cholesterol and blood pressure.[40] And, as we have previously seen, while inflammation is caused by factors such as obesity, junky diets, excess alcohol, and smoking, it is substantially lowered by physical activity.

How Much and What Kind of Exercise Are Best?

Some of us inherit genes that predispose us to cardiovascular diseases, but they by no means seal our fates. Instead, it is common knowledge that to prevent hypertension, coronary artery disease, and other problems, start by not smoking and by avoiding too many processed foods rich in sugar, saturated fats, and salt. Physical activity is also indispensable because the cardiovascular system never evolved to develop capacity and maintain itself in the absence of demand. Inactivity thus makes us vulnerable to high blood pressure and heart disease.

It is widely recognized that cardio exercise is best for the cardiovascular system. Extended periods of aerobic physical activity require the heart to pump high volumes of blood to every corner of the body, stimulating beneficial responses that keep blood pressure low and the heart strong. Cardio also combats other risk factors for cardiovascular disease, especially high inflammation and cholesterol.[41] Cardiorespiratory fitness, which benefits most strongly from aerobic exercise, is thus a powerful predictor of risk for cardiovascular diseases. One massive study of nearly ten thousand men found that individuals with good cardiorespiratory fitness had more than a fourfold lower risk for cardiovascular diseases than those with poor fitness, and those who improved their fitness cut their risk in half.[42] Not only does cardio prevent disease, but it can also help with treatment. People with hypertension, bad cholesterol levels, or full-blown coronary artery disease derive modest benefits from at least 150 minutes of physical activity a week and even more benefits from higher doses.[43] As we have seen before, shorter doses of high-intensity cardio appear to be as good as if not more effective than lengthier doses of low-intensity cardio.[44]

While cardio unquestionably invigorates and strengthens the cardiovascular system, lifting weights also improves cholesterol levels (raising HDLs and lowering LDLs) and lowers resting blood pressure (although not as much as cardio).[45] That said, doing only weights is apparently less protective than only cardio for the cardiovascular system.[46] My colleagues Rob Shave and Aaron Baggish and I have suggested this protection may arise from how the cardiovascular system trades off its ability to adapt to the contrasting challenges of weights versus cardio.[47] Professional athletes like runners who train solely for endurance maintain low blood pressure and develop large, elastic hearts that are better able to handle high volumes of blood flow but that don't cope well with high pressures needed to lift heavy weights. In contrast, resistance athletes like football linemen develop thicker and stiffer hearts that can manage high pressures but are less able to handle the high volumes necessary for cardio exercise. Thus, athletes who exclusively weight train without also doing some cardio appear to be at as much risk as sedentary individuals of developing chronic high blood pressure and cardiovascular disease. This risk is reflected in a massive study of the Finnish population including every athlete who competed in the Olympics between 1920 and 1965. Endurance athletes such as cross-country skiers had a stunning two-thirds lower risk of heart attacks than average Finns, while power athletes like weight lifters and wrestlers had one-third higher rates of heart attacks.[48] Bottom line: weight training isn't bad, but don't skip the cardio.

Respiratory Tract Infections and Other Contagions

As I edit these words in March 2020, COVID-19, the worst pandemic since the 1918 Spanish Flu, is overwhelming the globe, causing massive numbers of people to fall ill, many to die, and plunging the world into economic crisis. The virus is a stark reminder that contagious diseases have never ceased to pose a profound and terrifying threat to human health. Even though the majority of people who get COVID-19 experience only mild to moderate symptoms, it is many times deadlier than most viral infections of the respiratory tract, including influenza. According to the Centers for Disease Control and Prevention, in a typical year influenza kills about fifty thousand Americans, most of them elderly. Other infectious diseases like AIDS, hepatitis, and tuberculosis also take the lives of substantial numbers of people around the world annually.

You may be wondering what physical activity has to do with contagions like respiratory tract infections (RTIs). During epidemics like COVID-19, health officials urge us to wash our hands more often and more thoroughly, to practice social distancing, to cough into our elbows, and—trickiest of all—to stop touching our faces. These fundamental, sensible measures effectively help impede transmission of the virus. Other key, proven treatments include vaccines that teach our immune systems to protect us from particular viruses, and antiviral medicines. The final, complementary, but sometimes neglected method of protection is to bolster the immune system. And in this respect, regular physical activity, although no panacea, might be helpful.

What Is the Hypothesized Mismatch?

Just as viruses, bacteria, and other pathogens are constantly evolving to invade our bodies, evade our immune systems, and make more copies of themselves that we then sneeze, cough, or otherwise disperse to infect others, our immune systems have been simultaneously evolving to fight back. This evolutionary arms race has been going on for hundreds of millions of years, but ever since the origins of agriculture, humans have made ourselves vastly more vulnerable to contagious diseases like cholera, smallpox, and RTIs that are passed from one person to another. Hunter-gatherers live in small groups at low population densities and in temporary camps with no farm animals, but the development of agriculture and then industrialization enabled people to live permanently in villages, towns, and cities at extremely high population densities, often cheek by jowl with farm animals and other species like rats and mice. To make matters worse, sewers and clean water supplies were not constructed in most towns and cities until relatively recently, and public sanitation is still inadequate in many parts of the world. Contagious pathogens flourish in crowded, unhygienic conditions, and when they jump to humans from other species, they are especially dangerous because no one's immune system has encountered them before. So, while hunter-gatherers suffer from plenty of infectious illnesses, highly contagious epidemic diseases like COVID-19 are partly mismatches made possible by civilization, and that explains why social distancing and handwashing are key tools to fight them.[49]

Persistent lack of physical activity may be an additional, partial mismatch for the immune system. There are longstanding concerns that excessively demanding physical activities like running a marathon can compromise the immune system's capabilities, but several lines of evidence indicate that regular, moderate physical activity has the potential to reduce the risk of contracting certain contagious diseases, including RTIs.[50] In addition, exercise appears to slow the rate at which the immune system deteriorates as we age.[51] But exercise is no magic bullet. The immune system is labyrinthine with a multitude of different components that usually work in marvelous coordination but occasionally work at odds with one another. As allergies and autoimmune diseases like lupus attest, on rare but consequential occasions, immune responses that are meant to protect us can turn against our own bodies. In addition, pathogens are endlessly evolving new ways to bypass our immune defenses.

It is also unclear why physical inactivity might be a mismatch for the immune system apart from the generally negative effects of sedentariness on overall health and levels of stress (which, as we have seen, depresses the immune system). One possibility is that because heading off to the bush to hunt and gather potentially made our ancestors more likely to encounter pathogens, our immune systems evolved to compensate by ramping up our defenses when we are active. A related possible explanation stems from the stingy way our bodies use calories. The fatigue we experience when fighting a cold is a reminder that the immune system is often energetically costly. As a result, maybe our immune systems evolved to be less vigilant when they are less needed. For hunter-gatherers, unlike most industrial people, those times might have been when they were less physically active and thus less likely to be exposed to pathogens.

How Does Physical Activity Help?

How and to what extent physical activity may reduce the risk of certain communicable diseases including RTIs is hard to measure. One way to address this question is to compare the incidence of RTIs and other infectious diseases among individuals with differing levels of physical activity. Overall, such studies provide positive but not unqualified good news. In one investigation, researchers randomly assigned one hundred and fifteen physically inactive women in the Seattle area to remain sedentary or to walk for forty-five-minute sessions five times a week over the course of a year. At first, there was no difference between the groups, but after six months, the regular walkers suffered roughly one-half to one-third the rate of RTIs.[52] To test the effects of weight, researchers also asked more than one hundred women, some of them obese, to walk for five forty-five-minute sessions a week for twelve weeks during the winter when RTIs are most common. The women who walked, regardless of their weight, regularly experienced roughly half the number of days with RTIs.[53] Because stress also depresses immune function, another study followed more than one thousand Swedes for four months while collecting data on their levels of exercise, stress, and the incidence of RTIs. Compared to sedentary Swedes, both moderate and vigorous exercisers had a 15 to 18 percent reduced risk of contracting an RTI, with stronger reductions among those who reported they were feeling stressed.[54] Finally, a study of more than one hundred thousand nurses (two-thirds of them women) that controlled for smoking, weight, alcohol consumption, sex, and age found an inverse dose-response relationship between levels of physical activity and the risk of pneumonia, with a more than 30 percent reduction between the women (but not men) who were most and least active.[55] Despite these encouraging findings, not all studies report lower RTI rates among exercisers, in part because these sorts of trials are difficult to conduct.[56] We need more studies that follow large samples of people over extended time periods and which diagnose the incidence of infections accurately while also measuring physical activity levels and risk factors like stress. While more rigorous research is needed to test and quantify better how much exercise helps fight infectious diseases, there is no evidence that moderate levels of exercise increases anyone's risk.

Another research strategy is to test experimentally how different components of the immune system respond to varying doses and types of physical activity. The simplest way to do this is to draw blood or take saliva samples from people or laboratory animals before and after exercise to measure changes in the concentration of white blood cells, antibodies, and other agents of the immune system. A limitation of these studies is that they measure only the activity of the immune system, not clinical outcomes, but they generally find higher baseline levels of infection-fighting cells among individuals who exercise regularly and moderately, and lower levels immediately following intense, prolonged bouts of vigorous exercise.[57] In one elegant experiment, a group of thirteen young men were asked to pedal on a stationary bike until they reached exhaustion and then divided into two groups, one forbidden to exercise and another required to pedal moderately for a half-hour every day for two months, at which point both groups were given the same exhausting pedaling test. While the two months of moderate exercise led to higher levels of white blood cells in the exercised men, the bouts of acute exercise had the opposite effect, especially among the sedentary men.[58] These and other studies showing increased immune activity following moderate exercise but declines in white blood cell counts right after intense exercise have led to the hypothesis of a J-shaped relationship between exercise dose and immune function.[59] According to this idea, long-term physical inactivity depresses immune competence, moderate levels boost the immune system, and very high doses of physical activity temporarily compromise immune function, thus increasing vulnerability to infection, especially in unfit individuals.[60]

The widely held view that intense, prolonged physical activities like a full triathlon lead to an "open window" for infection is commonsensical, but more research is needed to establish how much exercise is too much and why. If you think of white blood cells and antibodies as soldiers battling foreign enemies, then *where* they are deployed may be more important than how many one measures in the bloodstream. A number of studies provide some support for this surveillance hypothesis. In particular, regular physical activity not only increases white blood cell counts but also appears to distribute preferentially certain cells from the bloodstream to the places they are most needed, including the vulnerable mucus-covered linings of the respiratory tract and gut.[61] Further, some of the most highly redeployed cells are those most effective at fighting viruses (these include natural killer cells and cytotoxic T cells).[62] Studies that compare sedentary and active people given vaccines also find that exercise helps us develop antibodies more rapidly and effectively.[63] Importantly, this exercise-based vaccine boost also occurs in the elderly. Indeed, while older individuals generally tend to have higher infection rates, recover more slowly, and respond less to vaccines, regular physical activity appears to slow the senescence of these aspects of immune function.[64]

Altogether, it appears that regular, moderate physical activity increases the immune system's capacities, but how much is optimal and for which contagious diseases is poorly understood. It bears repeating that the immune system is as complex and diverse as the multitude of hostile pathogens it evolved to fight. Efforts to quantify how well someone's immune system fights particular infectious diseases (COVID-19 among them) are thus clouded by variations between humans, germs, and many other factors including physical activity levels. In addition, we can't do controlled experiments on how well the immune system protects humans from potentially deadly diseases, and extrapolating studies of immune function in laboratory animals like mice to humans is complicated by differences between the immune systems of species and their enemies. Nonetheless, in one noteworthy study that could never be done in people, researchers gave a life-threatening strain of influenza to mice and then forced some of them to exercise for three days before the onset of symptoms. An impressive 82 percent of the mice who were exercised moderately (twenty to thirty minutes a day at a modest speed) survived, but 43 percent of the sedentary mice survived, and only 30 percent of the mice forced to exercise for two and a half hours a day survived.[65] For these mice, a little exercise was better than none, but too much was deadly, highlighting the vital importance of rest when fighting a serious infection.

How Much and What Kind of Exercise Are Best?

Almost all research on the effects of physical activity on the immune system have looked at cardio, and the few studies that have examined weight training find little to no effect (but also no harm).[66] What we don't yet know enough about is dose. As with so many benefits of physical activity, some is probably better than none for many aspects of immune function, but how much is too much and to what extent it temporarily depresses the immune system require further study. Recall from chapter 12 that the consensus view on this issue is the "open window" hypothesis. Because the immune system requires plentiful energy when it is fighting at full force, extreme bouts of exercise may reduce the calories needed to combat invading pathogens. However, the immune surveillance hypothesis posits that vigorous exercise can preferentially send needed cells to the places they do the most good. More research is needed, especially given the many factors affecting each individual and disease. There is no disagreement, however, that moderate levels of exercise are usually beneficial rather than detrimental, but once one is fighting any serious infection high levels of exercise are seriously inadvisable because the immune system needs all the energy it can get.

Chronic Musculoskeletal Conditions

Age is infamously unkind to muscles, bones, and joints. Among those fortunate enough to reach old age, it is not uncommon to be disabled by a trio of infirmities: muscle wasting (sarcopenia), bone loss (osteoporosis), and cartilage degeneration in joints (osteoarthritis). Weakened muscles fatigue elderly individuals when they climb stairs, carry groceries, walk, and do other basic tasks like rise from a chair. Loss of bone leads to painful collapsed vertebrae and fractures that inhibit activity. The nightmare scenario is a fractured neck of the thighbone from a seemingly trivial act like getting out of bed. Hip fractures can be the beginning of the end for some elderly by making them bedridden and thus vulnerable to additional, potentially mortal complications from inactivity like blood clots and pneumonia.[67] Finally, searing joint pain from arthritis diminishes mobility, which in turn hastens the aging process by engendering further physical disability and preventing seniors from doing what they want or need, sometimes leading to isolation and depression.

Fortunately, an evolutionary anthropological perspective highlights how and why aging doesn't necessarily have to become disastrous for muscles, bones, and joints.

What Is the Hypothesized Mismatch?

Some infirmity in old age is inescapable, but there is evidence that musculoskeletal diseases of aging are partly mismatches. As we already saw, grip strength tests indicate that hunter-gatherers age without losing as much muscle strength as average Westerners of the same age.[68] Although we have no good estimates of osteoporosis rates among hunter-gatherers, studies of bone quality and fracture rates from around the world indicate that osteoporosis rates have skyrocketed in postindustrial countries.[69] Today, the lifetime risk of osteoporosis is 40 to 50 percent among women and 13 to 22 percent among men, contributing to more than ten million fractures per year in developed nations.[70] Finally, my colleague Ian Wallace and I studied the skeletons of more than twenty-five hundred individuals who died over the age of fifty to show that osteoarthritis has been around for millions of years (even some Neanderthals were afflicted) but our chances of getting this disease at a given age have more than doubled since World War II.[71] Today, more than 25 percent of the U.S. adult population has been diagnosed with some form of osteoarthritis, most commonly in the knee.[72]

As always, the genes we inherited affect our susceptibility to sarcopenia, osteoporosis, and osteoarthritis. But because our genes haven't changed over the last few generations, the chief culprit of these mismatches must be environmental change. Modern processed diets and obesity are major causes, but given the basic functions of muscles and bones it is hardly surprising that physical inactivity is also to blame. The protective benefits of exercise, however, are distinct for each disease.

How Does Physical Activity Help?

Sarcopenia is the most obvious beneficiary of exercise. Because muscles are costly (right now, you are spending about one out of five calories simply maintaining your muscles[73]), they are the classic example of the "Use It or Lose It" principle of energy allocation. When we demand more from our muscles, especially contractions involving resistance, we activate genes that increase the size of fibers as well as repair and maintain muscle cells. As soon as we stop using them, muscles dwindle. Thus although aging affects hormone levels and nerve properties that inexorably diminish strength, staying physically active counters these declines. For today's retirees, staying active is a choice, but our elderly hunter-gatherer forebears had no alternative to lots of daily walking, carrying, digging, and climbing. In fact, as we saw, elderly hunter-gatherer grandparents are often more active than younger parents. Fortunately, the mechanisms by which physical activity, especially weight-bearing tasks that generate resistance, maintains as well as reverses muscle atrophy remain effective with advancing age. Even octogenarians can bulk up in a gym.[74]

Osteoporosis is a more complicated disease of disuse only partly prevented by exercise. It is a common misconception that bone is inert, not unlike the steel beams holding up a building. In reality bone is a dynamic tissue. We spend the first twenty to thirty years of life building up the bones in our skeleton, and thereafter gradually lose bone mass and density at a slow rate—as much as 1 percent per year.[75] That loss doesn't necessarily condemn us to osteoporosis, because under normal circumstances our bones are sufficiently built up to cope with this gradual loss without falling below the threshold at which they become too weak. Instead, we get osteoporosis only if we failed to develop enough peak bone mass when we were young, or if we lose bone too rapidly as we age.[76] When bones are insufficiently strong, vertebrae collapse, wrists snap, femurs fracture. One way to avoid osteoporosis is to develop as youngsters strong bones better able to withstand later losses. The other way to avoid the disease is to slow the rate of bone loss as we age. Age-related loss occurs in both men and women but is exacerbated in women after menopause, when there is a drop in levels of estrogen, which protects bones from being resorbed.[77] Although good nutrition, including lots of calcium and vitamin D, helps youngsters develop strong skeletons and prevents the elderly from resorbing their bones, the forces bones experience from physical activity are equally vital. In particular, weight-bearing activities that load the skeleton cause bone-growing cells to add more bone when we are young, and they prevent bone-resorbing cells from removing bone as we age.[78] Consequently, lifelong weight-bearing exercise helps prevent the disease.

Osteoarthritis is an enigmatic and poorly understood musculoskeletal disease despite afflicting millions of older people in industrialized countries. Because osteoarthritis occurs when the cartilage in joints wears away, many patients and doctors think it is a wear-and-tear consequence of aging. This view is wrong. Physical activities like running that load joints repeatedly and heavily do not cause higher rates of osteoarthritis and may sometimes be protective.[79] Indeed, if physical activity were a problem, we'd expect the disease to have become *less,* not *more,* common in today's more sedentary world. Instead, the disease occurs more commonly today as we age because of inflammation within joints that eats away cartilage. Sometimes, this inflammation is triggered by an accident that tears a meniscus or snaps a ligament. But most cases of the disease appear to be influenced by inflammation triggered by obesity and possibly also physical inactivity.[80]

How Much and What Kind of Exercise Are Best?

Because muscles, bones, and joints primarily function to generate and withstand forces, they maintain and repair themselves principally in response to high forces. There are no simple dose-response relationships between particular exercises and the vulnerability of muscles, bones, and joints to disease, but we can make a few generalizations.

Muscles benefit from all physical activities, but they respond most strongly to weight-bearing activities that require them to contract forcefully without changing length (isometric contractions) or as they lengthen (eccentric contractions). To prevent sarcopenia, do weights.

Bones also need weight-bearing exercises that apply forces of sufficient magnitude and rate to activate bone cells. Some of these forces occur from sudden impacts like a runner's body hitting the ground, but muscles generally create the highest forces.[81] Thus activities like jumping, running, and weight lifting that place demanding loads on bones help develop and maintain a strong skeleton much more than lower-impact activities like swimming or using an elliptical.[82]

Cartilage degeneration is probably countered by physical activity, but it is unknown how and to what extent different kinds of exercise help prevent osteoarthritis. Probably, the biggest benefit of physical activity is to prevent and reduce obesity, thus limiting inflammation as well as abnormally high pressures.[83] Regular loading from activities like walking and maybe even running might also increase the quantity and quality of cartilage in joints.[84] Finally, exercise, especially weights, strengthens the muscles around joints, reducing the likelihood they will be damaged from aberrant loads (like twisting a knee).[85] But everything has trade-offs. While physical activities generally help prevent osteoarthritis, some (especially relatively newfangled sports like downhill skiing) can make you more likely to seriously injure your joints, thus increasing your risk of the disease.

Cancer

Cancer scares me more than any other disease. Now the second-leading cause of death worldwide (killing about one in four), cancer is a sort of cellular Russian roulette that seems to strike indiscriminately, most often after the age of fifty. Over the last few decades, medical science has made impressive advances in understanding and treating various cancers, but a diagnosis too often remains a death sentence. Perhaps spectacular new treatments will be invented soon, but right now we need to pay more attention to *preventing* cancer. Prevention includes not only exercising but also recognizing that the disease is a form of evolution gone dreadfully wrong.

What Is the Hypothesized Mismatch?

Cancer isn't a single disease. Instead, it's an umbrella term for what happens when cells compete with each other in a kind of twisted unnatural selection within the body.[86] Think of your body as a giant ecosystem with nearly forty trillion cells from more than two hundred different cell lines. Normally, these cells cooperate harmoniously even as they acquire random mutations, nearly all harmless. Every once in a while, however, cells develop mutations that disrupt their function, and a tiny fraction of those mutations trigger cells to compete with each other. When such mutations occur, the cells become malignant. At this point, they divide uncontrollably, migrate throughout the body, and voraciously consume calories. If your immune system fails to kill these cancerous cells quickly enough, they overtake organs, disrupt their function, and starve other cells. The most common cancers occur in reproductive organs, intestines, skin, lungs, and marrow because cells in these tissues divide frequently and are exposed to external influences like radiation, toxins, and hormones that affect their likelihood of dividing or mutating.

As long as there is multicellular life, there will be cancer. And as more of us live to be older and have the chance to accrue harmful mutations, cancer rates will inevitably stay high. However, some cancers may be partly mismatches. Without the sophisticated technology available in modern hospitals, cancer is difficult to diagnose, but limited evidence suggests that cancer rates are lower in hunter-gatherer and nonindustrial populations.[87] The same was true of industrial populations until recently. In 1842, when Domenico Rigoni-Stern, chief physician of a hospital in Verona, Italy, published estimates of cancer rates in his hospital, he estimated that less than 1 percent of all the 150,673 deaths that occurred between 1760 and 1839 were from cancer.[88] Even if we account for doctors' inability to diagnose many cancers and the much younger age of death back then, these are at least ten times lower than contemporary cancer rates.[89] Further, wherever we look, the rates of many cancers are rising. For example, breast cancer rates in the U.K. doubled from 1921 to 2004.[90] According to one alarming estimate, there will be 27.5 million new cancer cases worldwide each year by 2040, a 62 percent increase from 2018.[91]

Because cancer isn't going away, we must figure out better ways not just to fight it but also to prevent or tame it. Fortunately, for several kinds of cancers, that includes physical activity.

How Does Physical Activity Help?

I am aware that people like me often sound like broken records. Fact after fact extolling the health benefits of physical activity can dampen the message's impact. Please don't react that way for cancer, however, because the cancer-fighting potential of exercise is underappreciated and insufficiently explored.

Let's start with the evidence. Numerous studies, many of them high quality, have examined the relationship between physical activity and cancer. One analysis pooled data from six prospective studies that, together, followed more than 650,000 elderly individuals for at least a decade.[92] Of the more than 116,000 deaths recorded, 25 percent were from cancer. When the researchers looked at the relationship between varying physical activity levels and cancer rates (controlling for sex, age, smoking, alcohol, and education), they found a clear dose-response relationship. Compared with those who were sedentary, modest exercisers had 13 to 20 percent lower cancer rates, and those who exercised moderately or more had 25 to 30 percent lower cancer rates. Other analyses—including one study of more than 1.4 million people—yield similar results.[93] Breast and colon cancers are most strongly affected by exercise. According to one estimate, three to four hours of moderate exercise a week is likely to reduce a woman's risk of breast cancer by 30 to 40 percent, and both men's and women's risk of colon cancer by 40 to 50 percent.[94]

How and why physical activity helps ward off cancer is only partly understood, but as an evolutionary perspective predicts, the mechanisms appear to be linked to energy. Life is fueled by energy whether you are an entire human being or just a cell within a body. Just as natural selection favors humans who acquire and then spend as many calories as possible on reproduction, the selection that drives cancer favors malignant cells that acquire as many calories as possible and then use them to create more copies of themselves. High levels of physical activity divert energy from cancerous cells in at least four possible ways.

(1) Reproductive hormones. Energy spent on physical activity is energy not spent on reproduction, a trade-off modulated by reproductive hormones like estrogen. Women who exercise moderately produce more than enough hormones to reproduce, but the bodies of sedentary women naturally shunt more energy toward reproduction, leading to 25 percent higher levels of estrogen.[95] Because reproductive hormones like estrogen induce cell division in breast tissue, inactivity increases the risk of breast cancer, while exercise has the opposite effect. Levels of estrogen, hence breast cancer, are also elevated by obesity and by having fewer pregnancies.[96]

(2) Sugar. Some cancer cells have a sweet tooth. In fact, many cancer cells tend to get their energy directly from sugars, which they burn anaerobically without oxygen. High levels of blood sugar from metabolic syndrome are thus associated with increased rates of cancer.[97] Exercise may thus help prevent and fight cancer by depriving cancer cells of ready energy. Furthermore, because high-intensity exercise inhibits anaerobic sugar metabolism, extremely vigorous exercise may be especially effective for preventing and fighting certain cancers.[98]

(3) Inflammation. Inflammation, which goes hand in hand with chronic positive energy balance and obesity, is a risk factor for many cancers. As we have already seen, inflammation causes various kinds of cellular damage, some of which are associated with mutations that can lead to cancer.[99] Physical activity thus counters cancer indirectly by helping prevent or reduce levels of inflammation either directly or indirectly.

(4) Antioxidants and immune function. Physical activity stimulates the body to spend energy on repair and maintenance systems to mop up the damage that exercise might cause in the first place. One of these investments is antioxidant production. These cleanup molecules counteract highly reactive atoms that cause many kinds

of damage including potentially cancerous mutations.[100] In addition, non-extreme levels of exercise boost the immune system, which plays a vital role in fighting cancer. An especially promising discovery is that vigorous exercise potently enhances the effectiveness of natural killer (NK) cells, the immune system's primary weapon that recognizes and destroys cancerous cells.[101]

How Much and What Kind of Exercise Are Best?

This question is poorly studied and difficult to answer given the incredible diversity of cancers and variation among individuals. Moderate to vigorous aerobic and resistance exercises have both been shown to lower the risk of certain cancers, especially colon and breast cancer, with higher doses generally associated with lower rates; exercise may also help patients undergoing treatment for cancer.[102]

Alzheimer's Disease

When my grandmother's short-term memory started to fail, we thought it was caused by the stress of taking care of my ailing grandfather. But after he died, her mind continued to decline slowly and relentlessly, memory by memory. At first she couldn't remember where she had put things, whom she had just spoken to, and what she had eaten for lunch. Then, as her Alzheimer's progressed, she started having trouble recognizing family members and friends and remembering basic words and key events in her life. Eventually, she lost her sense of both the present and the past. It was as if the disease had stolen her mind, leaving behind just her body.

What Is the Hypothesized Mismatch?

Alzheimer's is a complicated, poorly understood disease that must be in part an evolutionary mismatch. Studies of dementia in nonindustrial populations are limited, but conservative epidemiological studies that correct for differences in life expectancy indicate that Alzheimer's disease is about twenty times more common in industrial than nonindustrial populations.[103] And it's getting more common: worldwide prevalence of the disease is projected to increase fourfold in the first half of the twenty-first century.[104] Genes alone cannot explain this epidemic.

While Alzheimer's symptoms and progression are well known, its causes are not. The most common theory is that Alzheimer's results from plaques and tangles that smother nerve cells (neurons) near the surface of the brain, depriving the cells of nutrients, not unlike the way hair clogs a drain.[105] Treating these plaques and tangles, however, doesn't appear to reverse or prevent the disease, and many elderly people with plaques and tangles never develop Alzheimer's.[106] Mounting novel evidence suggests Alzheimer's is a kind of inflammatory autoimmune disease that initially affects cells in the brain known as astrocytes. Astrocytes, which number in the billions, normally regulate and protect neurons and their connections. When needed, astrocytes also produce toxin-like chemicals to defend the brain from infection. According to this theory, Alzheimer's occurs when astrocytes produce these toxins in the absence of infections, thus attacking other cells in the brain.[107]

One evolutionary explanation and preliminary support for this hypothesis comes from studies of Amazonian forager-farmers, the Tsimane (remember, they are the population without evidence of coronary heart disease). Although Westerners who carry the two copies of a gene called ApoE4 (a protein that transports fats in the bloodstream) are three to fifteen times *more* likely to get Alzheimer's in old age, elderly Tsimane with the same ApoE4 gene are *less* likely to show declines in cognitive performance if they suffer from many infections.[108] Alzheimer's may thus be an example of an evolutionary phenomenon called the hygiene hypothesis. According to this idea, ApoE4, which can be expressed by cells in the brain, might have evolved long ago to help protect the brain when infectious diseases were ubiquitous. Those of us today who live in bizarrely sterile environments without many germs and worms, face an increased chance that these formerly protective immune mechanisms turn against us. (The hygiene hypothesis also helps explain increased rates of allergies and many other autoimmune diseases.[109])

How Does Physical Activity Help?

Regardless of what causes Alzheimer's, if you are worried about the disease, then exercise. No effective drugs have yet been developed to treat Alzheimer's, and there is inconclusive evidence that keeping your mind sharp with mental games staves off dementia.[110] Exercise is by far the most effective known form of prevention and treatment. Further, the effects are impressive. An analysis of sixteen prospective studies including more than 160,000 individuals found that moderate levels of physical activity lowered the risk of Alzheimer's by 45 percent.[111] More physically intense activities may be associated with reduced risks for the disease.[112] Physical activity also slows the rate of cognitive and physical deterioration in Alzheimer's patients.[113]

How physical activity helps prevent and treat Alzheimer's is poorly known, but there is evidence for several evolved mechanisms. The most well supported is that physical activity—especially of longer duration but also more vigorous activities—causes the brain to produce a powerful molecule known as BDNF (brain-derived neurotrophic factor). BDNF first evolved to help mammals get energy during physical activity and at some point took on additional roles in the brain.[114] BDNF is a sort of growth tonic for the brain that nourishes and induces new brain cells, especially in regions involved in memory. But because we never evolved to be persistently sedentary, we never evolved a mechanism other than physical activity to produce high levels of BDNF. In a classic mismatch, absence of exercise deprives us of doses of BDNF that have been shown to improve memory and cognition and to maintain neuronal health that apparently helps prevent Alzheimer's.[115] One prospective study that followed more than two thousand individuals for decades found that women with the highest levels of BDNF have half the risk of developing Alzheimer's as those with the lowest levels.[116] Because BDNF helps astrocytes take care of brain cells and their connections, elevated levels of exercise-induced BDNF may help prevent the kind of astrocyte-induced damage hypothesized to underlie Alzheimer's.[117] Physical activity may also lower the risk of Alzheimer's by increasing blood flow to the brain, by suppressing inflammation, and by lowering damaging levels of oxidative stress.[118] Rodents that run on treadmills develop fewer plaques and tangles in their brains and have lower levels of inflammation associated with Alzheimer's.[119]

How Much and What Kind of Exercise Are Best?

Abundant evidence shows that physical activity is probably the single best way to lower the risk of Alzheimer's disease, but how much and what type are most effective is poorly known. One analysis of nineteen studies found that aerobic physical activity is most beneficial, but other reviews favor a mix of aerobic exercise, weights, and exercises that improve balance and coordination.[120] In addition, limited evidence suggests there may be a dose-response relationship between exercise intensity and risk.[121]

Mental Health: Depression and Anxiety

Exercise is no panacea for any disease, most especially those afflicting the mind, but even exerphobes admit some link between physical and mental health—an idea commonly and pithily expressed by quoting out of context the Roman poet Juvenal: "*Mens sana in corpore sano*" (a sound mind in a healthy body).[122] Yet using exercise to address mental health is rare. A 2018 study found that only 20 percent of doctors prescribed exercise to patients with anxiety or depression.[123] To some extent this attitude reflects recent extraordinary improvements in therapy and pharmaceuticals that have helped millions of people. But should we do more to explore and exploit the connections between physical and mental health? This topic is enormous, so let's focus briefly on just two common disorders: depression and anxiety.

What Is the Hypothesized Mismatch?

Only some of us will get cancer or heart disease, but everyone feels anxious or down in the dumps at times. While the ups and downs of daily moods are a normal part of being alive, they should not be confused with depressive and anxiety disorders, which are very different, serious clinical syndromes that affect one in five of us at some point. Depression takes many forms including major depressive disorder, which is defined as more than two weeks of extreme sadness, loss of pleasure in formerly engaging activities, diminished energy, altered appetite and sleep, poor concentration, low self-esteem, and general purposelessness. Unlike grief, depression tends to be persistent and characterized by feelings of low self-worth and guilt. There are also several forms of anxiety disorders. While some anxiety disorders are directed toward specific dreads (like speaking in public or acts of violence), generalized anxiety disorders involve chronic obsessive worries about nonspecific threats that are potential rather than actual. Depression and anxiety disorders are serious causes of disability and death.

We have made enormous progress in understanding depression and anxiety disorders, but as articulated by Randolph Nesse and others, an evolutionary perspective that considers these diseases as adaptations gone awry may help explain why we are so vulnerable and why they are so varied.[124] It is obviously adaptive for fear to lead us to avoid threats like poisonous snakes or being attacked by strangers. In anxiety disorder, however, these normal anxieties become irrational and uncontrolled. Likewise, it may sometimes be adaptive to be discouraged and unmotivated, hence disinclined to engage in behaviors unlikely to be successful like fighting someone who might kill us or wooing a lover who spurns us. In depression disorder, however, these low moods become directed persistently at ourselves, not the outside world. Just how and why these sorts of adaptive mechanisms turn pathological is poorly understood. As with all diseases, they involve interactions between genes and many complex environmental factors. But an evolutionary perspective prompts us to scrutinize the crucial role of the environment: Could these too be mismatch diseases? Are we vulnerable because we now face environmental factors, requiring less physical activity, that we never evolved to handle?

A first step in evaluating this hypothesis is to test if depression and anxiety disorders have become more common in modern, westernized societies. As ancient descriptions attest, these are hardly novel problems. Consider the despair of the prophet Elijah: "Then he himself went a day's journey into the wilderness and sat down under a broom tree. And he prayed that he might die, and said: 'It is enough! Now, Lord, take my life, for I am no better than my fathers' "

(1 Kings 19:4). We lack, however, reliable long-term data, especially in non-Western countries, and comparing diagnoses across cultures is complicated by different languages, contexts, perceptions, and beliefs. With these caveats in mind, some but not all studies suggest a trend toward higher rates of depression and anxiety disorders in societies undergoing modernization.[125] In addition, rates of depression and anxiety disorders have been rising recently in the United States and other developed countries. This epidemic is especially alarming among youngsters. When the psychologist Jean Twenge analyzed seventy years of survey data on almost eighty thousand American college and high school students, she found young adults were six to eight times more likely in 2007 to be suffering from major mental health disorders including depression than their peers in 1938.[126] Between 2009 and 2017, rates of depression rose 47 percent among those aged twelve to thirteen and by more than 60 percent among those aged fourteen to seventeen.[127]

It is appropriate to be skeptical about purported trends in mental health disorders given changes in the way these conditions are recognized and classified. But no one disagrees that these disorders are becoming more common, and they are strongly affected by rapidly changing social and physical environments. Our great-grandparents never confronted social media and 24/7 news cycles, not to mention as much obesity or physical inactivity. Not all these changes may cause depression, anxiety, or other mental health problems, but we owe it to ourselves and others to explore whether factors in our environment that we can change increase people's vulnerability to these conditions, and thus help prevent or treat them. Unsurprisingly, there is compelling evidence that implicates physical inactivity.

How Does Physical Activity Help?

Anyone doubting a connection between physical and mental health should consider an analysis of more than one million Americans showing that regular exercisers report 12 to 23 percent lower levels of mental health problems than sedentary people matched for sex, age, education, and income.[128] Dozens of more focused, high-quality analyses—many of them prospective randomized control studies—

confirm that exercise helps prevent and treat depression disorders and to a lesser extent generalized anxiety disorder.[129] To be sure, exercise is not a magic pill, but neither are drugs and psychotherapy, the most commonly used treatments. In fact, large-scale analyses of many trials find that exercise is at least as if not more effective than drugs and therapy.[130] Because physical activity has other benefits, it is hard to understand why more mental health clinicians and patients don't add exercise to their armamentarium.

How exercise alleviates depression and anxiety is less clear, and we should remember that some of the benefits of physical activity may arise from our physiology being poorly adapted to excess sedentariness. In this respect, exercise may be medicinal only because persistent inactivity increases our vulnerability to mental health disorders. While causal mechanisms are difficult to assess, here are several possible mechanisms, some of them already familiar.

First, physical activity has many direct effects on the brain. One is to flood the brain with mood-altering chemicals. As a reminder, exercise heightens the activity of transmitting molecules in the brain, notably dopamine, serotonin, and norepinephrine.[131] These neurotransmitters induce sensations of reward, well-being, arousal, and memory enhancement. Most pharmaceuticals such as SSRIs used to treat depression and anxiety manipulate levels of these neurotransmitters. Exercise also increases levels of other neurotransmitters, glutamate and GABA, that are often depleted in people with depression and anxiety.[132] Additional mood-enhancing molecules turned on by exercise include endogenous opioids such as endorphins and endocannabinoids that inhibit pain and produce positive moods.[133] Finally, as if this were not enough, remember that physical activity increases levels of BDNF and other growth factors that help maintain brain function. Altogether, regular physical activity alters brain chemistry, enhances electrical activity, and improves brain structure. Among other differences, the brains of more physically active people have enlarged memory regions, more cells, and increased blood supply.[134] These evolutionarily normal characteristics may reduce vulnerability to disease.

Additional possible mental health benefits from exercise are diverse and varied. As we have previously seen, regular physical activity lowers overall reactivity to stressful situations, keeping down chronic levels of cortisol, which has noxious effects on the brain.[135] Regular physical activity also improves sleep, gets people outside and in social groups, can distract us from obsessive negative thoughts, and involves doing something positive. Altogether, exercise can improve confidence and our belief in the ability to achieve our goals (self-efficacy). All of these effects are therapeutic.

How Much and What Kind of Exercise Are Best?

Whatever the mechanisms by which exercise boosts the brain—and they are many and varied—you'd have to be a flat-earther to discount exercise as a potential means to help prevent and treat mental health. To be sure, exercise is just one of many factors that affect the brain and the mind, and it is not a wonder drug that cures every ill. Exercise is also no substitute for other effective therapies, but most of the world doesn't have access to psychotherapy and medication, and only about 50 percent of patients undergoing antidepressant treatment get better.[136] Given these somber statistics, there needs to be wider recognition that regular physical inactivity is a mismatch that sometimes increases people's vulnerability to many diseases of the mind including dementia, depression, and anxiety. For similar reasons, exercise also benefits other neural and cognitive disorders from attention deficit hyperactivity disorder to Parkinson's. Exercise has additionally been shown to modestly but significantly improve memory, attention span, and various aspects of cognition including math and reading abilities.[137]

How much and what dose of exercise are beneficial for the brain, however, is unclear. If you want to maximize levels of BDNF in your brain as a form of prevention, cardio appears to be more effective than weights, especially if you do high-intensity workouts.[138] For treatment of depression and other mood disorders, varied results across studies hamper efforts to reach firm conclusions about doses and types. Some studies find that self-selected doses of exercise are more effective than specific prescribed doses, others show a benefit from high-intensity exercise over more moderate levels, and others find no difference between the effect of mild-, moderate-, and high-intensity levels of exercise on mood, well-being, and depression.[139] Most studies have focused on aerobic exercise, but the few that compare weights and cardio generally find them to be equally effective.[140] More research will provide better guidelines beyond what is already clear: move for the sake of your mind.

Epilogue

In 2019, I yet again drove with my students and colleagues up the twisty, slippery roads to the Pemja community in western Kenya where this book began—but this time without a treadmill. In most respects, the community has barely changed in the ten years since I first visited. Every day, I see women carrying enormous bundles of firewood and large yellow plastic barrels of water on their heads. As we pass by tiny fields on craggy hillsides, men and women are bent over, hoeing and harvesting corn and millet by hand without the help of any machines. Everywhere children are walking and running, going to and from school, tending cows and goats, and doing other chores. Apart from seeing kids playing soccer during recess in a dusty field using a ball bundled together from plastic bags, I observe no one doing anything remotely resembling exercise.

But things are beginning to change in Pemja if you look and ask. There is now a power line bringing electricity to a few parts of the community. The dirt roads are slightly improved. More students are attending high school. Most of the schoolchildren now wear plastic sandals instead of going barefoot. A handful of people have cell phones. The Industrial Revolution has not yet reached Pemja—not by a long shot—but it is slowly creeping up the roads from the nearby city of Eldoret as Kenya continues to transform. At some point, modernization will bring plumbing, vehicles, tractors, and other laborsaving devices. And when it does, people in Pemja will have to choose to do unnecessary, optional physical activity for the sake of their health—that is, exercise. They might even use treadmills. Chances are, however, that most of them will not exercise very much.

That avoidance makes sense because, as we have seen from the very start of this book, exercise is a fundamentally strange and unusual behavior from an evolutionary perspective. Thirteen chapters later, I hope you still agree. When all is said and done, exercise—despite its manifold benefits—requires overriding deep, natural instincts. So instead of shaming and blaming people who avoid exertion, we should help each other choose to exercise. But as the last few decades have shown, we won't succeed solely by medicalizing and commodifying exercise; instead, we should treat exercise the way we treat education by making it fun, social, emotionally worthwhile, and something that we willingly commit ourselves to do.

Finding new strategies to encourage and facilitate exercise should be a shared priority, not just to help each other but also to help our communities. The widespread, collective benefits of physical activity aren't always obvious until we face the consequences of its absence. Indeed, as I put the finishing touches on this book, the COVID-19 pandemic is sweeping across the world, causing millions of people to fall ill and many to die. At the moment we lack data on if or how much physical activity helps protect anyone from this particular contagion, but because everyone knows that exercise is generally salubrious, the outbreak has inspired me and many of my neighbors to exercise for our health, both mental and physical. All those efforts to stay active have a beneficial collective impact because, as this epidemic has made especially obvious, none of us is an island when it comes to health. Our well-being is interconnected.

Which brings me to a final point. Researching and writing this book has convinced me that a philosophy for how to use one's body is just as useful as a philosophy for how to live one's life. All of us get only one chance to enjoy a good life, and we don't want to die full of regret for having mislived it, and that includes having misused one's body. By following deep and ancient instincts to avoid the discomfort that comes with physical exertion, we increase the chances we will senesce faster and die younger, and we become more vulnerable to many diseases and chronic, disabling illnesses. We also miss out on the vigor, both physical and mental, that comes from being fit. To be sure, exercise is no magic pill that guarantees good health and a long life, and it is possible to live a reasonably long and healthy life without exercising. But thanks to our evolutionary history, lifelong physical activity dramatically increases the chances we will die healthy after seven or more decades.

So on cold, miserable mornings when I am exercised about exercise and struggle to head out the door for a run, I remind my brain, which thinks the rest of my body is a vehicle for moving it from place to place, that it really evolved to advise my body on when and how to move. Fortunately, that advice can be boiled down concisely and simply. *Make exercise necessary and fun. Do mostly cardio, but also some weights. Some is better than none. Keep it up as you age.*

Acknowledgments

I have so many people to thank I am worried I'll forget some of them. My deepest thanks go to my family, friends, colleagues, and students who kindly read all or most of the book, providing comments, critiques, and other forms of feedback. Of these, I am most deeply grateful to my wonderful, saintly wife, Tonia, who read the entire book from cover to cover, editing with a deft hand. She, too, has patiently endured my long obsession with physical activity, not to mention putting up with me as I researched and wrote the book. I also cannot thank enough my colleague and running buddy, Aaron Baggish, who not only read drafts of every chapter but also discussed many of the ideas in this book on countless early morning runs. Despite his efforts, I have retained a few semicolons. Other kind and helpful souls to whom I am indebted because they read all or large portions of the book and gave me invaluable feedback are Manuel Domínguez-Rodrigo, Alan Garber, Henry Gee, Steven Heymsfield, Eleanor Lieberman, Chris MacDonald, Barbara Natterson-Horowitz, and Tom Tippett. For reading and commenting on particular chapters and providing data and other insights, I am very thankful to Steve Austad, Paul Barriera, Bradley Cardinal, Mark Haykowsky, Nicholas Holowka, Carole Hooven, Tim Kistner, Mickey Mahaffey, Samuel Osher, Kieran O'Sullivan, Herman Pontzer, David Raichlen, Craig Rodgers, Ben Sibson, Jerome Siegel, Bo Waggoner, Ian Wallace, Peter Weyand, and Richard Wrangham.

I also thank many others who have helped through scientific collaborations, assistance in the field and lab, discussions, emails, and more. Listed alphabetically they are Brian Addison, Coren Apicella, Meir Barak, Francis

Berenbaum, Claude Bouchard, Dennis Bramble, Henrik Bunge, Ambrose Burfoot, Eamon Callison, Terence Capellini, Rachel Carmody, David Carrasco, David Carrier, Eric Castillo, Silvino Cubesare, Adam Daoud, Irene Davis, Sarah DeLeon, Maureen Devlin, Pierre D'Hemecourt, Peter Ellison, Carolyn Eng, David Felson, Paul Gompers, Michael Gurven, Brian Hare, Kristen Hawkes, Erin Hecht, Joe Henrich, Kim Hill, Dorothy Hintze, Michael Hintze, Jenny Hoffman, Mikko Ijäs, Josphine Jemutai, Joice Jepkirui, Mette Yun Johansen, Yana Kamberov, Erwan Le Corre, Kristi Lewton, Louis Liebenberg, Claire Lo, Zarin Machanda, Huian Mathre, Chris McDougall, Dave McGillivray, Jordan Metzl, Thomas Milani, Randolph Nesse, Lena Nordin, Robert Ojiambo, Paul Okutoyi, Erik Otarola-Castillo, Bente Pedersen, David Pilbeam, Steven Pinker, Yannis Pitsiladis, Mary Prendergast, Arnulfo Quimare, Michael Rainbow, Humberto Ramos Fernandez, Alonso Ramos Vaca, David Reich, Neil Roach, Campbell Rolian, Maryellen Ruvolo, Bob Sallis, Meshack Sang, Lee Saxby, Rob Shave, Freddy Sichting, Timothy Sigei, Martin Surbeck, Cliff Tabin, Adam Tenforde, Victoria Tobolsky, Ben Trumble, Madhu Venkadesan, Anna Warrener, William Werbel, Katherine Whitcome, Brian Wood, Gabriela Yañez, Andrew Yegian, and Katherine Zink.

Thanks also to the students, teachers, and everyone else who assisted, hosted, and helped us in Pemja, Kobujoi, and Eldoret, Kenya; in Lake Eyasi, Tanzania; and in the Sierra Tarahumara, Mexico. Special thanks to Manuel Domínguez-Rodrigo for providing me with a wonderful sanctuary in which to think and write in Madrid when I was writing part of the book.

I am very grateful to my agent, Max Brockman, who has unceasingly provided wise counsel and support from the time this book was just the germ of an idea to its conclusion. I also cannot thank enough my brilliant editor, Erroll McDonald, whose high standards and vision for this book helped me be a better writer. I also thank Laura Stickney and Rowan Cope for their editorial assistance and enthusiasm.

Last but not least, I thank my parents. As a child, I had no idea how inspirational it was to have a heroic mother who liberated a gym by running, and how fortunate I was that both my mother and my father jogged regularly and took me on innumerable hiking trips to the mountains every summer, sometimes for the entire summer. I was never good at sports, but thanks to my parents I grew up thinking it was necessary, normal, and fun to be physically active.

Notes

PROLOGUE

1. *Oxford English Dictionary* (2016).
2. Cregan-Reid, V. (2016), *Footnotes: How Running Makes Us Human* (London: Ebury Press).
3. Marathon des Sables, www.marathondessables.com.
4. No, I didn't make this up. Hats off to John Oliver, who used this example in his brilliant, funny critique of science journalism: www.youtube.com/watch?v=0Rnq1NpHdmw. The paper in question was Dolinsky, V. W., et al. (2012), "Improvements in skeletal muscle strength and cardiac function induced by resveratrol during exercise training contribute to enhanced exercise performance in rats," *Journal of Physiology* 590:2783–99. It's an excellent, careful study, but sadly, few news reports noted that the subjects were rats, not humans, nor bothered to report the study's findings properly.
5. Physical Activity Guidelines Advisory Committee (2018), *2018 Physical Activity Guidelines Advisory Committee Scientific Report* (Washington, D.C.: U.S. Department of Health and Human Services).
6. This often-quoted phrase comes from Theodosius Dobzhansky, who wrote a famous essay with the same title just after he retired: Dobzhansky, T. (1973), Nothing in biology makes sense except in the light of evolution, *American Biology Teacher* 35:125–29.

ONE ARE WE BORN TO REST OR RUN?

1. To the best of my knowledge, the first Westerner to describe the Tarahumara was an American adventurer, Frederick Schwatka, who in 1893 published *In the Land of Cave and Cliff Dwellers* (New York: Cassell). He was followed by the Norwegian Carl Lumholtz, whose 1902 photograph-rich monograph *Unknown Mexico* (New York: Charles Scribner's Sons) offers a stunning

glimpse into how the Tarahumara used to live. In 1935, the anthropologists W. C. Bennett and R. M. Zingg published a comprehensive monograph, *The Tarahumara: An Indian Tribe of Northern Mexico* (Chicago: University of Chicago Press), that remains an important source of information on the Tarahumara. Over the decades, other books and magazines, including *Runner's World,* continued to write about the Tarahumara, but most people today know about them from McDougall, C. (2009), *Born to Run: A Hidden Tribe, Superathletes, and the Greatest Race the World Has Never Seen* (New York: Alfred A. Knopf).

2. No one knows how old the *rarájipari* footrace is, but such traditions are ancient and widespread throughout the Americas, and are even depicted in ancient cave art. See Nabokov, P. (1981), *Indian Running: Native American History and Tradition* (Santa Barbara, Calif.: Capra Press).
3. Letsinger, A. C., et al. (2019), Alleles associated with physical activity levels are estimated to be older than anatomically modern humans, *PLOS ONE* 14:e0216155.
4. Tucker, R., Santos-Concejero, J., and Collins, M. (2013), The genetic basis for elite running performance, *British Journal of Sports Medicine* 47:545–49; Pitsiladis, Y., et al. (2013), Genomics of elite sporting performance: What little we know and necessary advances, *British Journal of Sports Medicine* 47:550–55.
5. For a terrific exploration of these challenges among Western athletes, see Hutchinson, A. (2018), *Endure: Mind, Body, and the Curiously Elastic Limits of Human Performance* (New York: William Morrow).
6. Lieberman, D. E., et al. (2020), Running in Tarahumara (Rarámuri) culture: Persistence hunting, footracing, dancing, work, and the fallacy of the athletic savage, *Current Anthropology* 6.
7. For a review of these and other disturbing stereotypes about athletics, see Coakley, J. (2015), *Sports in Society: Issues and Controversies,* 11th ed. (New York: McGraw-Hill).
8. To make matters worse, more than two-thirds of these Americans and Europeans were university undergraduates. Arnett, J. (2008), The neglected 95%: Why American psychology needs to become less American, *American Psychologist* 63:602–14.
9. Henrich, J., Heine, S. J., and Norenzayan, A. (2010), The weirdest people in the world?, *Behavioral and Brain Sciences* 33:61–83.
10. Schrire, C., ed. (1984), *Past and Present in Hunter Gatherer Studies* (Orlando, Fla.: Academic Press); Wilmsen, E. N. (1989), *Land Filled with Flies* (Chicago: University of Chicago Press).
11. The most comprehensive book is Marlowe, F. W. (2010), *The Hadza: Hunter-Gatherers of Tanzania* (Berkeley: University of California Press). A lovely book filled with beautiful photographs is Peterson, D., Baalow, R., and Cox, J. (2013), *Hadzabe: By the Light of a Million Fires* (Dar es Salaam, Tanzania: Mkuki na Nyota).
12. Schnorr, S. L., et al. (2014), Gut microbiome of the Hadza hunter-gatherers, *Nature Communications* 5:3654; Rampelli, S., et al. (2015), Metagenome sequencing of the Hadza hunter-gatherer gut microbiota, *Current Biology* 25:1682–93; Turroni, S., et al. (2016), Fecal metabolome of the Hadza hunter-gatherers: A host-microbiome integrative view, *Scientific Reports* 6:32826.
13. Lake Eyasi is a seasonal salt lake that dries up during the long, hot dry sea-

son. A few farmers of the Iraqw tribe live near the top of the lake, and the other denizens of the area are the Datoga, who primarily survive by herding goats and cows. The Datoga are encroaching on the Hadza's land, and their goats and cows are destroying the natural habitat, driving away wildlife, and making it increasingly difficult for the Hadza to hunt.

14. Raichlen, D. A., et al. (2017), Physical activity patterns and biomarkers of cardiovascular disease risk in hunter-gatherers, *American Journal of Human Biology* 29:e22919.
15. Marlowe (2010), *Hadza;* Pontzer, H., et al. (2015), Energy expenditure and activity among Hadza hunter-gatherers, *American Journal of Human Biology* 27:628–37.
16. Lee, R. B. (1979), *The !Kung San: Men, Women, and Work in a Foraging Society* (Cambridge, U.K.: Cambridge University Press).
17. Hill, K., et al. (1985), Men's time allocation to subsistence work among the Aché of eastern Paraguay, *Human Ecology* 13:29–47; Hurtado, A. M., and Hill, K. R. (1987), Early dry season subsistence ecology of Cuiva (Hiwi) foragers of Venezuela, *Human Ecology* 15:163–87.
18. Gurven, M., et al. (2013), Physical activity and modernization among Bolivian Amerindians, *PLOS ONE* 8:e55679.
19. Kelly, R. L. (2013), *The Lifeways of Hunter-Gatherers: The Foraging Spectrum,* 2nd ed. (Cambridge, U.K.: Cambridge University Press).
20. James, W. P. T., and Schofield, E. C. (1990), *Human Energy Requirements: A Manual for Planners and Nutritionists* (Oxford: Oxford University Press).
21. Leonard, W. R. (2008), Lifestyle, diet, and disease: Comparative perspectives on the determinants of chronic health risks, in *Evolution, Health, and Disease,* ed. S. C. Stearns and J. C. Koella (New York: Oxford University Press), 265–76.
22. Speakman, J. (1997), Factors influencing the daily energy expenditure of small mammals, *Proceedings of the Nutrition Society* 56:1119–36.
23. Hays, M., et al. (2005), Low physical activity levels of Homo sapiens among free-ranging mammals, *International Journal of Obesity* 29:151–56.
24. Church, T. S., et al. (2011), Trends over 5 decades in U.S. occupation-related physical activity and their associations with obesity, *PLOS ONE* 6:e19657.
25. Meijer, J. H., and Robbers, Y. (2014), Wheel running in the wild, *Proceedings of the Royal Society B* 281:20140210.
26. Mechikoff, R. A. (2014), *A History and Philosophy of Sport and Physical Education: From Ancient Civilization to the Modern World* (New York: McGraw-Hill).
27. Rice, E. A., Hutchinson, J. L., and Lee, M. (1958), *A Brief History of Physical Education* (New York: Ronald Press); Nieman, D. C. (1990), *Fitness and Sports Medicine: An Introduction* (Palo Alto, Calif.: Bull).
28. For a fabulous history of this topic, see McKenzie, S. (2013), *Getting Physical: The Rise of Fitness Culture in America* (Lawrence: University Press of Kansas).
29. Sargent, D. A. (1900), The place for physical training in the school and college curriculum, *American Physical Education Review* 5:1–7; Sargent, D. A. (1902), *Universal Test for Strength, Speed, and Endurance of the Human Body* (Cambridge, Mass.: Powell Press).
30. Vankim, N. A., and Nelson, T. F. (2013), Vigorous physical activity, mental health, perceived stress, and socializing among college students, *American Journal of Health Promotion* 28:7–15.

31. Physical Activity Guidelines Advisory Committee (2018), *2018 Physical Activity Guidelines Advisory Committee Scientific Report* (Washington, D.C.: U.S. Department of Health and Human Services).
32. For references and more detailed information on these and other estimates, see chapter 13.

TWO INACTIVITY: THE IMPORTANCE OF BEING LAZY

1. Harcourt, A. H., and Stewart, K. J. (2007), *Gorilla Society: Conflict, Compromise, and Cooperation Between the Sexes* (Hawthorne, N.Y.: Aldine de Gruyter).
2. Organ, C., et al. (2011), Phylogenetic rate shifts in feeding time during the evolution of *Homo, Proceedings of the National Academy of Sciences USA* 108:14555–59.
3. Goodall, J. (1986), *The Chimpanzees of Gombe: Patterns of Behavior* (Cambridge, Mass.: Harvard University Press); Pontzer, H., and Wrangham, R. W. (2004), Climbing and the daily energy cost of locomotion in wild chimpanzees: Implications for hominoid locomotor evolution, *Journal of Human Evolution* 46:317–35.
4. Pilbeam, D. R., and Lieberman, D. E. (2017), Reconstructing the last common ancestor of chimpanzees and humans, in *Chimpanzees and Human Evolution,* ed. M. N. Muller, R. W. Wrangham, and D. R. Pilbeam (Cambridge, Mass.: Harvard University Press), 22–141.
5. For every liter of O_2 you use, your body gets 5.1 kilocalories (kcals) from burning pure carbohydrates, and 4.7 kcal from burning pure fat. The ratio of O_2/CO_2 you expire signals how much fat versus carbohydrate you are using. If you are burning only carbohydrate, you produce exactly as much CO_2 as O_2, and if you are burning only fat, you produce 70 percent as much CO_2 as O_2. Most of the time, you burn a mixture of fat and carbohydrate, yielding an average of 4.8 kcal for every liter of oxygen.
6. Kilocalories are formally written Calories (with a capital C) to distinguish them from a calorie, the energy to raise one gram of water 1°C. Like the labels on our food, I will hereafter follow the convention of using the term "calories" for Calories.
7. Jones, W. P. T., and Schofield, E. C. (1990), *Human Energy Requirements: A Manual for Planners and Nutritionists* (Oxford: Oxford University Press).
8. I am sorry to report that an entire day of intense thinking amounts to only twenty to fifty extra calories, about what you get from eating half a dozen peanuts. See Messier, C. (2004), Glucose improvement of memory: A review, *European Journal of Pharmacology* 490:33–57.
9. Here is a more detailed explanation of the "doubly labeled water" method. Most water, H_2O, is made from hydrogen with a molecular weight of 1 (^1H, which has 1 proton) and oxygen with a molecular weight of 16 (^{16}O, which has 8 protons and 8 neutrons). It is also possible to make a harmless form of water using "heavy" hydrogen (^2H, which has an extra neutron) and oxygen (^{18}O, which has two extra neutrons). Regardless of what kind of hydrogen and oxygen is in the water, the H and O leave the body in different ways: the H and O leave as water when we urinate, breathe, and sweat, whereas the O also leaves the body when we breathe out carbon dioxide (CO_2). Because we can accurately and precisely measure ^2H and ^{18}O in urine, the difference in the rate at which

^2H versus ^{18}O remains in your urine over several days allows us to calculate how much CO_2 you breathed out, thus allowing a reasonably precise calculation of how much energy you spent. The method also allows us to calculate your fat-free body mass and how much water goes in and out of your body.

10. Pontzer, H., et al. (2012), Hunter-gatherer energetics and human obesity, *PLOS ONE* 7:e40503. This paper reports estimated BMRs; their estimated RMRs are probably about 10 percent higher: 1,169 calories for women and 1,430 calories for men.
11. Hadza men and women average 13 percent and 21 percent body fat, respectively; industrialized men and women average 23 percent and 38 percent body fat, respectively. Ibid.
12. White, M. (2012), *Atrocities: The 100 Deadliest Episodes in Human History* (New York: W. W. Norton).
13. For a very readable account, see Tucker, T. (2006), *The Great Starvation Experiment: The Heroic Men Who Starved So That Millions Could Live* (New York: Free Press). The two-volume monograph by Keys and colleagues is also fascinating: Keys, A., et al. (1950), *The Biology of Human Starvation* (Minneapolis: University of Minnesota Press).
14. At the start, the volunteers averaged 21.6 pounds of fat (14 percent of their body weight); after the starvation period, they averaged 6.5 pounds of fat (5.5 percent of their body weight).
15. Elia, M. (1992), Organ and tissue contribution to metabolic rate, in *Energy Metabolism: Tissue Determinants and Cellular Corollaries,* ed. J. M. Kinney and H. N. Ticker (New York: Raven Press), 61–77.
16. A 2015 experiment that put volunteers on the same starvation diet for just two weeks and measured organ size reductions using modern technology yielded comparable results. See Müller, M. J., et al. (2015), Metabolic adaptation to caloric restriction and subsequent refeeding: The Minnesota Starvation Experiment revisited, *American Journal of Clinical Nutrition* 102:807–19.
17. In science, a "theory" is not an untested idea (a hypothesis); a theory is a well-established, validated understanding of how the world works. Natural selection is as well validated as the theories of gravity and plate tectonics.
18. Marlowe, F. C., and Berbesque, J. C. (2009), Tubers as fallback foods and their impact on Hadza hunter-gatherers, *American Journal of Physical Anthropology* 140:751–58.
19. The data here come from Pontzer et al. (2012), Hunter-gatherer energetics and human obesity; Pontzer, H., et al. (2016), Metabolic acceleration and the evolution of human brain size and life history, *Nature* 533:390–92.
20. These data are calculated from Pontzer's measurements of their daily energy expenditure and estimates of their resting and basal metabolic rates. According to one study, basal metabolic rates are about 10 percent higher in humans than in chimpanzees after correcting for differences in body size. See Pontzer et al. (2016), Metabolic acceleration and the evolution of human brain size and life history.
21. Westerterp, K. R., and Speakman, J. R. (2008), Physical activity energy expenditure has not declined since the 1980s and matches energy expenditures of wild mammals, *International Journal of Obesity* 32:1256–63; Hayes, M., et al. (2005), Low physical activity levels of modern *Homo sapiens* among free-ranging mammals, *International Journal of Obesity* 29:151–56.

22. Pontzer, H., et al. (2010), Metabolic adaptation for low energy throughput in orangutans, *Proceedings of the National Academy of Sciences USA* 107:14048–52.
23. Taylor, C. R., and Rowntree, V. J. (1973), Running on two or on four legs: Which consumes more energy?, *Science* 179:186–87; Pontzer, H., Raichlen, D. A., and Sockol, M. D. (2009), The metabolic cost of walking in humans, chimpanzees, and early hominins, *Journal of Human Evolution* 56:43–54.
24. Aiello, L. C., and Key, C. (2002), Energetic consequences of being a *Homo erectus* female, *American Journal of Human Biology* 14:551–65.
25. Wrangham, R. W. (2009), *Catching Fire: How Cooking Made Us Human* (New York: Basic Books).
26. Webb, O. J., et al. (2011), A statistical summary of mall-based stair-climbing intervention, *Journal of Physical Activity and Health* 8:558–65.
27. Rosenthal, R. J., et al. (2017), Obesity in America, *Surgery for Obesity and Related Disorders* 13:1643–50.

THREE SITTING: IS IT THE NEW SMOKING?

1. Nash, O. (1940), *The Face Is Familiar* (Garden City, N.Y.: Garden City Publishing).
2. Levine, J. A. (2004), *Get Up! Why Your Chair Is Killing You and What You Can Do About It* (New York: St. Martin's Griffin). To test the claim that sitting is as bad as smoking, two Canadian doctors did the following back-of-the-envelope calculation. According to their analysis, patients who were physically inactive incurred on average three hundred dollars more in health-care costs than active individuals, whereas smokers ended up costing about five times more, between sixteen hundred and eighteen hundred dollars per year, than nonsmokers. Because the average smoker consumes sixteen cigarettes a day, these numbers suggest that physical inactivity is approximately as costly as three cigarettes a day, which equals a pack a week. See Khan, K., and Davis, J. (2010), A week of physical inactivity has similar health costs to smoking a packet of cigarettes, *British Journal of Sports Medicine* 44:345.
3. Rezende, L. F., et al. (2016), All-cause mortality attributable to sitting time: Analysis of 54 countries worldwide, *American Journal of Preventive Medicine* 51:253–63; Matthews, C. E. (2015), Mortality benefits for replacing sitting time with different physical activities, *Medicine and Science in Sports and Exercise* 47:1833–40. For counterevidence, see Pulsford, R. M., et al. (2015), Associations of sitting behaviours with all-cause mortality over a 16-year follow-up: The Whitehall II study, *International Journal of Epidemiology* 44:1909–16.
4. Kulinski, J. P. (2014), Association between cardiorespiratory fitness and accelerometer-derived physical activity and sedentary time in the general population, *Mayo Clinic Proceedings* 89:1063–71; Matthews (2015), Mortality benefits for replacing sitting time with different physical activities.
5. Although the internet is full of calculators that provide questionable values, many careful studies have measured the cost of standing versus sitting. A few recent studies include Júdice, P. B., et al. (2016), What is the metabolic and energy cost of sitting, standing, and sit/stand transitions?, *European Journal of Applied Physiology* 116:263–73; Fountain, C. J., et al. (2016), Metabolic and energy cost of sitting, standing, and a novel sitting/stepping protocol in recreationally active college students, *International Journal of Exercise Science*

9:223–29; Mansoubi, M., et al. (2015), Energy expenditure during common sitting and standing tasks: Examining the 1.5 MET definition of sedentary behavior, *BMC Public Health* 15:516–23; Miles-Chan, J., et al. (2013), Heterogeneity in the energy cost of posture maintenance during standing relative to sitting, *PLOS ONE* 8:e65827.

6. This kind of math, however, doesn't account for whether people who stand more compensate by eating more (just an apple a day) or fidgeting less.
7. Birds spend 16 to 25 percent more energy standing than sitting on the ground. See van Kampen, M. (1976), Activity and energy expenditure in laying hens: 3. The energy cost of eating and posture, *Journal of Agricultural Science* 87:85–88; Tickle, P. G., Nudds, R. L., and Codd, J. R. (2012), Barnacle geese achieve significant energy savings by changing posture, *PLOS ONE* 7:e46950. For cows and moose, see Vercoe, J. E. (1973), The energy cost of standing and lying in adult cattle, *British Journal of Nutrition* 30:207–10; Renecker, L. A., and Hudson, R. J. (1985), The seasonal energy expenditures and thermoregulatory responses of moose, *Canadian Journal of Zoology* 64:322–27.
8. A bent-hip and bent-knee posture requires either apes or crouching humans to constantly contract the hamstrings and quadriceps muscles to avoid collapsing onto the ground. See Sockol, M. D., Raichlen, D. A., and Pontzer, H. (2007), Chimpanzee locomotor energetics and the origin of human bipedalism, *Proceedings of the National Academy of Sciences USA* 104:12265–69.
9. Winter, D. A. (1995), Human balance and posture control during standing and walking, *Gait and Posture* 3:193–214.
10. Hunter-gatherers lack furniture not because they lack carpentry skills; instead, they don't make furniture because they move frequently, usually seven times a year, carrying everything they own. The benefits of furniture are outweighed by the costs of moving. Furniture started to become more common once farmers settled down into permanent dwellings.
11. Hewes, G. (1953), Worldwide distribution of certain postural habits, *American Anthropologist* 57:231–44.
12. For recent populations, see Nag, P. K., et al. (1986), EMG analysis of sitting work postures in women, *Applied Ergonomics* 17:195–97; Gurr, K., Straker, L., and Moore, P. (1998), Cultural hazards in the transfer of ergonomics technology, *International Journal of Industrial Ergonomics* 22:397–404. For *Homo erectus* and Neanderthals, see Trinkaus, E. (1975), Squatting among the Neandertals: A problem in the behavioral interpretation of skeletal morphology, *Journal of Archaeological Science* 2:327–51; Pontzer, H., et al. (2010), Locomotor anatomy and biomechanics of the Dmanisi hominins, *Journal of Human Evolution* 58:492–504. For early modern humans, see Pearson, O. M., et al. (2008), A description of the Omo I postcranial skeleton, including newly discovered fossils, *Journal of Human Evolution* 55:421–37; Rightmire, G. P., et al. (2006), Human foot bones from Klasies River main site, South Africa, *Journal of Human Evolution* 50:96–103.
13. Mays, S. (1998), *The Archaeology of Human Bones* (London: Routledge); Boulle, E. (1998), Evolution of two human skeletal markers of the squatting position: A diachronic study from antiquity to the modern age, *American Journal of Physical Anthropology* 115:50–56.
14. Ekholm, J., et al. (1985), Load on knee joint and knee muscular activity dur-

ing machine milking, *Ergonomics* 28:665–82; Eguchi, A. (2003), Influence of the difference in working postures during weeding on muscle activities of the lower back and the lower extremities, *Journal of Science Labour* 79:219–23; Nag et al. (1986), EMG analysis of sitting work postures in women; Miles-Chan, J. L., et al. (2014), Sitting comfortably versus lying down: Is there really a difference in energy expenditure?, *Clinical Nutrition* 33:175–78.

15. Castillo, E. R., et al. (2016), Physical fitness differences between rural and urban children from western Kenya, *American Journal of Human Biology* 28:514–23.
16. Mörl, F., and Bradl, I. (2013), Lumbar posture and muscular activity while sitting during office work, *Journal of Electromyography and Kinesiology* 23:362–68.
17. Rybcynski, W. (2016), *Now I Sit Me Down* (New York: Farrar, Straus and Giroux).
18. Aveling, J. H. (1879), *Posture in Gynecic and Obstetric Practice* (Philadelphia: Lindsay & Blakiston).
19. Prince, S. A., et al. (2008), A comparison of direct versus self-report measures for assessing physical activity in adults: A systematic review, *International Journal of Behavioral Nutrition and Physical Activity* 5:56–80.
20. To be precise, the device measures accelerations, the rate of change in velocity in the vertical direction, and usually also side-to-side and front-to-back accelerations. Because force equals mass times acceleration, such measurements of acceleration provide a good estimate of how much force is being generated to move a body. Of course, there can be problems. For example, if the accelerometer is worn on a hip, the device will fail to measure the activity of riding a bicycle.
21. Matthews, C. E., et al. (2008), Amount of time spent in sedentary behaviors in the United States, 2003–2004, *American Journal of Epidemiology* 167:875–81; Tudor-Locke, C., et al. (2011), Time spent in physical activity and sedentary behaviors on the working day: The American time use survey, *Journal of Occupational and Environmental Medicine* 53:1382–87; Evenson, K. R., Buchner, D. M., and Morland, K. B. (2012), Objective measurement of physical activity and sedentary behavior among US adults aged 60 years or older, *Preventing Chronic Disease* 9:E26; Martin, K. R. (2014), Changes in daily activity patterns with age in U.S. men and women: National Health and Nutrition Examination Survey 2003–04 and 2005–06, *Journal of the American Geriatric Society* 62:1263–71; Diaz, K. M. (2017), Patterns of sedentary behavior and mortality in U.S. middle-aged and older adults: A national cohort study, *Annals of Internal Medicine* 167:465–75.
22. Ng, S. W., and Popkin, B. (2012), Time use and physical activity: A shift away from movement across the globe, *Obesity Review* 13:659–80.
23. Raichlen, D. A., et al. (2017), Physical activity patterns and biomarkers of cardiovascular disease risk in hunter-gatherers, *American Journal of Human Biology* 29:e22919.
24. Gurven, M., et al. (2013), Physical activity and modernization among Bolivian Amerindians, *PLOS ONE* 8:e55679.
25. Katzmarzyk, P. T., Leonard, W. R., and Crawford, M. H. (1994), Resting metabolic rate and daily energy expenditure among two indigenous Siberian populations, *American Journal of Human Biology* 6:719–30; Leonard, W. R., Galloway, V. A., and Ivakine, E. (1997), Underestimation of daily energy

expenditure with the factorial method: Implications for anthropological research, *American Journal of Physical Anthropology* 103:443–54; Kashiwazaki, H., et al. (2009), Year-round high physical activity levels in agropastoralists of Bolivian Andes: Results from repeated measurements of DLW method in peak and slack seasons of agricultural activities, *American Journal of Human Biology* 21:337–45; Madimenos, F. C. (2011), Physical activity in an indigenous Ecuadorian forager-horticulturalist population as measured using accelerometry, *American Journal of Human Biology* 23:488–97; Christensen, D. L., et al. (2012), Cardiorespiratory fitness and physical activity in Luo, Kamba, and Maasai of rural Kenya, *American Journal of Human Biology* 24:723–29.

26. Raichlen, D. A., et al. (2020), Sitting, squatting, and the evolutionary biology of human inactivity, *Proceedings of the National Academy of Sciences USA* 117:7115–7121.
27. Classifying activity levels depends on the method being used. One standard convention is based on percentage of maximum heart rate: less than 40 percent is sedentary; 40 to 54 percent is light activity; 55 to 69 percent is moderate activity; 70 to 89 percent is vigorous; and more than 90 percent is high. Maximum heart rate is sometimes measured but usually estimated from age. A common formula for maximum heart rate is to subtract your age from the number 220, but for healthy adults a better equation is 208 – 0.7 • age. Another classification is based on measuring energy use from oxygen using METs (metabolic equivalents) in which 1 MET is the rate of energy used during quiet sitting, usually 3.5 milliliters of oxygen per kilogram of body weight per minute. By this convention, sedentary activity is 1 to 1.5 METs, light is 1.5 to 2.9 METs, moderate is 3 to 6 METs, and vigorous is greater than 6 METs. Activity levels can also be estimated from accelerometers using other equations. For maximum heart rate measurements, see Tanaka, H., Monahan, K. D., and Seals, D. R. (2001), Age-predicted maximal heart rate revisited, *Journal of the American College of Cardiology* 37:153–56. For estimating activity levels from accelerometers, see Freedson, P. S., Melanson, E., and Sirard, J. (1998), Calibration of the Computer Science and Applications Inc. accelerometer, *Medicine and Science in Sports and Exercise* 30:777–81; Matthews et al. (2008), Amount of time spent in sedentary behaviors in the United States, 2003–2004.
28. Evenson, K. R., Wen, F., and Herring, A. H. (2016), Associations of accelerometry-assessed and self-reported physical activity and sedentary behavior with all-cause and cardiovascular mortality among US adults, *American Journal of Epidemiology* 184:621–32.
29. Raichlen et al. (2017), Physical activity patterns and biomarkers of cardiovascular disease risk in hunter-gatherers.
30. I am grateful to Dr. Zarin Machanda for compiling these numbers.
31. Aggarwal, B. B., Krishnan, S., and Guha, S. (2011), *Inflammation, Lifestyle, and Chronic Diseases: The Silent Link* (Boca Raton, Fla.: CRC Press).
32. As an example, one worldwide best-selling book claimed that wheat and other foods with gluten cause inflammation of the brain. The data, however, indicate that unless you have celiac disease, eating wheat (especially whole wheat) or other grains will not cause your body, including your brain, to become inflamed unless you eat too much and become obese. For credible, peer-reviewed, evidence-based studies, see Lutsey, P. L., et al. (2007),

Whole grain intake and its cross-sectional association with obesity, insulin resistance, inflammation, diabetes, and subclinical CVD: The MESA Study, *British Journal of Nutrition* 98:397–405; Lefevre, M., and Jonnalagadda, S. (2012), Effect of whole grains on markers of subclinical inflammation, *Nutrition Review* 70:387–96; Vitaglione, P., et al. (2015), Whole-grain wheat consumption reduces inflammation in a randomized controlled trial on overweight and obese subjects with unhealthy dietary and lifestyle behaviors: Role of polyphenols bound to cereal dietary fiber, *American Journal of Clinical Nutrition* 101:251–61; Ampatzoglou, A., et al. (2015), Increased whole grain consumption does not affect blood biochemistry, body composition, or gut microbiology in healthy, low-habitual whole grain consumers, *Journal of Nutrition* 145:215–21.

33. For an excellent, readable review of how fat cells function and cause inflammation, see Tara, S. (2016), *The Secret Life of Fat: The Science Behind the Body's Least Understood Organ and What It Means for You* (New York: W. W. Norton).
34. Shen, W. (2009), Sexual dimorphism of adipose tissue distribution across the lifespan: A cross-sectional whole-body magnetic resonance imaging study, *Nutrition and Metabolism* 16:6–17; Hallgreen, C. E., and Hall, K. D. (2008), Allometric relationship between changes of visceral fat and total fat mass, *International Journal of Obesity* 32:845–52.
35. Weisberg, S. P., et al. (2003), Obesity is associated with macrophage accumulation in adipose tissue, *Journal of Clinical Investigation* 112:1796–808.
36. Levine, J. A., Schleusner, S. J., and Jensen, M. D. (2000), Energy expenditure of nonexercise activity, *American Journal of Clinical Nutrition* 72:1451–54.
37. Olsen, R. H., et al. (2000), Metabolic responses to reduced daily steps in healthy nonexercising men, *Journal of the American Medical Association* 299:1261–63.
38. Homer, A. R., et al. (2017), Regular activity breaks combined with physical activity improve postprandial plasma triglyceride, nonesterified fatty acid, and insulin responses in healthy, normal weight adults: A randomized crossover trial, *Journal of Clinical Lipidology* 11:1268–79; Peddie, M. C., et al. (2013), Breaking prolonged sitting reduces postprandial glycemia in healthy, normal-weight adults: A randomized crossover trial, *American Journal of Clinical Nutrition* 98:358–66.
39. Boden, G. (2008), Obesity and free fatty acids (FFA), *Endocrinology and Metabolism Clinics of North America* 37:635–46; de Vries, M. A., et al. (2014), Postprandial inflammation: Targeting glucose and lipids, *Advances in Experimental Medical Biology* 824:161–70.
40. Bruunsgaard, H. (2005), Physical activity and modulation of systemic low-level inflammation, *Journal of Leukocyte Biology* 78:819–35; Pedersen, B. K., and Febbraio, M. A. (2008), Muscle as an endocrine organ: Focus on muscle-derived interleukin-6, *Physiology Reviews* 88:1379–406; Pedersen, B. K., and Febbraio, M. A. (2012), Muscles, exercise, and obesity: Skeletal muscle as a secretory organ, *Nature Reviews Endocrinology* 8:457–65.
41. Petersen, A. M., and Pedersen, B. K. (2005), The anti-inflammatory effect of exercise, *Journal of Applied Physiology* 98:1154–62.
42. Fedewa, M. V., Hathaway, E. D., and Ward-Ritacco, C. L. (2017), Effect of exercise training on C reactive protein: A systematic review and meta-analysis of

randomised and non-randomised controlled trials, *British Journal of Sports Medicine* 51:670–76; Petersen and Pedersen (2005), Anti-inflammatory effect of exercise.

43. Another fascinating source of low-grade inflammation could be unusually sterile environments. Until recently, all humans grew up surrounded by dirt, germs, worms, and other pathogens that regularly challenged our immune systems. The anthropologist Thom McDade has shown that people who grow up in these sorts of evolutionarily "normal," unsterile environments have different inflammatory immune responses from those of us who grew up with high levels of sanitation. When humans who grew up in pathogen-rich conditions get an infection, their inflammatory response is sudden, strong, but short-lived. In contrast, the immune systems of people who grew up in highly hygienic environments with dishwashers, indoor plumbing, bleach, and lots of soap are different. When they get an infection, their inflammatory response tends to be much slower and less intense and lasts much longer. In other words, too much hygiene, especially when we are young, may predispose us to more chronic inflammation as we age. Then when we sit too much, we become more susceptible to persistent, low-grade inflammation. For more information, see McDade, T. W., et al. (2013), Do environments in infancy moderate the association between stress and inflammation in adulthood? Initial evidence from a birth cohort in the Philippines, *Brain, Behavior, and Immunity* 31:23–30.
44. Whitfield, G., Kelly, G. P., and Kohl, H. W. (2014), Sedentary and active: Self-reported sitting time among marathon and half-marathon participants, *Journal of Physical Activity and Health* 11:165–72.
45. Greer, A. E., et al. (2015), The effects of sedentary behavior on metabolic syndrome independent of physical activity and cardiorespiratory fitness, *Journal of Physical Activity and Health* 12:68–73.
46. Matthews, C. E., et al. (2012), Amount of time spent in sedentary behaviors and cause-specific mortality in US adults, *American Journal of Clinical Nutrition* 95:437–45.
47. One study using data from more than four million people estimated that for every two hours per day spent sitting, the risk of colon cancer increased 8 percent, as did some other cancers. Schmid, D., and Leitzmann, M. D. (2014), Television viewing and time spent sedentary in relation to cancer risk: A meta-analysis, *Journal of the National Cancer Institute* 106:dju098.
48. Healy, G. N., et al. (2011), Sedentary time and cardio-metabolic biomarkers in US adults: NHANES 2003–06, *European Heart Journal* 32:590–97.
49. Diaz, K. M., et al. (2017), Patterns of sedentary behavior and mortality in U.S. middle-aged and older adults: A national cohort study, *Annals of Internal Medicine* 167:465–75.
50. van Uffelen, J. G., et al. (2010), Occupational sitting and health risks: A systematic review, *American Journal of Preventive Medicine* 39:379–88.
51. Latouche, C., et al. (2013), Effects of breaking up prolonged sitting on skeletal muscle gene expression, *Journal of Applied Physiology* 14:453–56; Hamilton, M. T., Hamilton, D. G., and Zderic, T. W. (2014), Sedentary behavior as a mediator of type 2 diabetes, *Medicine and Sports Science* 60:11–26; Grøntved, A., and Hu, F. B. (2011), Television viewing and risk of type 2 dia-

betes, cardiovascular disease, and all-cause mortality: A meta-analysis, *Journal of the American Medical Association* 305:2448–55.

52. Healy et al. (2011), Sedentary time and cardio-metabolic biomarkers in US adults; Dunstan, D. W., et al. (2012), Breaking up prolonged sitting reduces postprandial glucose and insulin responses, *Diabetes Care* 35:976–83; Peddie, M. C. (2013), Breaking prolonged sitting reduces postprandial glycemia in healthy, normal-weight adults: A randomized crossover trial, *American Journal of Clinical Nutrition* 98:358–66; Duvivier, B. M. F. M., et al. (2013), Minimal intensity physical activity (standing and walking) of longer duration improves insulin action and plasma lipids more than shorter periods of moderate to vigorous exercise (cycling) in sedentary subjects when energy expenditure is comparable, *PLOS ONE* 8:e55542.
53. Takahashi, M. (2015), Effects of breaking sitting by standing and acute exercise on postprandial oxidative stress, *Asian Journal of Sports Medicine* 6:e24902.
54. Ansari, M. T. (2005), Traveler's thrombosis: A systematic review, *Journal of Travel Medicine* 12:142–54.
55. Ravussin, E., et al. (1986), Determinants of 24-hour energy expenditure in man: Methods and results using a respiratory chamber, *Journal of Clinical Investigation* 78:1568–78.
56. Koepp, G. A., Moore, G. K., and Levine, J. A. (2016), Chair-based fidgeting and energy expenditure, *BMJ Open Sport and Exercise Medicine* 2:e000152; Morishima, T. (2016), Prolonged sitting-induced leg endothelial dysfunction is prevented by fidgeting, *American Journal of Physiology: Heart and Circulatory Physiology* 311:H177–82.
57. Hagger-Johnson, G., et al. (2016), Sitting time, fidgeting, and all-cause mortality in the UK Women's Cohort Study, *American Journal of Preventive Medicine* 50:154–60.
58. Here are some studies of foraging societies to back this number up: the Aché of Paraguay sit 3.4 to 7.5 hours a day; the Hiwi of Venezuela sit 4 to 6 hours a day; the San sit for about 6.6 hours a day; and the Hadza sit at least 6.6 hours a day. Hill, K. R., et al. (1985), Men's time allocation to subsistence work among the Aché of eastern Paraguay, *Human Ecology* 13:29–47; Hurtado, A. M., and Hill, K. R. (1987), Early dry season subsistence ecology of Cuiva (Hiwi) foragers of Venezuela, *Human Ecology* 15:163–87; Leonard, W. R., and Robertson, M. L. (1997), Comparative primate energetics and hominid evolution, *American Journal of Physical Anthropology* 102:265–81; Raichlen et al. (2017), Physical activity patterns and biomarkers of cardiovascular disease risk in hunter-gatherers.
59. Raichlen, D. A., et al. (2020), Sitting, squatting, and the evolutionary biology of human inactivity, *Proceedings of the National Academy of Sciences USA* 117:7115–7121.
60. Møller, S. V., et al. (2016), Multi-wave cohort study of sedentary work and risk of ischemic heart disease, *Scandinavian Journal of Work and Environmental Health* 42:43–51.
61. Hayashi, R., et al. (2016), Occupational physical activity in relation to risk of cardiovascular mortality: The Japan Collaborative Cohort Study for Evaluation for Cancer Risk (JACC Study), *Preventive Medicine* 89:286–91.
62. van Uffelen et al. (2010), Occupational sitting and health risks.

63. Pynt, J., and Higgs, J. (2010), *A History of Seating, 3000 BC to 2000 AD: Function Versus Aesthetics* (Amherst, N.Y.: Cambria Press).
64. Åkerblom argued that the best way to maintain a lumbar curve when sitting is with a backrest that is nearly vertical at the base to support the lower back and that then angles backward to prevent the upper body from being curved forward into an ungainly slump. Additional refinements to the so-called Åkerblom curve include a slightly tilted seat that supports the underside of the thighs, armrests to make it easier to get in and out of the chair, an average seat height of eighteen inches (forty-six centimeters) off the ground, and moderate cushioning. See Åkerblom, B. (1948), *Standing and Sitting Posture: With Special Reference to the Construction of Chairs* (Stockholm: A.-B. Nordiska Bokhandeln). See also Andersson, B. J., et al. (1975), The sitting posture: An electromyographic and discometric study, *Orthopedic Clinics of North America* 6:105–20; Andersson, G. B. J., Jonsson, B., and Ortengren, R. (1974), Myoelectric activity in individual lumbar erector spinae muscles in sitting, *Scandinavian Journal of Rehabilitation Medicine,* Supplement, 3:91–108.
65. O'Sullivan, K., et al. (2012), What do physiotherapists consider to be the best sitting spinal posture?, *Manual Therapy* 17:432–37; O'Sullivan, K., et al. (2013), Perceptions of sitting posture among members of the community, both with and without non-specific chronic low back pain, *Manual Therapy* 18:551–56.
66. Hewes, G. (1953), Worldwide distribution of certain postural habits, *American Anthropologist* 57:231–44.
67. Back in the 1970s, researchers stuck enormous needles into people's backs to measure the pressure different postures generate in the disks between the bones of the spine. These painful experiments found that sitting almost tripled pressure within the disks of the lower back and that slouching postures that flatten the lower back generated even more pressure and stressed the tissues of the lower back, supposedly leading to degeneration and pain. Newer technologies that enable researchers to position tiny, more accurate sensors in the spine without causing injury show that previous measurements were inflated. Instead, sitting and standing generate similarly low pressures unlikely to cause injury. Further, dozens of experiments show that the supposedly ideal sitting postures many experts recommend—upright with a moderately curved lower back—actually *increase* back muscle activity and augment loading on the spine. For old studies, see Andersson, B. J., et al. (1975), The sitting posture: An electromyographic and discometric study, *Orthopedic Clinics of North America* 6:105–20; Nachemson, A., and Morris, J. (1964), *In vivo* measurements of intradiscal pressure. Discometry, a method for the determination of pressure in the lower lumbar discs. Lumbar discometry. Lumbar intradiscal pressure measurements *in vivo, Journal of Bone and Joint Surgery of America* 46:1077–92; Andersson, B. J., and Ortengren, R. (1974), Lumbar disc pressure and myoelectric back muscle activity during sitting. II. Studies on an office chair, *Scandinavian Journal of Rehabilitative Medicine* 6:115–21. For newer studies, see Claus, A., et al. (2008), Sitting versus standing: Does the intradiscal pressure cause disc degeneration or low back pain?, *Journal of Electromyography and Kinesiology* 18:550–58; Carcone, S. M., and Keir, P. J. (2007), Effects of backrest design on biomechanics and comfort during seated work, *Applied Ergonomics* 38:755–64; Lander, C., et al. (1987), The Balans chair and

its semi-kneeling position: An ergonomic comparison with the conventional sitting position, *Spine* 12:269–72; Curran, M., et al. (2015), Does using a chair backrest or reducing seated hip flexion influence trunk muscle activity and discomfort? A systematic review, *Human Factors* 57:1115–48.

68. Christensen, S. T., and Hartvigsen, J. (2008), Spinal curves and health: A systematic critical review of the epidemiological literature dealing with associations between sagittal spinal curves and health, *Journal of Manipulative and Physiological Therapeutics* 31:690–714; Roffey, D. M., et al. (2010), Causal assessment of awkward occupational postures and low back pain: Results of a systematic review, *Spine Journal* 10:89–99.
69. Roffey et al. (2010), Causal assessment of awkward occupational postures and low back pain; Kwon, B. K., et al. (2011), Systematic review: Occupational physical activity and low back pain, *Occupational Medicine* 61:541–48.
70. Driessen, M. T., et al. (2010), The effectiveness of physical and organisational ergonomic interventions on low back pain and neck pain: A systematic review, *Occupational and Environmental Medicine* 67:277–85; O'Sullivan, K., et al. (2012), The effect of dynamic sitting on the prevention and management of low back pain and low back discomfort: A systematic review, *Ergonomics* 55:898–908; O'Keeffe, M., et al. (2013), Specific flexion-related low back pain and sitting: Comparison of seated discomfort on two different chairs, *Ergonomics* 56:650–58; O'Keeffe, M., et al. (2016), Comparative effectiveness of conservative interventions for nonspecific chronic spinal pain: Physical, behavioral/psychologically informed, or combined? A systematic review and meta-analysis, *Journal of Pain* 17:755–74.
71. Lahad, A. (1994), The effectiveness of four interventions for the prevention of low back pain, *Journal of the American Medical Association* 272:1286–91; Tveito, T. H., Hysing, M., and Eriksen, H. R. (2004), Low back pain interventions at the workplace: A systematic literature review, *Occupational Medicine* 54:3–13; van Poppel, M. N., Hooftman, W. E., and Koes, B. W. (2004), An update of a systematic review of controlled clinical trials on the primary prevention of back pain at the workplace, *Occupational Medicine* 54:345–52; Bigos, S. J., et al. (2009), High-quality controlled trials on preventing episodes of back problems: Systematic literature review in working-age adults, *Spine Journal* 9:147–68; Moon, H. J., et al. (2013), Effect of lumbar stabilization and dynamic lumbar strengthening exercises in patients with chronic low back pain, *Annals of Rehabilitative Medicine* 37:110–17; Steele, J., et al. (2013), A randomized controlled trial of limited range of motion lumbar extension exercise in chronic low back pain, *Spine* 38:1245–52; Lee, J. S., and Kang, S. J. (2016), The effects of strength exercise and walking on lumbar function, pain level, and body composition in chronic back pain patients, *Journal of Exercise Rehabilitation* 12:463–70.

FOUR SLEEP: WHY STRESS THWARTS REST

1. Woolf, V. (1925), *The Common Reader* (London: Hogarth Press).
2. Lockley, S. W., and Foster, R. G. (2012), *Sleep: A Very Short Introduction* (New York: Oxford University Press).
3. Soldatos, C. R., et al. (2005), How do individuals sleep around the world? Results from a single-day survey in ten countries, *Sleep Medicine* 6:5–13.

4. For two popular, widely read accounts, see Huffington, A. (2016), *The Sleep Revolution: Transforming Your Life One Night at a Time* (New York: Harmony Books); Walker, M. (2017), *Why We Sleep: Unlocking the Power of Sleep and Dreams* (New York: Simon & Schuster). For data on the effects of sleep deprivation, see Mitler, M. M., et al. (1988), Catastrophes, sleep, and public policy: Consensus report, *Sleep* 11:100–109.
5. For an excellent account of one such experiment, see Boese, A. (2007), *Elephants on Acid, and Other Bizarre Experiments* (New York: Harvest Books).
6. Sharma, S., and Kavuru, M. (2010), Sleep and metabolism: An overview, *International Journal of Endocrinology* 2010:270832; Van Cauter, E., and Copinschi, G. (2000), Interrelationships between growth hormone and sleep, *Growth Hormone and IGF Research* 10:S57–S62.
7. Capellini, I., et al. (2010), Ecological constraints on mammalian sleep architecture, in *The Evolution of Sleep: Phylogenetic and Functional Perspectives,* ed. P. McNamara, R. A. Barton, and C. L. Nunn (Cambridge, U.K.: Cambridge University Press), 12–33.
8. Schacter, D. L. (2001), *The Seven Sins of Memory: How the Mind Forgets and Remembers* (Boston: Houghton Mifflin).
9. Our zebra, for example, needs to remember it was a human who shot her sister, not what kind of grass she was eating, what color shirt he wore, or whether it was sunny.
10. Stickgold, R., and Walker, M. P. (2013), Sleep-dependent memory triage: Evolving generalization through selective processing, *Nature Neuroscience* 16:139–45.
11. Although sleep helps us store and sort memories, does it explain how much we sleep? I also figure things out when I take a long shower, go for a walk, or otherwise relax, and it is not clear that more sleep leads to more memory consolidation. Elephants, the biggest-brained land animal and one of the smartest, sleep only two hours a day in the wild, whereas much less intelligent brown bats sleep twenty hours a day. See Gravett, N., et al. (2017), Inactivity/sleep in two wild free-roaming African elephant matriarchs—does large body size make elephants the shortest mammalian sleepers?, *PLOS ONE* 12:e0171903.
12. The most damaging are free radicals, molecules with unpaired electrons that react readily with other molecules, causing all kinds of cellular damage.
13. Mander, B. A., et al. (2015), β-amyloid disrupts human NREM slow waves and related hippocampus-dependent memory consolidation, *Nature Neuroscience* 18:1051–57.
14. The body's main energy store are adenosine molecules with three phosphates attached: ATP (adenosine triphosphate). When ATPs break down, releasing energy, adenosine molecules slowly accumulate in the brain, helping make you drowsy. Caffeine keeps you awake by binding to the receptors in the brain that normally bind to adenosine, blocking their effect.
15. The brain is bathed in cerebrospinal fluid that keeps blood away from brain cells. This is because direct contact with blood destroys neurons (as, for example, during a stroke). In addition, by keeping blood away from direct contact with the brain, the blood-brain barrier prevents infectious agents and toxins in the bloodstream from getting into the brain.
16. Xie, L., et al. (2013), Sleep drives metabolite clearance from the adult brain, *Science* 342:373–77.

17. Suntsova, N., et al. (2002), Sleep-waking discharge patterns of median preoptic nucleus neurons in rats, *Journal of Physiology* 543:666–77.
18. American Automobile Association Foundation for Traffic Safety (2014), *Prevalence of Motor Vehicle Crashes Involving Drowsy Drivers, United States, 2009–2013* (Washington, D.C.: AAA Foundation for Traffic Safety).
19. Mascetti, G. G. (2016), Unihemispheric sleep and asymmetrical sleep: Behavioral, neurophysiological, and functional perspectives, *Nature and Science of Sleep* 8:221–38.
20. Capellini (2010), Ecological constraints on mammalian sleep architecture.
21. To be honest, we are not sure how much chimpanzees are asleep during this time. See Samson, D., and Nunn, C. L. (2015), Sleep intensity and the evolution of human cognition, *Evolutionary Anthropology* 24:225–37.
22. Ford, E. S., Cunningham, T. J., and Croft, J. B. (2015), Trends in self-reported sleep duration among US adults from 1985 to 2012, *Sleep* 38:829–32; Groeger, J. A., Zijlstra, F. R. H., and Dijk, D. J. (2004), Sleep quantity, sleep difficulties, and their perceived consequences in a representative sample of some 2000 British adults, *Journal of Sleep Research* 13:359–71; Luckhaupt, S. E., Tak, S., and Calvert, G. M. (2010), The prevalence of short sleep duration by industry and occupation in the National Health Interview Survey, *Sleep* 33:149–59; Ram, S., et al. (2010), Prevalence and impact of sleep disorders and sleep habits in the United States, *Sleep and Breathing* 14:63–70. Note that worldwide the percentage of people who report not getting the minimum of seven to eight hours they supposedly require is slightly lower, about one in four. See Soldatos et al. (2005), How do individuals sleep around the world?
23. In one careful study that compared self-reported with sensor-measured sleep time in more than six hundred people, the average individual who reported six and a half hours of sleep in fact slept only five hours, and those who reported seven and a half hours actually slept seven hours. See Lauderdale, D. S., et al. (2008), Self-reported and measured sleep duration: How similar are they?, *Epidemiology* 19:838–45.
24. Lauderdale, D. S., et al. (2006), Objectively measured sleep characteristics among early-middle-aged adults: The CARDIA study, *American Journal of Epidemiology* 164:5–16; Blackwell, T., et al. (2011), Factors that may influence the classification of sleep-wake by wrist actigraphy: The MrOS Sleep Study, *Journal of Clinical Sleep Medicine* 7:357–67; Natale, V., et al. (2014), The role of actigraphy in the assessment of primary insomnia: A retrospective study, *Sleep Medicine* 15:111–15; Lehnkering, H., and Siegmund, R. (2007), Influence of chronotype, season, and sex of subject on sleep behavior of young adults, *Chronobiology International* 24:875–88; Robillard, R., et al. (2014), Sleep-wake cycle in young and older persons with a lifetime history of mood disorders, *PLOS ONE* 9:e87763; Heeren, M., et al. (2014), Active at night, sleepy all day: Sleep disturbances in patients with hepatitis C virus infection, *Journal of Hepatology* 60:732–40.
25. See Walker (2017), *Why We Sleep*.
26. Evans, D. S., et al. (2011), Habitual sleep/wake patterns in the Old Order Amish: Heritability and association with non-genetic factors, *Sleep* 34:661–69; Knutson, K. L. (2014), Sleep duration, quality, and timing and their associations with age in a community without electricity in Haiti, *American Journal of Human Biology* 26:80–86; Samson, D. R., et al. (2017), Segmented sleep

in a nonelectric, small-scale agricultural society in Madagascar, *American Journal of Human Biology* 29:e22979.

27. Contrary to these findings, however, a comparison of two populations of former foragers in Argentina, the Toba, found that during winter people in communities without electricity slept more than one hour longer than those with lights. See de la Iglesia, H. O., et al. (2015), Access to electric light is associated with shorter sleep duration in a traditionally hunter-gatherer community, *Journal of Biological Rhythms* 30:342–50.
28. Youngstedt, S. D., et al. (2016), Has adult sleep duration declined over the last 50+ years?, *Sleep Medicine Reviews* 28:69–85.
29. In his 2017 best seller, *Why We Sleep,* the sleep expert Matthew Walker dismissed these data by claiming that even if the measurements are correct, hunter-gatherers *ought* to sleep more. As evidence, he points to their life expectancy of fifty-eight and susceptibility to infectious diseases, both of which he supposes would be ameliorated if they had more sleep. These criticisms, however, are problematic. If you correct for high rates of infant mortality, most hunter-gatherers live at least seven decades, and their susceptibility to infectious diseases is actually considerably lower than that of both farmers and industrialized people who lack access to modern health care. Walker also incorrectly suggests that hunter-gatherers are highly stressed, hence sleep deprived, because they are calorie deprived. For data on hunter-gatherer life expectancy and causes of death, see Gurven, M., and Kaplan, H. (2007), Longevity among hunter-gatherers: A cross-cultural examination, *Population and Development Review* 33:321–65.
30. Kripke, D. F., et al. (2002), Mortality associated with sleep duration and insomnia, *Archives of General Psychiatry* 59:131–36. The study was actually a follow-up to a 1964 study that received little attention: Hammond, E. C. (1964), Some preliminary findings on physical complaints from a prospective study of 1,064,004 men and women, *American Journal of Public Health* 54:11–23.
31. Tamakoshi, A., and Ohno, Y. (2004), Self-reported sleep duration as a predictor of all-cause mortality: Results from the JACC study, Japan, *Sleep* 27:51–54; Youngstedt, S. D., and Kripke, D. F. (2004), Long sleep and mortality: Rationale for sleep restriction, *Sleep Medicine Reviews* 8:159–74; Bliwise, D. L., and Young, T. B. (2007), The parable of parabola: What the U-shaped curve can and cannot tell us about sleep, *Sleep* 30:1614–15; Ferrie, J. E., et al. (2007), A prospective study of change in sleep duration; associations with mortality in the Whitehall II cohort, *Sleep* 30:1659–66; Hublin, C., et al. (2007), Sleep and mortality: A population-based 22-year follow-up study, *Sleep* 30:1245–53; Shankar, A., et al. (2008), Sleep duration and coronary heart disease mortality among Chinese adults in Singapore: A population-based cohort study, *American Journal of Epidemiology* 168:1367–73; Stranges, S., et al. (2008), Correlates of short and long sleep duration: A cross-cultural comparison between the United Kingdom and the United States: The Whitehall II study and the Western New York Health Study, *American Journal of Epidemiology* 168:1353–64.
32. Lopez-Minguez, J., et al. (2015), Circadian system heritability as assessed by wrist temperature: A twin study, *Chronobiology International* 32:71–80; Jones, S. E., et al. (2019), Genome-wide association analyses of chronotype in

697,828 individuals provides insights into circadian rhythms, *Nature Communications* 10:343.

33. Ekirch, A. R. (2005), *At Day's Close: Night in Times Past* (New York: W. W. Norton).
34. Samson et al. (2017), Segmented sleep in a nonelectric, small-scale agricultural society in Madagascar; Worthman, C. M. (2002), After dark: The evolutionary ecology of human sleep, in *Perspectives in Evolutionary Medicine*, ed. W. R. Trevathan, E. O. Smith, and J. J. McKenna (Oxford: Oxford University Press), 291–313; Worthman, C. M., and Melby, M. K. (2002), Towards a comparative developmental ecology of human sleep, in *Adolescent Sleep Patterns: Biological, Social, and Psychological Influences*, ed. M. A. Carskadon (Cambridge, U.K.: Cambridge University Press), 69–117.
35. Randler, C. (2014), Sleep, sleep timing, and chronotype in animal behaviour, *Behaviour* 94:161–66.
36. The study needs to be replicated, however, to correct for people's tendency to spend about an hour a night moving in their sleep, making it seem as if they were awake. See Samson, D. R., et al. (2017), Chronotype variation drives night-time sentinel-like behaviour in hunter-gatherers, *Proceedings of the Royal Society of Science B: Biological Science* 28:20170967.
37. Snyder, F. (1966), Toward an evolutionary theory of dreaming, *American Journal of Psychiatry* 123:121–36; Nunn, C. L., Samson, D. R., and Krystal, A. D. (2016), Shining evolutionary light on human sleep and sleep disorders, *Evolutionary Medicine and Public Health* 2016:227–43.
38. Van Meijl, T. (2013), Māori collective sleeping as cultural resistance, in *Sleep Around the World: Anthropological Perspectives*, ed. K. Glaskin and R. Chenhall (London: Palgrave Macmillan), 133–49.
39. Lohmann, R. I. (2013), Sleeping among the Asabano: Surprises in intimacy and sociality at the margins of consciousness, in Glaskin and Chenhall, *Sleep Around the World*, 21–44; Musharbash, Y. (2013), Embodied meaning: Sleeping arrangements in Central Australia, in ibid., 45–60.
40. Worthman and Melby (2002), Towards a comparative developmental ecology of human sleep.
41. Reiss, B. (2017), *Wild Nights: How Taming Sleep Created Our Restless World* (New York: Basic Books).
42. Ekirch (2005), *At Day's Close.*
43. Worthman (2008), After dark.
44. Alexeyeff, K. (2013), Sleeping safe: Perceptions of risk and value in Western and Pacific infant co-sleeping, in Glaskin and Chenhall, *Sleep Around the World*, 113–32.
45. Ferber, R. (1986), *Solve Your Child's Sleep Problems* (New York: Fireside Books).
46. McKenna, J. J., Ball, H. L., and Gettler, L. T. (2007), Mother-infant cosleeping, breastfeeding, and sudden infant death syndrome: What biological anthropology has discovered about normal infant sleep and pediatric sleep medicine, *Yearbook of Physical Anthropology* 45:133–61.
47. McKenna, J. J., and McDade, T. (2006), Why babies should never sleep alone: A review of the co-sleeping controversy in relation to SIDS, bedsharing, and breast feeding, *Paediatric Respiratory Reviews* 6:134–52; Fleming, P., Blair, P., and McKenna, J. J. (2006), New knowledge, new insights, new recommendations, *Archives of Diseases in Childhood* 91:799–801.

48. For diagnosable insomnia, see Ohayon, M. M., and Reynolds, C. F., III (2009), Epidemiological and clinical relevance of insomnia diagnosis algorithms according to the Diagnostic and Statistical Manual of Disorders (DSM-IV) and the International Classification of Sleep Disorders (ICSD), *Sleep Medicine* 10:952–60. For self-reported claims on insomnia, see Centers for Disease Control and Prevention, 1 in 3 adults don't get enough sleep, press release, Feb. 18, 2016, www.cdc.gov.
49. Mai, E., and Buysse, D. J. (2008), Insomnia: Prevalence, impact, pathogenesis, differential diagnosis, and evaluation, *Sleep Medicine Clinics* 3:167–74.
50. Chong, Y., Fryer, C. D., and Gu, Q. (2013), Prescription sleep aid use among adults: United States, 2005–2010, *National Center for Health Statistics Data Brief* 127:1–8.
51. Borbély, A. A. (1982), A two process model of sleep regulation, *Human Neurobiology* 1:195–204.
52. Czeisler, C. A., et al. (1999), Stability, precision, and near-24-hour period of the human circadian pacemaker, *Science* 284:2177–81.
53. Yes, this is a reference to Harry Potter. See also McNamara, P., and Auerbach, S. (2010), Evolutionary medicine of sleep disorders: Toward a science of sleep duration, in McNamara, Barton, and Nunn, *Evolution of Sleep*, 107–22.
54. Murphy, P. J., and Campbell, S. S. (1997), Nighttime drop in body temperature: A physiological trigger for sleep onset?, *Sleep* 20:505–11; Uchida, S., et al. (2012), Exercise effects on sleep physiology, *Frontiers in Neurology* 3:48; Youngstedt, S. D. (2005), Effects of exercise on sleep, *Clinics in Sports Medicine* 24:355–65.
55. Kubitz, K. A., et al. (1996), The effects of acute and chronic exercise on sleep: A meta-analytic review, *Sports Medicine* 21:277–91; Youngstedt, S. D., O'Connor, P. J., and Dishman, R. K. (1997), The effects of acute exercise on sleep: A quantitative synthesis, *Sleep* 20:203–14; Singh, N. A., Clements, K. M., and Fiatarone, M. A. (1997), A randomized controlled trial of the effect of exercise on sleep, *Sleep* 20:95–101; Dishman, R. K., et al. (2015), Decline in cardiorespiratory fitness and odds of incident sleep complaints, *Medicine and Science in Sports and Exercise* 47:960–66; Dolezal, B. A., et al. (2017), Interrelationship between sleep and exercise: A systematic review, *Advances in Preventive Medicine* 2017:1364387.
56. Loprinzi, P. D., and Cardinal, B. J. (2011), Association between objectively-measured physical activity and sleep, NHANES 2005–2006, *Mental Health and Physical Activity* 4:65–69.
57. Fowler, P. M., et al. (2017), Greater effect of east versus west travel on jet lag, sleep, and team sport performance, *Medicine and Science in Sports and Exercise* 49:2548–61.
58. Gao, B., et al. (2019), Lack of sleep and sports injuries in adolescents: A systematic review and meta-analysis, *Journal of Pediatric Orthopedics* 39:e324–e333.
59. Hartescu, I., Morgan, K., and Stevinson, C. D. (2015), Increased physical activity improves sleep and mood outcomes in inactive people with insomnia: A randomized controlled trial, *Journal of Sleep Research* 24:526–34; Hartescu, I., and Morgan, K. (2019), Regular physical activity and insomnia: An international perspective, *Journal of Sleep Research* 28:e12745; Inoue, S., et al. (2013), Does habitual physical activity prevent insomnia? A cross-sectional

and longitudinal study of elderly Japanese, *Journal of Aging and Physical Activity* 21:119–39; Skarpsno, E. S., et al. (2018), Objectively measured occupational and leisure-time physical activity: Cross-sectional associations with sleep problems, *Scandinavian Journal of Work and Environmental Health* 44:202–11.

60. For an excellent popular account of stress and its many cortisol-mediated effects on the body, see Sapolsky, R. M. (2004), *Why Zebras Don't Get Ulcers: An Updated Guide to Stress, Stress-Related Diseases, and Coping,* 3rd ed. (San Francisco: W. H. Freeman).
61. Insomnia is defined as having persistent trouble falling or staying asleep even in ideal circumstances. See Leproult, R., et al. (1997), Sleep loss results in an elevation of cortisol levels the next evening, *Sleep* 20:865–70; Spiegel, K., Leproult, R., and Van Cauter, E. (1999), Impact of sleep debt on metabolic and endocrine function, *Lancet* 354:1435–39; Ohayon, M. M., et al. (2010), Using difficulty resuming sleep to define nocturnal awakenings, *Sleep Medicine* 11:236–41.
62. Spiegel, K., et al. (2004), Sleep curtailment in healthy young men is associated with decreased leptin levels: Elevated ghrelin levels and increased hunger and appetite, *Annals of Internal Medicine* 141:846–50.
63. Vgontzas, A. N., et al. (2004), Adverse effects of modest sleep restriction on sleepiness, performance, and inflammatory cytokines, *Journal of Clinical Endocrinology and Metabolism* 89:2119–26.
64. Shi, T., et al. (2019), Does insomnia predict a high risk of cancer? A systematic review and meta-analysis of cohort studies, *Journal of Sleep Research* 2019:e12876.
65. Colrain, I. M., Nicholas, C. L., and Baker, F. C. (2014), Alcohol and the sleeping brain, *Handbook of Clinical Neurology* 125:415–31.
66. Chong, Fryer, and Gu (2013), Prescription sleep aid use among adults.
67. Kripke, D. F., Langer, R. D., and Kline, L. E. (2012), Hypnotics' association with mortality or cancer: A matched cohort study, *BMJ Open* 2:e000850.
68. Kripke, D. F. (2016), Hypnotic drug risks of mortality, infection, depression, and cancer: But lack of benefit, *F1000Research* 5:918.
69. Huedo-Medina, T. B., et al. (2012), Effectiveness of non-benzodiazepine hypnotics in treatment of adult insomnia: Meta-analysis of data submitted to the Food and Drug Administration, *British Medical Journal* 345:e8343.
70. The quotation appears in Huffington (2016), *Sleep Revolution,* 48.
71. Buysse, D. J. (2014), Sleep health: Can we define it? Does it matter?, *Sleep* 37:9–17.
72. To go a step further, sleep also involves some physical activity because sleepers generally toss and turn about an hour every night, helping prevent bedsores and other problems that arise from being too inert. There are also a few pathological forms of somnambulant physical activity including sleepwalking and restless leg syndrome.

FIVE SPEED: NEITHER TORTOISE NOR HARE

1. Bourliere, F. (1964), *The Natural History of Mammals,* 3rd ed. (New York: Alfred A. Knopf).
2. A typical bull can run twenty-two miles an hour, whereas most untrained

but fit humans can run fifteen miles an hour. Elite sprinters can run about as fast as a bull.

3. Fiske-Harrison, A., et al. (2014), *Fiesta: How to Survive the Bulls of Pamplona* (London: Mephisto Press).
4. Salo, A. L., et al. (2010), Elite sprinting: Are athletes individually step frequency or step length reliant?, *Medicine and Science in Sport and Exercise* 43:1055–62.
5. According to the sprinting expert Peter Weyand, Bolt's peak stride frequency was 2.1 to 2.2 strides per second, slightly below other sprinters who could manage 2.3 to 2.4 strides per second and in the case of shorter sprinters as much as 2.5 strides per second.
6. Garland, T., Jr. (1983), The relation between maximal running speed and body mass in terrestrial mammals, *Journal of Zoology* 199:157–70.
7. In a famous experiment, the Harvard biologist C. Richard Taylor claimed that a cheetah can sustain its impressive speed for only four minutes before it overheats and must stop. However, this laboratory experiment was conducted on a cheetah running far below its maximum running speed (try getting a treadmill going that fast with a dangerous, speedy predator on it), and careful studies that implanted temperature sensors in wild cheetahs in South Africa found that they stop running well before they overheat. See Hetem, R. S., et al. (2013), Cheetah do not abandon hunts because they overheat, *Biology Letters* 9:20130472; Taylor, C. R., and Rowntree, V. J. (1973), Temperature regulation and heat balance in running cheetahs: A strategy for sprinters?, *American Journal of Physiology* 224:848–51.
8. Wilson, A. M., et al. (2018), Biomechanics of predator-prey arms race in lion, zebra, cheetah, and impala, *Nature* 554:183–88.
9. Turning slows down runners considerably by requiring them to apply a sideways force to redirect their body into the direction of the turn. Turning also makes animals less stable because to go sideways they need to put their feet to the side instead of underneath their center of mass. In both respects, humans have a disadvantage. First, a running human with only one leg on the ground at a time is more likely to fall than the quadruped that has at least two legs on the ground, one on the inside that can help support the body. Humans also lack traction. Athletes today use shoes with cleats or spikes, but the cheetah has huge retractable claws that act like cleats to give it traction when changing direction. See Jindrich, D. L., Besier, T. F., and Lloyd, D. G. (2006), A hypothesis for the function of braking forces during running turns, *Journal of Biomechanics* 39:1611–20.
10. Rubenson, J., et al. (2004), Gait selection in the ostrich: Mechanical and metabolic characteristics of walking and running with and without an aerial phase, *Proceedings of the Royal Society B: Biological Science* 271:1091–99. Humans versus ostriches: Jindrich, D. L., et al. (2007), Mechanics of cutting maneuvers by ostriches (*Struthio camelus*), *Journal of Experimental Biology* 210:1378–90. In case you are wondering, *Tyrannosaurus rex* was probably not a fast runner; new estimates suggest it had a maximum speed of twelve miles per hour. See Sellers, W. I., et al. (2017), Investigating the running abilities of *Tyrannosaurus rex* using stress-constrained multibody dynamic analysis, *PeerJ* 5:e3420.

11. Gambaryan, P., and Hardin, H. (1974), *How Mammals Run: Anatomical Adaptations* (New York: Wiley); Galis, F., et al. (2014), Fast running restricts evolutionary change of the vertebral column in mammals, *Proceedings of the National Academy of Sciences USA* 111:11401–6.
12. See Castillo, E. R., and Lieberman, D. E. (2018), Shock attenuation in the human lumbar spine during walking and running, *Journal of Experimental Biology* 221:jeb177949. I've been focusing here on the energetic disadvantages of being bipedal, but being upright causes a host of other problems, one of which is that our heads jiggle dreadfully. A human runner's head bounces up and down, but the head of a galloping dog or horse stays impressively still even though the rest of its body is moving. The quadruped's head looks like a missile mounted on its body. This missile-like stability is crucial for running because the reflexes that stabilize the eyeballs to keep an animal's vision from blurring are not fast enough to overcome rapid displacements. Experiments show that when the head rotates too fast, animals (including humans) can no longer effectively focus on whatever they are looking at including obstacles in their way. Overly bouncy heads are thus potentially perilous for runners. Quadrupeds solve this problem by having a relatively horizontal neck that runs from the front of the thorax to the back of the head. Thanks to this cantilevered configuration, when the animal's body falls or rises, it raises or lowers its neck to keep its head still. Some animals have gone a step further and added a highly elastic structure (the nuchal ligament) that passively stabilizes the head without requiring muscles to do much work. Humans, however, run like pogo sticks because we have short, little, vertical necks that attach to the middle of the base of the skull. Although we have evolved some special mechanisms to keep our heads from pitching too much, there is nothing we can do to stop our heads from bouncing up and down when we run. For more information, see chapter 9 in Lieberman, D. E. (2011), *The Evolution of the Human Head* (Cambridge, Mass.: Harvard University Press).
13. Weyand, P. G., Lin, J. E., and Bundle, M. W. (2006), Sprint performance-duration relationships are set by the fractional duration of external force application, *American Journal of Physiology: Regulatory, Integrative, and Comparative Physiology* 290:R758–R765; Weyand, P. G., et al. (2010), The biological limits to running speed are imposed from the ground up, *Journal of Applied Physiology* 108:950–61; Bundle, M. W., and Weyand, P. G. (2012), Sprint exercise performance: Does metabolic power matter?, *Exercise Sports Science Reviews* 40:174–82.
14. We can also break down protein as a source of fuel, but this is much less common and occurs only when fat and sugars are in short supply. Also, in case you were wondering, ADP sometimes can be broken down to an AMP (adenosine *mono*phosphate), releasing yet more energy. This happens less often, though.
15. Gillen, C. M. (2014), *The Hidden Mechanics of Exercise* (Cambridge, Mass.: Harvard University Press).
16. See McArdle, W. D., Katch, F. I., and Katch, V. L. (2007), *Exercise Physiology: Energy, Nutrition, and Human Performance* (Philadelphia: Lippincott Williams & Wilkins), 143; Gastin, P. B. (2001), Energy system interaction and relative contribution during maximal exercise, *Sports Medicine Journal* 10:725–41.

17. Although this process is normally very slow, muscle cells can supply an enzyme (creatine phosphatase) that speeds up that transfer by five hundred–fold.
18. Boobis, L., Williams, C., and Wootton, S. (1982), Human muscle metabolism during brief maximal exercise, *Journal of Physiology* 338:21–22; Nevill, M. E., et al. (1989), Effect of training on muscle metabolism during treadmill sprinting, *Journal of Applied Physiology* 67:2376–82. Note that intensive training and consuming foods like meat that are rich in creatine can increase these reserves only modestly. See Koch, A. J., Pereira, R., and Machado, M. (2014), The creatine kinase response to resistance training, *Journal of Musculoskeletal and Neuronal Interactions* 14:68–77.
19. To be precise, glycolysis is a ten-step process that requires two ATPs per sugar, yielding four ATPs, hence a net of two ATPs.
20. Bogdanis, G. C., et al. (1995), Recovery of power output and muscle metabolites following 30 s of maximal sprint cycling in man, *Journal of Physiology* 482:467–80.
21. Mazzeo, R. S., et al. (1986), Disposal of blood [1-13C]lactate in humans during rest and exercise, *Journal of Applied Physiology* 60:232–41; Robergs, R. A., Ghiasvand, F., and Parker, D. (2004), Biochemistry of exercise-induced metabolic acidosis, *American Journal of Physiology: Regulatory, Integrative, and Comparative Physiology* 287:R502–R516.
22. This process is also known as the citric acid cycle and involves a second step called oxidative phosphorylation, but I am lumping them together to keep things simple.
23. Maughan, R. J. (2000), Physiology and biochemistry of middle distance and long distance running, in *Handbook of Sports Science and Medicine: Running*, ed. J. A. Hawley (Oxford: Blackwell), 14–27.
24. That limit, a topic for chapter 9, is strongly affected by your genes and your environment, including how you train.
25. To be precise, about 20 percent of the ATP we used came from our immediate but limited creatine phosphate supplies, about 50 percent came from short-term glycolysis, and only about 30 percent came from our long-term aerobic energy systems. See Bogdanis et al. (1995), Recovery of power output and muscle metabolites following 30 s of maximal sprint cycling in man.
26. Steinmetz, P. R. H., et al. (2012), Independent evolution of striated muscles in cnidarians and bilaterians, *Nature* 482:231–34.
27. The thin filaments are known as actin, the thick filaments as myosin, and the projections on the myosin filaments that pull on the actin filaments are known as myosin heads. When an ATP binds with a myosin head, it shortens the sarcomere by only six nanometers, about fifteen thousand times less than the thickness of a piece of paper. Because a typical large muscle contains at least a million muscle fibers, and each fiber has about five thousand parallel sarcomeres, billions of them contract together in a typical muscle to generate enormous forces at the cost of billions of ATP. Interestingly, no ATP is required for a myosin head to bind to an actin protein, but ATP causes each myosin head to ratchet and then release itself from the actin. As a result, when ATP runs out after death, muscles go into rigor mortis in the absence of ATP to release the myosin heads from the actin.
28. The red tinge, which is only evident if you stain or cook them, comes

from having a higher percentage of myoglobin molecules. Myoglobin is a hemoglobin-like protein in your blood that functions to transport oxygen within muscle cells to your mitochondria. Muscles that use a lot of aerobic energy need more myoglobin than those that rely on glycolysis.

29. McArdle, W. D., Katch, F. I., and Katch, V. L. (2000), *Essentials of Exercise Physiology*, 2nd ed. (Baltimore: Lippincott Williams & Wilkins).
30. Costill, D. L. (1976), Skeletal muscle enzymes and fiber composition in male and female track athletes, *Journal of Applied Physiology* 40:149–54.
31. Zierath, J. R., and Hawley, J. A. (2004), Skeletal muscle fiber type: Influence on contractile and metabolic properties, *PLOS Biology* 2:e348.
32. Handsfield, G. G., et al. (2017), Adding muscle where you need it: Nonuniform hypertrophy patterns in elite sprinters, *Scandinavian Journal of Medicine and Science in Sports* 27:1050–60.
33. Van De Graaff, K. M. (1977), Motor units and fiber types of primary ankle extensors of the skunk (*Mephitis mephitis*), *Journal of Neurophysiology* 40:1424–31; Rodríguez-Barbudo, M. V., et al. (1984), Histochemical and morphometric examination of the cranial tibial muscle of dogs with varying aptitudes (greyhound, German shepherd, and fox terrier), *Zentralblatt für Veterinarmedizin: Reihe C* 13:300–312; Williams, T. M., et al. (1997), Skeletal muscle histology and biochemistry of an elite sprinter, the African cheetah, *Journal of Comprehensive Physiology B* 167:527–35.
34. Costa, A. M., et al. (2012), Genetic inheritance effects on endurance and muscle strength: An update, *Sports Medicine* 42:449–58.
35. Interestingly, there are different genes that appear to affect people whose maximum aerobic capacity (VO_2 max) responds poorly to training (so-called nonresponders) and those who respond strongly (high responders). See Bouchard, C., et al. (2011), Genomic predictors of the maximal O_2 uptake response to standardized exercise training programs, *Journal of Applied Physiology* 110:1160–70.
36. For an excellent review, see Epstein, D. (2013), *The Sports Gene: Inside the Science of Extraordinary Athletic Performance* (New York: Current Books).
37. Yang, N., et al. (2003), ACTN3 genotype is associated with human elite athletic performance, *American Journal of Human Genetics* 73:627–31; Berman, Y., and North, K. N. (2010), A gene for speed: The emerging role of alpha-actinin-3 in muscle metabolism, *Physiology* 25:250–59.
38. Moran, C. N., et al. (2007), Association analysis of the ACTN3 R577X polymorphism and complex quantitative body composition and performance phenotypes in adolescent Greeks, *European Journal of Human Genetics* 15:88–93.
39. Pitsiladis, Y., et al. (2013), Genomics of elite sporting performance: What little we know and necessary advances, *British Journal of Sports Medicine* 47:550–55; Tucker, R., Santos-Concejero, J., and Collins, M. (2013), The genetic basis for elite running performance, *British Journal of Sports Medicine* 47:545–49.
40. Bray, M. S., et al. (2009), The human gene map for performance and health-related fitness phenotypes: The 2006–2007 update, *Medicine and Science in Sports and Exercise* 41:35–73; Guth, L. M., and Roth, S. M. (2013), Genetic influence on athletic performance, *Current Opinions in Pediatrics* 25:653–58.
41. Simoneau, J. A., and Bouchard, C. (1995), Genetic determinism of fiber type proportion in human skeletal muscle, *FASEB Journal* 9:1091–95.

42. Ama, P. F., et al. (1986), Skeletal muscle characteristics in sedentary black and Caucasian males, *Journal of Applied Physiology* 61:1758–61.
43. Wood, A. R., et al. (2014), Defining the role of common variation in the genomic and biological architecture of adult human height, *Nature Genetics* 46:1173–86.
44. Price, A. (2014), *Year of the Dunk: A Modest Defiance of Gravity* (New York: Crown).
45. Carling, C., et al. (2016), Match-to-match variability in high-speed running activity in a professional soccer team, *Journal of Sports Science* 34:2215–23.
46. Van Damme, R., et al. (2002), Performance constraints in decathletes, *Nature* 415:755–56. Another example is a study that measured general athletic abilities in Australian professional soccer players which found that endurance abilities such as fifteen-hundred-meter running speed barely compromised power-based abilities such as how far the players could jump. See Wilson, R. S. (2014), Does individual quality mask the detection of performance trade-offs? A test using analyses of human physical performance, *Journal of Experimental Biology* 217:545–51.
47. Wilson, R. S., James, R. S., and Van Damme, R. (2002), Trade-offs between speed and endurance in the frog *Xenopus laevis:* A multi-level approach, *Journal of Experimental Biology* 205:1145–52; Garland, T., and Else, P. L. (1987), Seasonal, sexual, and individual variation in endurance and activity metabolism in lizards, *American Journal of Physiology* 252:R439–R449; Garland, T. (1988), Genetic basis of activity metabolism: I. Inheritance of speed, stamina, and antipredator displays in the garter snake *Thamnophis sirtalis, Evolution* 42:335–50; Schaffer, H. B., Austin, C. C., and Huey, R. B. (1989), The consequences of metamorphosis on salamander (*Ambystoma*) locomotor performance, *Physiological Zoology* 64:212–31.
48. Elaine Morgan's 1982 book, *The Aquatic Ape: A Theory of Human Evolution* (London: Souvenir Press), popularized the notion that humans evolved to swim, and despite many critiques by scientists this pseudoscientific idea remains popular on the internet, often in the guise of a conspiracy theory. While freshwater and marine resources have sometimes been important in human evolution, there is little evidence that humans were selected to swim well, and many claimed adaptations for swimming such as downward-facing nostrils are disproven just-so stories. For all these features, there are better-tested, more compelling adaptive hypotheses. In addition, even the best Olympic swimmers don't swim very fast or well. Whereas a sea lion or dolphin can swim twenty-five miles an hour, the fastest human swimmers can barely paddle six miles an hour. For more information, see Langdon, J. H. (1997), Umbrella hypotheses and parsimony in human evolution: A critique of the Aquatic Ape Hypothesis, *Journal of Human Evolution* 33:479–94; Gee, H. (2013), *The Accidental Species: Misunderstandings of Human Evolution* (Chicago: University of Chicago Press).
49. Apicella, C. L. (2014), Upper-body strength predicts hunting reputation and reproductive success in Hadza hunter-gatherers, *Evolution and Human Behavior* 35:508–18; Walker, R., and Hill, K. (2003), Modeling growth and senescence in physical performance among the Aché of eastern Paraguay, *American Journal of Human Biology* 15:196–208.
50. This is because when the muscle is lengthened, the sarcomeres are stretched,

thus reducing the overlap between the myosin and the actin filaments. In this elongated state, they are less able to produce force and have to work harder.

51. Staron, R. S. (1991), Strength and skeletal muscle adaptations in heavy-resistance-trained women after detraining and retraining, *Journal of Applied Physiology* 70:631–40; Staron, R. S., et al. (1994), Skeletal muscle adaptations during early phase of heavy-resistance training in men and women, *Journal of Applied Physiology* 76:1247–55; Häkkinen, K., et al. (1998), Changes in muscle morphology, electromyographic activity, and force production characteristics during progressive strength training in young and older men, *Journals of Gerontology Series A: Biological Sciences and Medical Sciences* 53:B415–B423.
52. Handsfield et al. (2017), Adding muscle where you need it.
53. Seynnes, O. R., de Boer, M., and Narici, M. V. (2007), Early skeletal muscle hypertrophy and architectural changes in response to high-intensity resistance training, *Journal of Applied Physiology* 102:368–73. This phenomenon is known as the size principle. As a muscle is recruited to generate force, it first activates the smaller slow-twitch (type I) fibers and then activates larger, more forceful fast-twitch (type II) fibers. Because maximal force production such as lifting a very heavy weight recruits all types of muscle fibers, they all respond, but as the activity is repeated, the higher force, type II muscle fibers are recruited more readily, causing them to respond the most. See Gorassini, Y. S. B. (2002), Intrinsic activation of human motoneurons: Reduction of motor unit recruitment thresholds by repeated contractions, *Journal of Neurophysiology* 87:1859–66. See also Abe, T., Kumagai, K., and Brechue, W. F. (2000), Fascicle length of leg muscles is greater in sprinters than distance runners, *Medicine and Science in Sports and Exercise* 32:1125–29; Andersen, J. L., and Aargaard, P. (2010), Effects of strength training on muscle fiber types and size; consequences for athletes training for high-intensity sport, *Scandinavian Journal of Medicine and Science in Sports* 20(S2): 32–38.
54. MacInnis, M. J., and Gibala, M. J. (2017), Physiological adaptations to interval training and the role of exercise intensity, *Journal of Physiology* 595:2915–30.

SIX STRENGTH: FROM BRAWNY TO SCRAWNY

1. Obituary of Charles Atlas, *New York Times*, Dec. 24, 1972, 40.
2. Most of these details come from Gaines, C. (1982), *Yours in Perfect Manhood: Charles Atlas* (New York: Simon & Schuster).
3. For a history of physical culture in America, I recommend Black, J. (2013), *Making the American Body* (Lincoln: University of Nebraska Press).
4. Eaton, S. B., Shostak, M., and Konner, M. (1988), *The Paleolithic Prescription: A Program of Diet and Exercise and a Design for Living* (New York: Harper & Row).
5. The Paleo Diet is trademarked by Loren Cordain.
6. Paleo dieters are advised to eat lots of meat (grass fed of course), vegetables, and fruits, and to avoid dairy, grains, legumes, and other modern foods unavailable to cavemen. Beyond its inaccurate view of what cavemen actually ate, the biggest problem with this diet is the flawed assumption that ancestral diets are necessarily healthy. Because natural selection cares only about reproductive success, we evolved to eat foods that first and foremost promote reproduction, not health. Many of the foods that hunter-gatherers and paleo

dieters eat such as red meat have questionable health effects, and not all modern foods are unhealthy. For more, see Lieberman, D. E. (2013), *The Story of the Human Body: Evolution, Health, and Disease* (New York: Pantheon). Also, Zuk, M. (2013), *Paleofantasy: What Evolution Really Tells Us About Sex, Diet, and How We Live* (New York: W. W. Norton).

7. Le Corre, E. (2019), *The Practice of Natural Movement: Reclaim Power, Health, and Freedom* (Las Vegas, Nev.: Victory Belt).
8. Sisson, M. (2012), *The New Primal Blueprint: Reprogram Your Genes for Effortless Weight Loss, Vibrant Health, and Boundless Energy*, 2nd ed. (Oxnard, Calif.: Primal Blueprint), 46–50.
9. Truswell, A. S., and Hanson, J. D. L. (1976), Medical research among the !Kung, in *Kalahari Hunter-Gatherers: Studies of the !Kung San and Their Neighbors*, ed. R. B. Lee and I. Devore (Cambridge, Mass.: Harvard University Press), 166–94. The quotation is from page 170.
10. O'Keefe, J. H., et al. (2011), Exercise like a hunter-gatherer: A prescription for organic physical fitness, *Progress in Cardiovascular Diseases* 53:471–79; Durant, J. (2013), *The Paleo Manifesto: Living Wild in the Manmade World* (New York: Harmony Books).
11. Ratey, J. J., and Manning, R. (2014), *Go Wild: Free Your Body and Mind from the Afflictions of Civilization* (New York: Little, Brown), 8.
12. In metric: Men: 162 centimeters, fifty-three kilograms; women 150 centimeters, forty-six kilograms. Marlowe, F. W. (2010), *The Hadza: Hunter-Gatherers of Tanzania* (Berkeley: University of California Press); Hiernaux, J., and Hartong, D. B. (1980), Physical measurements of the Hadza, *Annals of Human Biology* 7:339–46.
13. Adult grip strength among male Hadza averages thirty-three kilograms; females average twenty-one kilograms. Mathiowetz, V., et al. (1985), Grip and pinch strength: Normative data for adults, *Archives of Physical Medicine and Rehabilitation* 66:69–74; Günther, C. M., et al. (2008), Grip strength in healthy Caucasian adults: Reference values, *Journal of Hand Surgery of America* 33:558–65; Leyk, D., et al. (2007), Hand-grip strength of young men, women, and highly trained female athletes, *European Journal of Applied Physiology* 99:415–21.
14. Walker, R., and Hill, K. (2003), Modeling growth and senescence in physical performance among the Aché of eastern Paraguay, *American Journal of Human Biology* 15:196–208.
15. Blurton-Jones, N., and Marlowe, F. W. (2002), Selection for delayed maturity: Does it take 20 years to learn to hunt and gather?, *Human Nature* 13:199–238.
16. Evans, W. J. (1995), Effects of exercise on body composition and functional capacity of the elderly, *Journals of Gerontology Series A: Biological Sciences and Medical Sciences* 50:147–50; Phillips, S. M. (2007), Resistance exercise: Good for more than just Grandma and Grandpa's muscles, *Applied Physiology, Nutrition, and Metabolism* 32:1198–205.
17. Spenst, L. F., Martin, A. D., and Drinkwater, D. T. (1993), Muscle mass of competitive male athletes, *Journal of Sports Science* 11:3–8.
18. Some muscle men and women think they need to eat enormous amounts of protein to supply their growing, strapping muscles. Careful studies, however, find that the added muscle mass power lifters pack onto their bodies requires only about an additional 20 percent more protein than needed by

elite endurance athletes like distance runners. Apparently, a two-hundred-pound bodybuilder gets little benefit from consuming more than about four to five ounces of extra protein. Further, the body cannot store excess protein but instead must break it down and excrete it. Consequently, protein overconsumption can potentially lead to a variety of troubles, especially for the kidneys. See Lemon, P. W., et al. (1992), Protein requirements and muscle mass/strength changes during intensive training in novice bodybuilders, *Journal of Applied Physiology* 73:767–75; Phillips, S. M. (2004), Protein requirements and supplementation in strength sports, *Nutrition* 20:689–95; Hoffman, J. R., et al. (2006), Effect of protein intake on strength, body composition, and endocrine changes in strength/power athletes, *Journal of the International Society of Sports Nutrition* 3:12–18; Pesta, D. H., and Samuel, V. T. (2014), A high-protein diet for reducing body fat: Mechanisms and possible caveats, *Nutrition and Metabolism* 11:53.

19. For more details on du Chaillu's life and influence, see Conniff, R. (2011), *The Species Seekers: Heroes, Fools, and the Mad Pursuit of Life on Earth* (New York: W. W. Norton).
20. Paul du Chaillu (1867), *Stories of the Gorilla Country* (New York: Harper).
21. Peterson, D., and Goodall, J. (2000), *Visions of Caliban: On Chimpanzees and People* (Athens: University of Georgia Press).
22. Wrangham, R. W., and Peterson, D. (1996), *Demonic Males: Apes and the Origins of Human Violence* (Boston: Houghton Mifflin).
23. Bauman, J. E. (1923), The strength of the chimpanzee and orang, *Scientific Monthly* 16:432–39; Bauman, J. E. (1926), Observations on the strength of the chimpanzee and its implications, *Journal of Mammalogy* 7:1–9.
24. Finch, G. (1943), The bodily strength of chimpanzees, *Journal of Mammalogy* 24:224–28.
25. Edwards, W. E. (1965), *Study of Monkey, Ape, and Human Morphology and Physiology Relating to Strength and Endurance Phase IX: The Strength Testing of Five Chimpanzee and Seven Human Subjects* (Fort Belvoir, Va.: Defense Technical Information Center).
26. Scholz, M. N., et al. (2006), Vertical jumping performance of bonobo (*Pan paniscus*) suggests superior muscle properties, *Proceedings of the Royal Society B: Biological Sciences* 273:2177–84.
27. The key difference in muscle architecture between the species is that chimpanzees have a higher percentage of fast-twitch fibers, which produce more force, and their fibers are also longer, generating more velocity per unit of force. The result is more force and power. See O'Neill, M. C., et al. (2017), Chimpanzee super strength and human skeletal muscle evolution, *Proceedings of the National Academy of Sciences, USA* 114:7343–48.
28. For a history of thinking about the Neanderthals, see Trinkaus, E., and Shipman, P. (1993), *The Neanderthals: Changing the Image of Mankind* (New York: Alfred A. Knopf).
29. King, W. (1864), The reputed fossil man of the Neanderthal, *Quarterly Journal of Science* 1:88–97. The quotation is from page 96.
30. Among the Inuit, body fat percentages are about 12 to 15 percent in males and 19 to 26 percent in females. See Churchill, S. E. (2014), *Thin on the Ground: Neanderthal Biology, Archeology, and Ecology* (Ames, Iowa: John Wiley & Sons).
31. One classic study showed that the racket arm professional tennis players

use to whack millions upon millions of balls can be one-third thicker than the arm used just for tossing. Because muscles generate the major forces that bones must resist, it makes sense to infer that archaic humans such as Neanderthals were very strong. A caveat to this inference, however, is that exercise doesn't affect bones as simply as muscles. If I were to work out my upper body in the gym for the next year, my biceps and triceps would bulk up noticeably, but my arm bones would thicken almost imperceptibly. Unlike muscles, bones primarily grow bigger in response to loads only during youth. Thus, if cavemen like Neanderthals grew thicker bones because they were stronger and more active, they must have been super-active before adulthood. This is perplexing because modern hunter-gatherer children reportedly don't work very hard—much less than children who grow up on farms. See Pearson, O. M., and Lieberman, D. E. (2004), The aging of Wolff's "law": Ontogeny and responses of mechanical loading to cortical bone, *Yearbook of Physical Anthropology* 29:63–99; Lee, R. B. (1979), *The !Kung San: Men, Women, and Work in a Foraging Society* (Cambridge, U.K.: Cambridge University Press); Kramer, K. L. (2011), The evolution of human parental care and recruitment of juvenile help, *Trends in Ecology and Evolution* 26:533–40; Kramer, K. L. (2005), *Maya Children: Helpers at the Farm* (Cambridge, Mass.: Harvard University Press).

32. Bones can get thicker from loading, but the upper face never experiences high loads, even when chewing hard food. So the only other way facial bones can get extra thick is from hormones. One possibility is growth hormone, but this hormone also causes giantism, which certainly doesn't apply to the short and stocky Neanderthals. Growth hormone also doesn't augment muscle mass. See Lange, K. H. (2002), GH administration changes myosin heavy chain isoforms in skeletal muscle but does not augment muscle strength or hypertrophy, either alone or combined with resistance exercise training in healthy elderly men, *Journal of Clinical Endocrinology and Metabolism* 87:513–23.
33. Penton-Voak, I. S., and Chen, J. Y. (2004), High salivary testosterone is linked to masculine male facial appearance in humans, *Evolution and Human Behavior* 25:229–41; Verdonck, A. M., et al. (1999), Effect of low-dose testosterone treatment on craniofacial growth in boys with delayed puberty, *European Journal of Orthodontics* 21:137–43.
34. Cieri, R. L., et al. (2014), Craniofacial feminization, social tolerance, and the origins of behavioral modernity, *Current Anthropology* 55:419–43.
35. Bhasin, S., et al. (1996), The effects of supraphysiologic doses of testosterone on muscle size and strength in normal men, *New England Journal of Medicine* 335:1–7.
36. Fox, P. (2016), Teen girl uses "crazy strength" to lift burning car off dad, *USA Today*, Jan. 12, 2016, www.usatoday.com.
37. Walker, A. (2008), The strength of great apes and the speed of humans, *Current Anthropology* 50:229–34.
38. Haykowsky, M. J., et al. (2001), Left ventricular wall stress during leg-press exercise performed with a brief Valsalva maneuver, *Chest* 119:150–54.
39. One benefit of having to produce force over a wide range of motions is that muscles have to work harder when they are stretched or contracted. As long as a nerve stimulates a muscle to fire, the heads that stick off the thick myo-

sin molecules repeatedly grab onto actin filaments, cock, release, and then grab again as if they were playing tug-of-war. Although millions of myosin heads are repeatedly doing their thing at any one time, the length of the muscle at any given time determines just how much force it can produce to resist an opposing load. A muscle produces its maximum force when firing isometrically near its resting length. When a muscle shortens, the filaments overlap, limiting the opportunity for myosin heads to pull against actin filaments. And when a muscle lengthens, the filaments also reduce their overlap, hence the opportunity for the myosin heads to grab onto actin filaments.

40. For an excellent review, see Herzog, W., et al. (2008), Mysteries of muscle contraction, *Journal of Applied Biomechanics* 24:1–13.
41. Fridén, J., and Lieber, R. L. (1992), Structural and mechanical basis of exercise-induced muscle injury, *Medicine and Science in Sports and Exercise* 24:521–30; Schoenfeld, B. J. (2012), Does exercise-induced muscle damage play a role in skeletal muscle hypertrophy?, *Journal of Strength and Conditioning Research* 26:1441–53.
42. MacDougall, J. D., et al. (1984), Muscle fiber number in biceps brachii in bodybuilders and control subjects, *Journal of Applied Physiology: Respiratory, Environmental, and Exercise Physiology* 57:1399–403.
43. MacDougall, J. D., et al. (1977), Biochemical adaptation of human skeletal muscle to heavy resistance training and immobilization, *Journal of Applied Physiology: Respiratory, Environmental, and Exercise Physiology* 43:700–703; Damas, F., Libardi, C. A., and Ugrinowitsch, C. (2018), The development of skeletal muscle hypertrophy through resistance training: The role of muscle damage and muscle protein synthesis, *European Journal of Applied Physiology* 118:485–500.
44. French, D. (2016), Adaptations to anaerobic training programs, in *Essentials of Strength Training and Conditioning*, 4th ed., ed. G. G. Haff and N. T. Triplett (Champaign, Ill.: Human Kinetics), 87–113.
45. Rana, S. R., et al. (2008), Comparison of early phase adaptations for traditional strength and endurance, and low velocity resistance training programs in college-aged women, *Journal of Strength and Conditioning Research* 22:119–27.
46. Tinker, D. B., Harlow, H. J., and Beck, T. D. (1998), Protein use and muscle-fiber changes in free-ranging, hibernating black bears, *Physiological Zoology* 71:414–24; Hershey, J. D., et al. (2008), Minimal seasonal alterations in the skeletal muscle of captive brown bears, *Physiological and Biochemical Zoology* 81:138–47.
47. Evans, W. J. (2010), Skeletal muscle loss: Cachexia, sarcopenia, and inactivity, *American Journal of Clinical Nutrition* 91:1123S–1127S.
48. de Boer, M. D., et al. (2007), Time course of muscular, neural, and tendinous adaptations to 23 day unilateral lower-limb suspension in young men, *Journal of Physiology* 583:1079–91.
49. LeBlanc, A., et al. (1985), Muscle volume, MRI relaxation times (T2), and body composition after spaceflight, *Journal of Applied Physiology* 89:2158–64; Edgerton, V. R., et al. (1995), Human fiber size and enzymatic properties after 5 and 11 days of spaceflight, *Journal of Applied Physiology* 78:1733–39; Akima, H., et al. (2000), Effect of short-duration spaceflight on thigh and leg muscle volume, *Medicine and Science in Sports and Exercise* 32:1743–47.

50. Akima, H., et al. (2001), Muscle function in 164 men and women aged 20–84 yr., *Medicine and Science in Sports and Exercise* 33:220–26; Purves-Smith, F. M., Sgarioto, N., and Hepple, R. T. (2014), Fiber typing in aging muscle, *Exercise Sport Science Reviews* 42:45–52.
51. Dodds, R. M., et al. (2014), Grip strength across the life course: Normative data from twelve British studies, *PLOS ONE* 9:e113637; Dodds, R. M., et al. (2016), Global variation in grip strength: A systematic review and meta-analysis of normative data, *Age and Ageing* 45:209–16.
52. Jette, A., and Branch, L. (1981), The Framingham Disability Study: II. Physical disability among the aging, *American Journal of Public Health* 71:1211–16.
53. Walker and Hill (2003), Modeling growth and senescence in physical performance among the Aché of eastern Paraguay; Blurton-Jones and Marlowe (2002), Selection for delayed maturity; Apicella, C. L. (2014), Upper-body strength predicts hunting reputation and reproductive success in Hadza hunter-gatherers, *Evolution and Human Behavior* 35:508–18.
54. Beaudart, C., et al. (2017), Nutrition and physical activity in the prevention and treatment of sarcopenia: Systematic review, *Osteoporosis International* 28:1817–33; Lozano-Montoya, I. (2017), Nonpharmacological interventions to treat physical frailty and sarcopenia in older patients: A systematic overview—the SENATOR Project ONTOP Series, *Clinical Interventions in Aging* 12:721–40.
55. Fiatarone, M. A., et al. (1990), High-intensity strength training in nonagenarians: Effects on skeletal muscle, *Journal of the American Medical Association* 263:3029–34.
56. Donges, C. E., and Duffield, R. (2012), Effects of resistance or aerobic exercise training on total and regional body composition in sedentary overweight middle-aged adults, *Applied Physiology, Nutrition, and Metabolism* 37:499–509; Mann, S., Beedie, C., and Jimenez, A. (2014), Differential effects of aerobic exercise, resistance training, and combined exercise modalities on cholesterol and the lipid profile: Review, synthesis, and recommendations, *Sports Medicine* 44:211–21.
57. Phillips, S. M., et al. (1997), Mixed muscle protein synthesis and breakdown after resistance exercise in humans, *American Journal of Physiology* 273:E99–E107; McBride, J. M. (2016), Biomechanics of resistance exercise, in Haff and Triplett, *Essentials of Strength Training and Conditioning*, 19–42.
58. If you want to learn the science behind how to train like Batman, check out Zehr, E. P. (2008), *Becoming Batman: The Possibility of a Superhero* (Baltimore: Johns Hopkins University Press).
59. Haskell, W. L., et al. (2007), Physical activity and public health: Updated recommendation for adults from the American College of Sports Medicine and the American Heart Association, *Medicine and Science in Sports and Exercise* 39:1423–34; Nelson, M. E., et al. (2007), Physical activity and public health in older adults: Recommendation from the American College of Sports Medicine and the American Heart Association, *Medicine and Science in Sports and Exercise* 39:1435–45.

SEVEN FIGHTING AND SPORTS: FROM FANGS TO FOOTBALL

1. Wrangham, R. W., and Peterson, D. (1996), *Demonic Males: Apes and the Origins of Human Violence* (Boston: Houghton Mifflin).

2. Wrangham, R. W., Wilson, M. L., and Muller, M. N. (2006), Comparative rates of violence in chimpanzees and humans, *Primates* 47:14–26.
3. To continue this logic, fights can have major effects on evolution even if they happen every few generations. The fact that I have never had a major fight in my life doesn't mean my body wasn't influenced by strong selection on those of my ancestors who assuredly sometimes fought. Had they lost rather than won, I might not be here. Further, our bodies are replete with inherited features that accrued over thousands of generations regardless of whether they still benefit us.
4. Oates, J. C. (1987), On boxing, *Ontario Review*.
5. Lystad, R. P., Kobi, G., and Wilson, J. (2014), The epidemiology of injuries in mixed martial arts: A systematic review and meta-analysis, *Orthopaedic Journal of Sports Medicine* 2:2325967113518492.
6. For articulations of this theory, see Fry, D. R. (2006), *The Human Potential for Peace: An Anthropological Challenge to Assumptions About War and Violence* (Oxford: Oxford University Press); Hrdy, S. B. (2009), *Mothers and Others: The Evolutionary Origins of Mutual Understanding* (Cambridge, Mass.: Harvard University Press); van Schaik, C. P. (2016), *The Primate Origins of Human Nature* (Hoboken, N.J.: Wiley and Sons).
7. Allen, M. W., and Jones, T. L. (2014), *Violence and Warfare Among Hunter-Gatherers* (London: Taylor and Francis).
8. Daly, M., and Wilson, M. (1988), *Homicide* (New Brunswick, N.J.: Transaction); Wrangham and Peterson (1996), *Demonic Males*.
9. Pinker, S. (2011), *The Better Angels of Our Nature: Why Violence Has Declined* (New York: Penguin).
10. Wrangham, R. W. (2019), *The Goodness Paradox: The Strange Relationship Between Goodness and Violence in Human Evolution* (New York: Pantheon).
11. Morganteen, J. (2009), Victim's face mauled in Stamford chimpanzee attack, *Stamford Advocate*, Feb. 18, 2009; Newman, A., and O'Connor, A. (2009), Woman mauled by chimp is still in critical condition, *New York Times*, Feb. 18, 2009.
12. Churchill, S. E., et al. (2009), Shanidar 3 Neandertal rib puncture wound and paleolithic weaponry, *Journal of Human Evolution* 57:163–78; Murphy, W. A., Jr., et al. (2003), The iceman: Discovery and imaging, *Radiology* 226:614–29.
13. Lahr, M. M., et al. (2016), Inter-group violence among early Holocene hunter-gatherers of West Turkana, Kenya, *Nature* 529:394–98.
14. There is some debate over this site. See Stojanowski, C. M., et al. (2016), Contesting the massacre at Nataruk, *Nature* 539:E8–E11, and the reply by Lahr and colleagues.
15. Thomas, E. M. (1986), *The Harmless People*, 2nd ed. (New York: Vintage).
16. Land degradation, forced resettlement, and alcohol have led to widespread social problems among some hunter-gatherer societies, but for reviews of evidence for violence prior to these problems, see Lee, R. B. (1979), *The !Kung San: Men, Women, and Work in a Foraging Society* (Cambridge, U.K.: Cambridge University Press); Keeley, L. H. (1996), *War Before Civilization* (New York: Oxford University Press); Wrangham and Peterson (1996), *Demonic Males;* Boehm, C. (1999), *Hierarchy in the Forest* (Cambridge, Mass.: Harvard University Press); Gighlieri, M. (1999), *The Dark Side of Man: Tracing the Ori-*

gins of Male Violence (Reading, Mass.: Perseus Books); Allen and Jones (2014), *Violence and Warfare Among Hunter-Gatherers.*

17. Darwin, C. R. (1871), *The Descent of Man and Selection in Relation to Sex* (London: J. Murray).
18. Dart, R. A. (1953), The predatory transition from ape to man, *International Anthropological and Linguistic Review* 1:201–17.
19. Ardrey, R. (1961), *African Genesis: A Personal Investigation into the Animal Origins and Nature of Man* (New York: Atheneum Press).
20. Vrba, E. (1975), Some evidence of the chronology and palaeoecology of Sterkfontein, Swartkrans, and Kromdraai from the fossil Bovidae, *Nature* 254:301–4; Brain, C. K. (1981), *The Hunters of the Hunted: An Introduction to African Cave Taphonomy* (Chicago: University of Chicago Press).
21. Lovejoy, C. O. (1981), The origin of man, *Science* 211:341–50.
22. Lovejoy, C. O. (2009), Reexamining human origins in light of *Ardipithecus ramidus, Science* 326:74e1–74e8.
23. Grabowski, M., et al. (2015), Body mass estimates of hominin fossils and the evolution of human body size, *Journal of Human Evolution* 85:75–93.
24. Smith, R. J., and Jungers, W. L. (1997), Body mass in comparative primatology, *Journal of Human Evolution* 32:523–59.
25. Having big, weapon-like canines can also help males fight, but body size dimorphism is most predictive of male-male competition. See Plavcan, J. M. (2012), Sexual size dimorphism, canine dimorphism, and male-male competition in primates: Where do humans fit in?, *Human Nature* 23:45–67; Plavcan, J. M. (2000), Inferring social behavior from sexual dimorphism in the fossil record, *Journal of Human Evolution* 39:327–44.
26. Grabowski et al. (2015), Body mass estimates of hominin fossils and the evolution of human body size.
27. Keeley (1996), *War Before Civilization.*
28. Kaplan, H., et al. (2000), A theory of human life history evolution: Diet, intelligence, and longevity, *Evolutionary Anthropology* 9:156–85.
29. Isaac, G. L. (1978), The food-sharing behavior of protohuman hominids, *Scientific American* 238:90–108; Tanner, N. M., and Zilhman, A. (1976), Women in evolution: Innovation and selection in human origins, *Signs* 1:585–608.
30. Illner, K., et al. (2000), Metabolically active components of fat free mass and resting energy expenditure in nonobese adults, *American Journal of Physiology* 278:E308–E315; Lassek, W. D., and Gaulin, S. J. C. (2009), Costs and benefits of fat-free muscle mass in men: Relationship to mating success, dietary requirements, and native immunity, *Evolution and Human Behavior* 30:322–28.
31. Malina, R. M., and Bouchard, C. (1991), *Growth, Maturation, and Physical Activity* (Champaign, Ill.: Human Kinetics); Bribiescas, R. G. (2006), *Men: Evolutionary and Life History* (Cambridge, Mass.: Harvard University Press).
32. Watson, D. M. (1998), Kangaroos at play: Play behaviour in the Macropodoidea, in *Animal Play: Evolutionary, Comparative, and Ecological Perspectives*, ed. M. Beckoff and J. A. Byers (Cambridge, U.K.: Cambridge University Press), 61–95.
33. Cieri, R. L., et al. (2014), Craniofacial feminization, social tolerance, and the origins of behavioral modernity, *Current Anthropology* 55:419–33.

34. Barrett, R. L., and Harris, E. F. (1993), Anabolic steroids and cranio-facial growth in the rat, *Angle Orthodontist* 63:289–98; Verdonck, A., et al. (1999), Effect of low-dose testosterone treatment on craniofacial growth in boys with delayed puberty, *European Journal of Orthodontics* 21:137–43; Penton-Voak, I. S., and Chen, J. Y. (2004), High salivary testosterone is linked to masculine male facial appearance in humans, *Evolution and Human Behavior* 25:229–41; Schaefer, K., et al. (2005), Visualizing facial shape regression upon 2nd to 4th digit ratio and testosterone, *Collegium Anthropologicum* 29:415–19. Large browridges can also develop from excess quantities of growth hormone, which is produced by the pituitary gland, but we can rule out high GH levels for early human browridge enlargement, because this hormone acts throughout the body, causing gigantism, which also leads to high stature and especially large hands and feet.
35. Shuey, D. L., Sadler, T. W., and Lauder, J. M. (1992), Serotonin as a regulator of craniofacial morphogenesis: Site specific malformations following exposure to serotonin uptake inhibitors, *Teratology* 46:367–78; Pirinen, S. (1995), Endocrine regulation of craniofacial growth, *Acta Odontologica Scandinavica* 53:179–85; Byrd, K. E., and Sheskin, T. A. (2001), Effects of post-natal serotonin levels on craniofacial complex, *Journal of Dental Research* 80:1730–35.
36. Other hormonal changes may also include lower levels of cortisol and increased levels of the neurotransmitter serotonin. See Dugatkin, L., and Trut, L. (2017), *How to Tame a Fox (and Build a Dog)* (Chicago: University of Chicago Press).
37. Wilson, M. L., et al. (2014), Lethal aggression in *Pan* is better explained by adaptive strategies than human impacts, *Nature* 513:414–17.
38. Hare, B., Wobber, V., and Wrangham, R. W. (2012), The self-domestication hypothesis: Evolution of bonobo psychology is due to selection against aggression, *Animal Behaviour* 83:573–85. See also Sannen, A., et al. (2003), Urinary testosterone metabolite levels in bonobos: A comparison with chimpanzees in relation to social system, *Behaviour* 140:683–96; McIntyre, M. H., et al. (2009), Bonobos have a more human-like second-to-fourth finger length ratio (2D: 4D) than chimpanzees: A hypothesized indication of lower prenatal androgens, *Journal of Human Evolution* 56:361–65; Wobber, V., et al. (2010), Differential changes in steroid hormones before competition in bonobos and chimpanzees, *Proceedings of the National Academy of Sciences USA* 107:12457–62; Wobber, V., et al. (2013), Different ontogenetic patterns of testosterone production reflect divergent male reproductive strategies in chimpanzees and bonobos, *Physiology and Behavior* 116–17:44–53; Stimpson, C. D., et al. (2016), Differential serotonergic innervation of the amygdala in bonobos and chimpanzees, *Social Cognitive and Affective Neuroscience* 11:413–22.
39. Lieberman, D. E., et al. (2007), A geometric morphometric analysis of heterochrony in the cranium of chimpanzees and bonobos, *Journal of Human Evolution* 52:647–62; Shea, B. T. (1983), Paedomorphosis and neoteny in the pygmy chimpanzee, *Science* 222:521–22.
40. For a review, see Wrangham (2019), *Goodness Paradox.*
41. Lee (1979), *The !Kung San.*
42. Among other things, this explains why U.S. senators are not allowed to

carry firearms when they deliberate. In 1902, when Senator John McLaurin denounced his fellow senator from South Carolina, Ben Tillman, for "a willful, malicious, and deliberate lie," Tillman promptly punched McLaurin in the jaw, sparking a tumultuous fistfight among dozens of the senators present. Imagine what might have happened if Tillman had been toting a gun.

43. Hoplologists claim their field was inspired by the English adventurer Richard Burton, who published in 1884 his classic monograph, *The Book of the Sword* (London: Chatto & Windus). Ethnographies that include descriptions of fights are too numerous to list but include Chagnon, N. A. (2013), *Noble Savages: My Life Among Two Dangerous Tribes—the Yanomamo and the Anthropologists* (New York: Simon & Schuster); Daly, M., and Wilson, M. (1999), An evolutionary psychological perspective on homicide, in *Homicide: A Sourcebook of Social Research,* ed. M. D. Smith and M. A. Zahn (Thousand Oaks, Calif.: Sage), 58–71.
44. Shepherd, J. P., et al. (1990), Pattern, severity, and aetiology of injuries in victims of assault, *Journal of the Royal Society of Medicine* 83:75–78; Brink, O., Vesterby, A., and Jensen, J. (1998), Pattern of injuries due to interpersonal violence, *Injury* 29:705–9.
45. These hypotheses have been hotly disputed as "just so" stories because human faces and hands must also have been selected for other functions like using tools, speech, and chewing. See Morgan, M. H., and Carrier, D. R. (2013), Protective buttressing of the human fist and the evolution of hominin hands, *Journal of Experimental Biology* 216:236–44; Carrier, D. R., and Morgan, M. H. (2015), Protective buttressing of the hominin face, *Biological Reviews* 90:330–46; King, R. (2013), Fists of fury: At what point did human fists part company with the rest of the hominid lineage?, *Journal of Experimental Biology* 216:2361–62; Nickle, D. C., and Goncharoff, L. M. (2013), Human fist evolution: A critique, *Journal of Experimental Biology* 216:2359–60.
46. Briffa, M., et al. (2013), Analysis of contest data, in *Animal Contests,* ed. I. C. W. Hardy and M. Briffa (Cambridge, U.K.: Cambridge University Press), 47–85; Kanehisa, H., et al. (1998), Body composition and isokinetic strength of professional sumo wrestlers, *European Journal of Applied Physiology and Occupational Physiology* 77:352–59; García-Pallarés, J., et al. (2011), Stronger wrestlers more likely to win: Physical fitness factors to predict male Olympic wrestling performance, *European Journal of Applied Physiology* 111:1747–58.
47. Briffa, M., and Lane, S. M. (2017), The role of skill in animal contests: A neglected component of fighting ability, *Proceedings of the Royal Society B* 284:20171596.
48. For a thorough review, see Green, T. A. (2001), *Martial Arts of the World* (Santa Barbara, Calif.: ABC-CLIO).
49. Biologists quantify these characteristics as resource holding potential. See Parker, G. A. (1974), Assessment strategy and the evolution of animal conflicts, *Journal of Theoretical Biology* 47:223–43.
50. Sell, A., et al. (2009), Human adaptations for the visual assessment of strength and fighting ability from the body and face, *Proceedings of the Royal Society B* 276:575–84; Kasumovic, M. M., Blake, K., and Denson, T. F. (2017), Using knowledge from human research to improve understanding of contest theory and contest dynamics, *Proceedings of the Royal Society B* 284:2182.

51. Apparently, the actors were supposed to spend several days shooting what would be a five-minute duel featuring Indiana Jones's whip versus the assassin's sword. However, Ford was suffering from dysentery and asked the director, Steven Spielberg, if he couldn't just shoot the guy.
52. Pruetz, J. D. (2015), New evidence on the tool-assisted hunting exhibited by chimpanzees (*Pan troglodytes verus*) in a savannah habitat at Fongoli, Senegal, *Royal Society Open Science* 2:140507.
53. Harmand, S., et al. (2015), 3.3-million-year-old stone tools from Lomekwi 3, West Turkana, Kenya, *Nature* 521:310–15; McPherron, S., et al. (2010), Evidence for stone-tool-assisted consumption of animal tissues before 3.39 million years ago at Dikika, Ethiopia. *Nature* 466:857–60; Semaw, S., et al. (1997), 2.5-million-year-old stone tools from Gona, Ethiopia, *Nature* 385:333–36; Toth, N., Schick, K., and Semaw, S. (2009), The Oldowan: The tool making of early hominins and chimpanzees compared, *Annual Review of Anthropology* 38:289–305.
54. Shea, J. J. (2016), *Tools in Human Evolution: Behavioral Differences Among Technological Primates* (Cambridge, U.K.: Cambridge University Press); Zink, K. D., and Lieberman, D. E. (2016), Impact of meat and Lower Palaeolithic food processing techniques on chewing in humans, *Nature* 531:500–503.
55. Keeley, L. H., and Toth, N. (1981), Microwear polishes on early stone tools from Koobi Fora, Kenya, *Nature* 293:464–65.
56. Wilkins, J., et al. (2012), Evidence for early hafted hunting technology, *Science* 338:942–46; Wilkins, J., Schoville, B. J., and Brown, K. S. (2014), An experimental investigation of the functional hypothesis and evolutionary advantage of stone-tipped spears, *PLOS ONE* 9:e104514.
57. Churchill, S. E. (2014), *Thin on the Ground: Neanderthal Biology, Archeology, and Ecology* (Ames, Iowa: John Wiley & Sons); Gaudzinski-Windheuser, S., et al. (2018), Evidence for close-range hunting by last interglacial Neanderthals, *Nature Ecology and Evolution* 2:1087–92. See also Berger, T. D., and Trinkaus, E. (1995), Patterns of trauma among the Neandertals, *Journal of Archaeological Science* 22:841–52.
58. Goodall, J. (1986), *The Chimpanzees of Gombe: Patterns of Behavior* (Cambridge, Mass.: Harvard University Press); Westergaard, G. C., et al. (2000), A comparative study of aimed throwing by monkeys and humans, *Neuropsychologia* 38:1511–17.
59. Fleisig, G. S., et al. (1995), Kinetics of baseball pitching with implications about injury mechanisms, *American Journal of Sports Medicine* 23:233–39; Hirashima, M., et al. (2002), Sequential muscle activity and its functional role in the upper extremity and trunk during overarm throwing, *Journal of Sports Science* 20:301–10.
60. Roach, N. T., and Lieberman, D. E. (2014), Upper body contributions to power generation during rapid, overhand throwing in humans, *Journal of Experimental Biology* 217:2139–49.
61. Pappas, A. M., Zawacki, R. M., and Sullivan, T. J. (1985), Biomechanics of baseball pitching: A preliminary report, *American Journal of Sports Medicine* 13:216–22.
62. Roach, N. T., et al. (2013), Elastic energy storage in the shoulder and the evolution of high-speed throwing in *Homo*, *Nature* 498:483–86.
63. Brown, K. S., et al. (2012), An early and enduring advanced technology origi-

nating 71,000 years ago in South Africa, *Nature* 491:590–93; Shea (2016), *Tools in Human Evolution.*

64. Henrich, J. (2017), *The Secret of Our Success: How Culture Is Driving Human Evolution, Domesticating Our Species, and Making Us Smarter* (Princeton, N.J.: Princeton University Press).
65. Wrangham, R. W. (2009), *Catching Fire: How Cooking Made Us Human* (New York: Basic Books).
66. Fuller, N. J., Laskey, M. A., and Elia, M. (1992), Assessment of the composition of major body regions by dual-energy X-ray absorptiometry (DEXA), with special reference to limb muscle mass, *Clinical Physiology* 12:253–66; Gallagher, D., et al. (1997), Appendicular skeletal muscle mass: Effects of age, gender, and ethnicity, *Journal of Applied Physiology* 83:229–39; Abe, T., Kearns, C. F., and Fukunaga, T. (2003), Sex differences in whole body skeletal muscle mass measured by magnetic resonance imaging and its distribution in young Japanese adults, *British Journal of Sports Medicine* 37:436–40.
67. Boehm, C. H. (1999), *Hierarchy in the Forest: The Evolution of Egalitarian Behavior* (Cambridge, Mass.: Harvard University Press).
68. Fagen, R. M. (1981), *Animal Play Behavior* (New York: Oxford University Press); Palagi, E., et al. (2004), Immediate and delayed benefits of play behaviour: New evidence from chimpanzees (*Pan troglodytes*), *Ethology* 110:949–62; Nunes, S., et al. (2004), Functions and consequences of play behaviour in juvenile Belding's ground squirrels, *Animal Behaviour* 68:27–37.
69. Fagen (1981), *Animal Play Behavior;* Pellis, S. M., Pellis, V. C., and Bell, H. C. (2010), The function of play in the development of the social brain, *American Journal of Play* 2:278–96.
70. Poliakoff, M. B. (1987), *Combat Sports in the Ancient World: Competition, Violence, and Culture* (New Haven, Conn.: Yale University Press); McComb, D. G. (2004), *Sport in World History* (New York: Taylor and Francis).
71. Homer, *The Iliad,* trans. Robert Fagles (1990) (New York: Penguin), bk. 23, lines 818–19.
72. For data on the effects of hunting and fighting on reproductive success in small-scale societies, see Marlowe, F. W. (2001), Male contribution to diet and female reproductive success among foragers, *Current Anthropology* 42:755–59; Smith, E. A. (2004), Why do good hunters have higher reproductive success?, *Human Nature* 15:343–64; Gurven, M., and von Rueden, C. (2006), Hunting, social status, and biological fitness, *Biodemography and Social Biology* 53:81–99; Apicella, C. L. (2014), Upper-body strength predicts hunting reputation and reproductive success in Hadza hunter-gatherers, *Evolution and Human Behavior* 35:508–18; Glowacki, L., and Wrangham, R. W. (2015), Warfare and reproductive success in a tribal population, *Proceedings of the National Academy of Sciences USA* 112:348–53.

    For studies of the relationship between athletics and reproductive success, see De Block, A., and Dewitte, S. (2009), Darwinism and the cultural evolution of sports, *Perspectives in Biology and Medicine* 52:1–16; Puts, D. A. (2010), Beauty and the beast: Mechanisms of sexual selection in humans, *Evolution and Human Behavior* 31:157–75; Lombardo, M. P. (2012), On the evolution of sport, *Evolutionary Psychology* 10:1–28.
73. Rousseau, J.-J., *Émile,* trans. Allan Bloom (New York: Basic Books), 119.
74. Harvard Athletics Mission Statement, www.gocrimson.com.

EIGHT WALKING: ALL IN A DAY'S WALK

1. Spottiswoode, C. N., Begg, K. S., and Begg, C. M. (2016), Reciprocal signaling in honeyguide-human mutualism, *Science* 353:387–89.
2. To be precise, the average daily travel distances are 14.1 kilometers for male hunter-gatherers and 9.5 kilometers for female hunter-gatherers, but these distances do not include the thousands more steps taken in and around camp. Marlowe, F. W. (2005), Hunter-gatherers and human evolution, *Evolutionary Anthropology* 14:54–67.
3. Althoff, T., et al. (2017), Large-scale physical activity data reveal worldwide activity inequality, *Nature* 547:336–39.
4. Palinski-Wade, E. (2015), *Walking the Weight Off for Dummies* (Hoboken, N.J.: John Wiley and Sons).
5. Tudor-Locke, C., and Bassett, D. R., Jr. (2004), How many steps/day are enough? Preliminary pedometer indices for public health, *Sports Medicine* 34:1–8. Quotation from page 1.
6. Cloud, J. (2009), The myth about exercise, *Time,* Aug. 9, 2009.
7. In terms of absolute time, humans start walking later than most animals, but the delay has nothing to do with the fact that we walk on two versus four legs. With the exception of highly vulnerable quadrupeds like zebras, which risk being eaten by hyenas and lions and can walk within a day of birth, the age at which most animals start walking is primarily determined by how far along the brain has finished developing. Primates like macaques with smaller brains start walking at two months, bigger-brained chimpanzee infants require about six months to start walking, and humans start walking just at the age, one year, predicted by our brain development. See Garwicz, M., Christensson, M., and Psouni, E. (2011), A unifying model for timing of walking onset in humans and other mammals, *Proceedings of the National Academy of Sciences USA* 106:21889–93.
8. Donelan, J. M., Kram, R., and Kuo, A. D. (2002), Mechanical work for step-to-step transitions is a major determinant of the metabolic cost of human walking, *Journal of Experimental Biology* 205:3717–27; Marsh, R. L., et al. (2004), Partitioning the energetics of walking and running: Swinging the limbs is expensive, *Science* 303:80–83.
9. Biewener, A. A., and Patek, S. N. (2018), *Animal Locomotion,* 2nd ed. (Oxford: Oxford University Press).
10. Thompson, N. E., et al. (2015), Surprising trunk rotational capabilities in chimpanzees and implications for bipedal walking proficiency in early hominins, *Nature Communications* 6:8416.
11. Holowka, N. B., et al. (2019), Foot callus thickness does not trade off protection for tactile sensitivity during walking, *Nature* 571:261–64.
12. Tan, U. (2005), Unertan syndrome quadrupedality, primitive language, and severe mental retardation: A new theory on the evolution of human mind, *NeuroQuantology* 4:250–55; Ozcelik, T., et al. (2008), Mutations in the very low-density lipoprotein receptor VLDLR cause cerebellar hypoplasia and quadrupedal locomotion in humans, *Proceedings of the National Academy of Sciences USA* 105:4232–36.
13. Türkmen, S., et al. (2009), CA8 mutations cause a novel syndrome characterized by ataxia and mild mental retardation with predisposition to quadru-

pedal gait, *PLOS Genetics* 5:e1000487; Shapiro, L. J., et al. (2014), Human quadrupeds, primate quadrupedalism, and Uner Tan syndrome, *PLOS ONE* 9:e101758.

14. There are two reasons to be confident that the missing link resembled a knuckle-walking chimpanzee that also sometimes climbed trees. The first is that humans and chimpanzees are more closely related to each other than either are to gorillas, but chimpanzees and gorillas are extremely similar in many respects (especially after controlling for size differences) including their peculiar knuckle-walking form of locomotion. Unless all the many similarities, including knuckle walking, between chimpanzees and gorillas evolved independently (a statistical impossibility), their last common ancestor must have been very much like a chimpanzee (or a gorilla, which is less likely because big apes are rare). The second reason is that early hominin fossils, apart from being bipeds, most resemble chimpanzees in countless features from head to toe. For a detailed review, see Pilbeam, D. R., and Lieberman, D. E. (2017), Reconstructing the last common ancestor of humans and chimpanzees, in *Chimpanzees and Human Evolution*, ed. M. N. Muller, R. W. Wrangham, and D. R. Pilbeam (Cambridge, Mass.: Harvard University Press), 22–141.
15. Taylor, C. R., and Rowntree, V. J. (1973), Running on two or four legs: Which consumes more energy?, *Science* 179:186–87.
16. Sockol, M. D., Pontzer, H., and Raichlen, D. A. (2007), Chimpanzee locomotor energetics and the origin of human bipedalism, *Proceedings of the National Academy of Sciences USA* 104:12265–69.
17. For chimpanzees, see Pontzer, H., Raichlen, D. A., and Sockol, M. D. (2009), The metabolic cost of walking in humans, chimpanzees, and early hominins, *Journal of Human Evolution* 56:43–54. For data on other animals, see Rubenson, J., et al. (2007), Reappraisal of the comparative cost of human locomotion using gait-specific allometric analyses, *Journal of Experimental Biology* 210:3513–24.
18. Apes also have to spend extra energy to stabilize their forward-leaning trunks and their permanently hunched shoulders. The shoulders of typical quadrupeds like dogs are positioned along the side of the body, but ape shoulders are located high on their backs to help them climb but demand yet more muscular effort to keep stable. See Larson, S. G., and Stern, J. T., Jr. (1987), EMG of chimpanzee shoulder muscles during knuckle-walking: Problems of terrestrial locomotion in a suspensory adapted primate, *Journal of Zoology* 212:629–55.
19. My scenario is, of course, a hypothesis, but it is supported by evidence from the three oldest genera of hominins: *Sahelanthropus* (7 million years old), *Orrorin* (6 million years old), and *Ardipithecus* (5.8 to 4.3 million years old). All of these hominins are bipeds but otherwise look very much like apes, especially chimpanzees. Keep in mind, also, that there might have been other benefits to upright posture such as feeding, carrying, and possibly even fighting, but none make as much sense to me as saving energy. After all, chimps and gorillas can stand upright to carry things, but they spend three times as many calories as we do to do so. For more details, see Pilbeam and Lieberman (2017), Reconstructing the last common ancestor of humans and chimpanzees.

20. In case you are interested in the math, I weigh sixty-eight kilograms, and the average cost of human walking is 0.08mlO_2/kg/m; 1 liter of O_2 yields 5 kcals; chimpanzee walking is 2.15 times more costly per kilogram; running a marathon costs approximately 2,600 kcals.
21. Huang, T. W., and Kuo, A. D. (2014), Mechanics and energetics of load carriage during human walking, *Journal of Experimental Biology* 217:605–13.
22. Maloiy, G. M., et al. (1986), Energetic cost of carrying loads: Have African women discovered an economic way?, *Nature* 319:668–69; Heglund, N. C., et al. (1995), Energy-saving gait mechanics with head-supported loads, *Nature* 375:52–54; Lloyd, R., et al. (2010), Comparison of the physiological consequences of head-loading and back-loading for African and European women, *European Journal of Applied Physiology* 109:607–16.
23. Bastien, G. J., et al. (2005), Energetics of load carrying in Nepalese porters, *Science* 308:1755; Minetti, A., Formenti, F., and Ardigò, L. (2006), Himalayan porter's specialization: Metabolic power, economy, efficiency, and skill, *Proceedings of the Royal Society B* 273:2791–97.
24. Castillo, E. R., et al. (2014), Effects of pole compliance and step frequency on the biomechanics and economy of pole carrying during human walking, *Journal of Applied Physiology* 117:507–17.
25. Knapik, J., Harman, E., and Reynolds, K. (1996), Load carriage using packs: A review of physiological, biomechanical, and medical aspects, *Applied Ergonomics* 27:207–16; Stuempfle, K. J., Drury, D. G., and Wilson, A. L. (2004), Effect of load position on physiological and perceptual responses during load carriage with an internal frame backpack, *Ergonomics* 47:784–89; Abe, D., Muraki, S., and Yasukouchi, A. (2008), Ergonomic effects of load carriage on the upper and lower back on metabolic energy cost of walking, *Applied Ergonomics* 39:392–98.
26. Petersen, A. M., Leet, T. L., and Brownson, R. C. (2005), Correlates of physical activity among pregnant women in the United States, *Medicine and Science in Sports and Exercise* 37:1748–53.
27. Shostak, M. (1981), *Nisa: The Life and Words of a !Kung Woman* (New York: Vintage). Quotation is from page 178.
28. Whitcome, K. K., Shapiro, L. J., and Lieberman, D. E. (2007), Fetal load and the evolution of lumbar lordosis in bipedal hominins, *Nature* 450:1075–78.
29. Wall-Scheffler, C. M., Geiger, K., and Steudel-Numbers, K. L. (2007), Infant carrying: The role of increased locomotor costs in early tool development, *American Journal of Physical Anthropology* 133:841–46; Watson, J. C., et al. (2008), The energetic costs of load-carrying and the evolution of bipedalism, *Journal of Human Evolution* 54:675–83; Junqueira, L. D., et al. (2015), Effects of transporting an infant on the posture of women during walking and standing still, *Gait and Posture* 41:841–46.
30. Hall, C., et al. (2004), Energy expenditure of walking and running: Comparison with prediction equations, *Medicine and Science in Sports and Exercise* 36:2128–34.
31. Wishnofsky, M. (1958), Caloric equivalents of gained or lost weight, *American Journal of Clinical Nutrition* 6:542–46. For a much better analysis, see Hall, K. D., et al. (2011), Quantification of the effect of energy imbalance on bodyweight, *Lancet* 378:826–37.
32. The notion behind fat-burning zones is that the harder you work, the more

energy you burn, but a greater proportion of that energy is carbohydrate (glycogen). When sitting quietly, you burn only fat but not much. When you walk, jog, run, and then sprint, you burn more energy, but with greater intensity a higher percentage is glycogen, so at maximum capacity you burn only glycogen. The fat-burning zone is thus a low-to-moderate level of activity that burns about as much fat as glycogen, but it's sort of a fiction because there is a wide range of variation between individuals of moderate activity levels in which you may burn a lower percentage of but maximum amount of fat. How much time you exercise is also important. See Carey, D. G. (2009), Quantifying differences in the "fat burning" zone and the aerobic zone: Implications for training, *Journal of Strength and Conditioning Research* 23:2090–95.

33. Technically, the dose was 0, 4, 8, and 12 kcal/kg week. Swift, D. L., et al. (2014), The role of exercise and physical activity in weight loss and maintenance, *Progress in Cardiovascular Disease* 56:441–47; Ross, R., and Janssen, I. (2001), Physical activity, total and regional obesity: Dose-response considerations, *Medicine and Science in Sports and Exercise* 33:S521–S527; Morss, G. M., et al. (2004), Dose Response to Exercise in Women aged 45–75 yr (DREW): Design and rationale, *Medicine and Science in Sports and Exercise* 36:336–44.
34. Kraus, W. E., et al. (2002), Effects of the amount and intensity of exercise on plasma lipoproteins, *New England Journal of Medicine* 347:1483–92; Sigal, R. J., et al. (2007), Effects of aerobic training, resistance training, or both on glycemic control in type 2 diabetes, *Annals of Internal Medicine* 147:357–69; Church, T. S., et al. (2010), Exercise without weight loss does not reduce C-reactive protein: The INFLAME study, *Medicine and Science in Sports and Exercise* 42:708–16.
35. Ellison, P. T. (2001), *On Fertile Ground: A Natural History of Human Reproduction* (Cambridge, Mass.: Harvard University Press).
36. Church, T. S., et al. (2009), Changes in weight, waist circumference, and compensatory responses with different doses of exercise among sedentary, overweight postmenopausal women, *PLOS ONE* 4:e4515.
37. Thomas, D. M., et al. (2012), Why do individuals not lose more weight from an exercise intervention at a defined dose? An energy balance analysis, *Obesity Review* 13:835–47. See also Gomersall, S. R., et al. (2013), The ActivityStat hypothesis: The concept, the evidence, and the methodologies, *Sports Medicine* 43:135–49; Willis, E. A., et al. (2014), Nonexercise energy expenditure and physical activity in the Midwest Exercise Trial 2, *Medicine and Science in Sports and Exercise* 46:2286–94; Gomersall, S. R., et al. (2016), Testing the activitystat hypothesis: A randomised controlled trial, *BMC Public Health* 16:900; Liguori, G., et al. (2017), Impact of prescribed exercise on physical activity compensation in young adults, *Journal of Strength and Conditioning Research* 31:503–8.
38. For a review, see Thomas et al. (2012), Why do individuals not lose more weight from an exercise intervention at a defined dose? For the half-marathon study, see Westerterp, K. R., et al. (1992), Long-term effect of physical activity on energy balance and body composition, *British Journal of Nutrition* 68:21–30.
39. Foster, G. D., et al. (1997), What is a reasonable weight loss? Patients' expectations and evaluations of obesity treatment outcomes, *Journal of Consulting*

and Clinical Psychology 65:79–85; Linde, J. A., et al. (2012), Are unrealistic weight loss goals associated with outcomes for overweight women?, *Obesity* 12:569–76.

40. Raichlen, D. A., et al. (2017), Physical activity patterns and biomarkers of cardiovascular disease risk in hunter-gatherers, *American Journal of Human Biology* 29:e22919.
41. Average weekly time spent watching live TV in the United States from 4th quarter 2013 to 2nd quarter 2018, Statista, www.statista.com.
42. Flack, K. D., et al. (2018), Energy compensation in response to aerobic exercise training in overweight adults, *American Journal of Physiology: Regulatory, Integrative, and Comparative Physiology* 315:R619–R626.
43. Ross, R., et al. (2000), Reduction in obesity and related comorbid conditions after diet-induced weight loss or exercise-induced weight loss in men, *Annals of Internal Medicine* 133:92–103.
44. Pontzer, H., et al. (2012), Hunter-gatherer energetics and human obesity, *PLOS ONE* 7:e40503.
45. Pontzer, H., et al. (2016), Constrained total energy expenditure and metabolic adaptation to physical activity in adult humans, *Current Biology* 26:410–17.
46. A similar phenomenon may occur in severely obese individuals whose basal metabolic rate falls after they lose massive amounts of weight through diet and exercise. Note, however, that such compensatory downshifts are not always observed in less extreme dieters. Compare for example the following two studies: Ross et al. (2000), Reduction in obesity and related comorbid conditions after diet-induced weight loss or exercise-induced weight loss in men; and Johannsen, D. L., et al. (2012), Metabolic slowing with massive weight loss despite preservation of fat-free mass, *Journal of Clinical Endocrinology and Metabolism* 97:2489–96.
47. This estimate comes from Physical Activity Levels (PALs), total energy expenditure divided by basal metabolic rate. According to published data, PALs are 2.03 and 1.78 for male and female Hadza, respectively, and 1.48 and 1.66 for male and female Westerners, respectively. Data from Pontzer et al. (2012), Hunter-gatherer energetics and human obesity; Pontzer, H., et al. (2016), Metabolic acceleration and the evolution of human brain size and life history, *Nature* 533:390–92.
48. This elevated metabolic rate is termed the excess post-exercise oxygen consumption but typically requires sustained moderate- to high-intensity activity. See Speakman, J. R., and Selman, C. (2003), Physical activity and metabolic rate, *Proceedings of the Nutrition Society* 62:621–34; LaForgia, J., Withers, R. T., and Gore, C. J. (2006), Effects of exercise intensity and duration on the excess post-exercise oxygen consumption, *Journal of Sports Science* 24:1247–64.
49. Donnelly, J. E., et al. (2009), Appropriate physical activity intervention strategies for weight loss and prevention of weight regain for adults, *Medicine and Science in Sports and Exercise* 41:459–71.
50. Pavlou, K. N., Krey, Z., and Steffee, W. P. (1989), Exercise as an adjunct to weight loss and maintenance in moderately obese subjects, *American Journal of Clinical Nutrition* 49:1115–23.
51. Andersen, R. E., et al. (1999), Effects of lifestyle activity vs structured aerobic exercise in obese women: A randomized trial, *Journal of the American*

Medical Association 281:335–40; Jakicic, J. M., et al. (1999), Effects of intermittent exercise and use of home exercise equipment on adherence, weight loss, and fitness in overweight women: A randomized trial, *Journal of the American Medical Association* 282:1554–60; Jakicic, J. M., et al. (2003), Effect of exercise duration and intensity on weight loss in overweight, sedentary women: A randomized trial, *Journal of the American Medical Association* 290:1323–30.

52. Tudor-Locke, C. (2003), *Manpo-Kei: The Art and Science of Step Counting: How to Be Naturally Active and Lose Weight* (Vancouver, B.C.: Trafford).
53. Butte, N. F., and King, J. C. (2005), Energy requirements during pregnancy and lactation, *Public Health and Nutrition* 8:1010–27.
54. A widespread notion, the aquatic ape hypothesis (AAH) claims that humans evolved to swim. AAH supporters point to features like hairlessness, downward-facing nostrils, and subcutaneous fat as evidence that our ancestors were selected to wade, dive, and swim, but all of these features have been better explained as adaptations for other functions. For a review of the theory and its critiques, see Gee, H. (2015), *The Accidental Species: Misunderstandings of Human Evolution* (Chicago: University of Chicago Press). To these critiques I would add that considering humans good at swimming is risible: even the best human swimmers are slow, inefficient, and poorly able to maneuver compared with mammals that clearly were adapted for swimming like otters, seals, and beavers. You can walk faster than the world's fastest swimmers and spend about one-fifth the amount of energy. Finally, if there is any place in the world I would most avoid swimming, it would be the crocodile-infested lakes and rivers of Africa. See Di Prampero, P. E. (1986), The energy cost of human locomotion on land and in water, *International Journal of Sports Medicine* 7:55–72.

NINE RUNNING AND DANCING: JUMPING FROM ONE LEG TO THE OTHER

1. Carrier, D. R. (1984), The energetic paradox of human running and hominid evolution, *Current Anthropology* 25:483–95.
2. Taylor, C. R., Schmidt-Nielsen, K., and Raab, J. L. (1970), Scaling of energetic cost of running to body size in mammals, *American Journal of Physiology* 219:1104–7.
3. Dennis Bramble was Carrier's undergraduate mentor and then his colleague in Salt Lake City. They had published an important study on how humans breathe while running in 1983. See Bramble, D. M., and Carrier, D. R. (1983), Running and breathing in mammals, *Science* 219:251–56.
4. Bramble, D. M., and Lieberman, D. E. (2004), Endurance running and the evolution of *Homo, Nature* 432:345–52.
5. As horse enthusiasts know, canters are essentially slow gallops (technically, a three-beat gait instead of a galloping four-beat gait, both different from the two-beat gait of a trot).
6. Alexander, R. M., Jayes, A. S., and Ker, R. F. (1980), Estimates of energy cost for quadrupedal running gaits, *Journal of Zoology* 190:155–92; Heglund, N. C., and Taylor, C. R. (1988), Speed, stride frequency, and energy cost per stride: How do they change with body size and gait?, *Journal of Experimental*

Biology 138:301–18; Hoyt, D. F., and Taylor, C. R. (1981), Gait and the energetics of locomotion in horses, *Nature* 292:239–40; Minetti, A. E. (2003), Physiology: Efficiency of equine express postal systems, *Nature* 426:785–86.

7. The maximum recommended speed for long rides is a medium trot, which is about five to six miles per hour for ponies and six to eight miles per hour for horses. For comparison, world-class marathoners who complete in just over two hours are running thirteen miles per hour, and amateurs who run a 3:30, which is considered respectable but not fast, are running seven and a half miles per hour. Loving, N. S. (1997), *Go the Distance: The Complete Resource for Endurance Horses* (North Pomfret, Vt.: Trafalgar Press). See also Tips and hints for endurance riding, Old Dominion Equestrian Endurance Organization, www.olddominionrides.org.
8. Holekamp, K. E., Boydston, E. E., and Smale, E. (2000), Group travel in social carnivores, in *On the Move: How and Why Animals Travel in Groups*, ed. S. Boinski and P. Garber (Chicago: University of Chicago Press), 587–627; Pennycuick, C. J. (1979), Energy costs of locomotion and the concept of "foraging radius," in *Serengeti: Dynamics of an Ecosystem*, ed. A. R. E. Sinclair and M. Norton-Griffiths (Chicago: University of Chicago Press), 164–84.
9. Dill, D. B., Bock, A. V., and Edwards, H. T. (1933), Mechanism for dissipating heat in man and dog, *American Journal of Physiology* 104:36–43.
10. Number of marathon finishers in the United States from 2004 to 2016, Statista, www.statista.com.
11. Rubenson, J., et al. (2007), Reappraisal of the comparative cost of human locomotion using gait-specific allometric analyses, *Journal of Experimental Biology* 210:3513–24.
12. Alexander, R. M., et al. (1979), Allometry of the limb bones of mammals from shrews (*Sorex*) to elephant (*Loxodonta*), *Journal of Zoology* 3:305–14.
13. Ker, R. F., et al. (1987), The spring in the arch of the human foot, *Nature* 325:147–49.
14. Bramble, D. M., and Jenkins, F. A. J., Jr. (1993), Mammalian locomotor-respiratory integration: Implications for diaphragmatic and pulmonary design, *Science* 262:235–40.
15. Kamberov, Y. G., et al. (2018), Comparative evidence for the independent evolution of hair and sweat gland traits in primates, *Journal of Human Evolution* 125:99–105.
16. Lieberman, D. E. (2015), Human locomotion and heat loss: An evolutionary perspective, *Comprehensive Physiology* 5:99–117.
17. Shave, R. E., et al. (2019), Selection of endurance capabilities and the trade-off between pressure and volume in the evolution of the human heart, *Proceedings of the National Academy of Sciences USA* 116:19905–10. See also Hellsten, Y., and Nyberg, M. (2015), Cardiovascular adaptations to exercise training, *Comprehensive Physiology* 6:1–32.
18. Lieberman, D. E. (2011), *The Evolution of the Human Head* (Cambridge, Mass.: Harvard University Press).
19. O'Neill, M. C., et al. (2017), Chimpanzee super strength and human skeletal muscle evolution, *Proceedings of the National Academy of Sciences USA* 114:7343–48.
20. Ama, P. F., et al. (1986), Skeletal muscle characteristics in sedentary black

and Caucasian males, *Journal of Applied Physiology* 61:1758–61; Hamel, P., et al. (1986), Heredity and muscle adaptation to endurance training, *Medicine and Science in Sports and Exercise* 18:690–96.

21. Lieberman, D. E. (2006), The human gluteus maximus and its role in running, *Journal of Experimental Biology* 209:2143–55.
22. During the aerial phase of every stride when running, one's legs scissor in opposite directions, causing the body to yaw like a boat to one side and then the other. By pumping the arms opposite to the legs, we counter some of this angular momentum, and because arms weigh much less than legs, we also rotate our torsos. Arm swinging combined with torso rotation helps us run in a straight line, but running humans need to rotate their torsos independently of both their hips and their heads, something stiff-backed apes cannot accomplish. See Hinrichs, R. N. (1990), Upper extremity function in distance running, in *Biomechanics of Distance Running*, ed. P. R. Cavanagh (Champaign, Ill.: Human Kinetics), 107–33; Thompson, N. E., et al. (2015), Surprising trunk rotational capabilities in chimpanzees and implications for bipedal walking proficiency in early hominins, *Nature Communications* 6:8416.
23. A running dog or horse flexes and extends its nearly horizontal neck to keep its head stable like a missile. These adjustments help its eyes focus on whatever lies ahead, including obstacles or prey. Bipedal humans who bounce up and down like pogo sticks with vertical necks can't prevent their heads from also bouncing up and down. Instead, every time a running human's body lands, its head wants to pitch forward at the same time its arms want to fall downward. Both stay still, however, because just before the feet hit the ground, we fire a pencil-thin muscle in the shoulder that links the arm to the head via a sheet of tissue (the nuchal ligament) that runs along the midline of the back of the skull. By firing this muscle just before each foot strike, the arm and head keep each other still. Compared with apes, humans also have extra-sensitive organs of balance in our inner ears to help sense and cope with the various kinds of jiggling. See Lieberman (2011), *Evolution of the Human Head*.
24. O'Connell, J. F., Hawkes, K., and Blurton-Jones, N. G. (1988), Hadza scavenging: Implications for Plio-Pleistocene hominid subsistence, *Current Anthropology* 29:356–63.
25. Braun, D. R., et al. (2010), Early hominin diet included diverse terrestrial and aquatic animals 1.95 Ma in East Turkana, Kenya, *Proceedings of the National Academy of Sciences* 107:10002–7; Egeland, C. P. M., Domínguez-Rodrigo, M., and Barba, M. (2011), The hunting-versus-scavenging debate, in *Deconstructing Olduvai: A Taphonomic Study of the Bed I Sites*, ed. M. Domínguez-Rodrigo (Dordrecht, Netherlands: Springer), 11–22.
26. Lombard, M. (2005), Evidence of hunting and hafting during the Middle Stone Age at Sibidu Cave, KwaZulu-Natal, South Africa: A multianalytical approach, *Journal of Human Evolution* 48:279–300; Wilkins, J., et al. (2012), Evidence for early hafted hunting technology, *Science* 338:942–46.
27. For Africa, see Schapera, I. (1930), *The Khoisan Peoples of South Africa: Bushmen and Hottentots* (London: Routledge and Kegan Paul); Heinz, H. J., and Lee, M. (1978), *Namkwa: Life Among the Bushmen* (London: Jonathan Cape).

For Asia, see Shah, H. M. (1900), *Aboriginal Tribes of India and Pakistan: The Bhils and Kolhis* (Karachi: Mashoor Offset Press). For Australia, see Sollas, W. J. (1924), *Ancient Hunters, and Their Modern Representatives* (New York: Macmillan); McCarthy, F. D. (1957), *Australian Aborigines: Their Life and Culture* (Melbourne: Colorgravure); Tindale, N. B. (1974), *Aboriginal Tribes of Australia: Their Terrain, Environmental Controls, Distribution, Limits, and Proper Names* (Berkeley: University of California Press); Bliege-Bird, R., and Bird, D. (2008), Why women hunt: Risk and contemporary foraging in a western desert aboriginal community, *Current Anthropology* 49:655–93. For the Americas, see Lowie, R. H. (1924), Notes on Shoshonean ethnography, *Anthropological Papers of the American Museum of Natural History* 20:185–314; Kroeber, A. L. (1925), *Handbook of the Indians of California* (Washington, D.C.: Bureau of American Ethnology); Nabokov, P. (1981), *Indian Running: Native American History and Tradition* (Santa Fe, N.M.: Ancient City Press).

28. Liebenberg, L. (2006), Persistence hunting by modern hunter-gatherers, *Current Anthropology* 47:1017–25.
29. Lieberman, D. E., et al. (2020), Running in Tarahumara (Rarámuri) culture, *Current Anthropology* 6 (forthcoming).
30. Liebenberg, L. (1990), *The Art of Tracking: The Origin of Science* (Cape Town: David Philip).
31. Ijäs, M. (2017), *Fragments of the Hunt: Persistence Hunting, Tracking, and Prehistoric Art* (Helsinki: Aalto University Press).
32. Among the Tarahumara, these hunts usually took between two and six hours over distances of twelve to thirty-six kilometers.
33. Tindale (1974), *Aboriginal Tribes of Australia*, 106.
34. Kraske, R. (2005), *Marooned: The Strange but True Adventures of Alexander Selkirk, the Real Robinson Crusoe* (New York: Clarion Books).
35. Nabokov (1981), *Indian Running;* Lieberman et al. (2020), Running in Tarahumara (Rarámuri) culture.
36. Burfoot, A. (2016), *First Ladies of Running* (New York: Rodale).
37. A telling example of women's running is recounted in the book *Nisa*, about a San woman whose life experiences were transcribed and published by Marjorie Shostak. On pages 101–2, Nisa describes how she ran a kudu down as a youngster:

> Another day, when I was already fairly big, I went with some of my friends and with my younger brother away from the village and into the bush. While we were walking, I saw the tracks of a baby kudu in the sand. I called out, "Hey, Everyone! Come here! Come look at these kudu tracks." The others came over and we all looked at them.
>
> We started to follow the tracks and walked and walked and after a while, we saw a little kudu lying quietly in the grass, dead asleep. I jumped up and tried to grab it. It cried out, "Ehnnn . . . ehnnn . . ." I hadn't really caught it well and it freed itself and ran away. We all ran, chasing after it, and we ran and ran. But I ran so fast that they all dropped behind and I was alone, chasing it, running as fast as I could. Finally, I was able to grab it, I jumped on it and killed it. . . .
>
> I gave the animal to my cousin and he carried it. On the way back, one of the other girls spotted a small steenbok and she and her older brother

ran after it. They chased it and finally her brother killed it. That day we brought a lot of meat back to the village and everyone had plenty to eat.

Then on page 93, Nisa describes how she ran to scavenge:

I remember another time, when I was the first to notice a dead wildebeest, one recently killed by lions, lying in the bush. Mother and I had gone gathering and were walking along, she in one direction and I a short distance away. That's when I saw the wildebeest. . . . She stayed with the animal while I ran back, but we had gone deep into the mongongo groves and soon I got tired. I stopped to rest. Then I got up and started to run again, following along our tracks, ran and then rested and then ran until I finally got back to the village.

It was hot and everyone was resting in the shade. . . . My father and my older brother and everyone in the village followed me [back to the wildebeest]. When we arrived, they skinned the animal, cut the meat into strips and carried it on branches back to the village.

38. Lieberman, D. E., et al. (2010), Foot strike patterns and collision forces in habitually barefoot versus shod runners, *Nature* 463:531–35.
39. van Gent, R. N., et al. (2007), Incidence and determinants of lower extremity running injuries in long distance runners: A systematic review, *British Journal of Sports Medicine* 41:469–80.
40. There are too many studies to cite, but here are a few that summarize the evidence: van Mechelen, W. (1992), Running injuries: A review of the epidemiological literature, *Sports Medicine* 14:320–35; Rauh, M. J., et al. (2006), Epidemiology of musculoskeletal injuries among high school cross-country runners, *American Journal of Epidemiology* 163:151–59; van Gent et al. (2007), Incidence and determinants of lower extremity running injuries in long distance runners; Tenforde, A. S., et al. (2011), Overuse injuries in high school runners: Lifetime prevalence and prevention strategies, *Physical Medicine and Rehabilitation* 3:125–31; Videbæk, S., et al. (2015), Incidence of running-related injuries per 1000 h of running in different types of runners: A systematic review and meta-analysis, *Sports Medicine* 45:1017–26.
41. Kluitenberg, B., et al. (2015), What are the differences in injury proportions between different populations of runners? A systematic review and meta-analysis, *Sports Medicine* 45:1143–61.
42. Daoud, A. I., et al. (2012), Foot strike and injury rates in endurance runners: A retrospective study, *Medicine and Science in Sports and Exercise* 44:1325–34.
43. Buist, I., et al. (2010), Predictors of running-related injuries in novice runners enrolled in a systematic training program: A prospective cohort study, *American Journal of Sports Medicine* 38:273–80.
44. Alentorn-Geli, E., et al. (2017), The association of recreational and competitive running with hip and knee osteoarthritis: A systematic review and meta-analysis, *Journal of Orthopaedic and Sports Physical Therapy* 47:373–90; Miller, R. H. (2017), Joint loading in runners does not initiate knee osteoarthritis, *Exercise and Sport Sciences Reviews* 45:87–95.
45. Wallace, I. J., et al. (2017), Knee osteoarthritis has doubled in prevalence since the mid-20th century, *Proceedings of the National Academy of Sciences USA* 114:9332–36.

46. Truth be told, the only study to test this rule did not find it reduced injury rates. More research is needed. See Buist, I., et al. (2008), No effect of a graded training program on the number of running-related injuries in novice runners: A randomized controlled trial, *American Journal of Sports Medicine* 36:33–39.
47. Ferber, R., et al. (2015), Strengthening of the hip and core versus knee muscles for the treatment of patellofemoral pain: A multicenter randomized controlled trial, *Journal of Athletic Training* 50:366–77.
48. Dierks, T. A., et al. (2008), Proximal and distal influences on hip and knee kinematics in runners with patellofemoral pain during a prolonged run, *Journal of Orthopedic and Sports Physical Therapy* 38:448–56.
49. Nigg, B. M., et al. (2015), Running shoes and running injuries: Mythbusting and a proposal for two new paradigms: "Preferred movement path" and "comfort filter," *British Journal of Sports Medicine* 49:1290–94; Nigg, B. M., et al. (2017), The preferred movement path paradigm: Influence of running shoes on joint movement, *Medicine and Science in Sports and Exercise* 49:1641–48.
50. Counterarguments to this hypothesis include evidence that appreciable numbers of people still get injured despite presumably running as they wish in shoes they found comfortable. Further, if there were any truly natural way to run and kind of footwear, it would be barefoot. In addition, there is little evidence that someone's habitual gait is his optimal gait, especially in terms of avoiding injury. Just as there are better ways to swim or wrestle, why wouldn't there be better ways to run? Most important, an evolutionary perspective reminds us that everything we do involves trade-offs, and running is no exception. For example, in terms of form, studies show that landing on the ball of the foot puts less stress on the knee but more on the ankle. See Kulmala, J. P., et al. (2013), Forefoot strikers exhibit lower running-induced knee loading than rearfoot strikers, *Medicine and Science in Sports and Exercise* 45:2306–13; Stearne, S. M., et al. (2014), Joint kinetics in rearfoot versus forefoot running: Implications of switching technique, *Medicine and Science in Sports and Exercise* 46:1578–87.
51. Henrich, J. (2017), *The Secret of Our Success: How Culture Is Driving Human Evolution, Domesticating Our Species, and Making Us Smarter* (Princeton, N.J.: Princeton University Press).
52. For more information, see Lieberman, D. E., et al. (2015), Effects of stride frequency and foot position at landing on braking force, hip torque, impact peak force, and the metabolic cost of running in humans, *Journal of Experimental Biology* 218:3406–14; Cavanagh, P. R., and Kram, R. (1989), Stride length in distance running: Velocity, body dimensions, and added mass effects, *Medicine and Science in Sports and Exercise* 21:467–79; Heiderscheit, B. C., et al. (2011), Effects of step rate manipulation on joint mechanics during running, *Medicine and Science in Sports and Exercise* 43:296–302; Almeida, M. O., Davis, I. S., Lopes, A. D. (2015), Biomechanical differences of foot-strike patterns during running: A systematic review with meta-analysis, *Journal of Orthopedic and Sports Physical Therapy* 45:738–55.
53. Lieberman, D. E. (2014), Strike type variation among Tarahumara Indians in minimal sandals versus conventional running shoes, *Journal of Sport*

and Health Science 3:86–94; Lieberman, D. E., et al. (2015), Variation in foot strike patterns among habitually barefoot and shod runners in Kenya, *PLOS ONE* 10:e0131354.

54. Holowka, N. B., et al. (2019), Foot callus thickness does not trade off protection for tactile sensitivity during walking, *Nature* 571:261–64.
55. Thomas, E. M. (1989), *The Harmless People* (New York: Vintage).
56. Marshall, L. J. (1976), *The !Kung of Nyae Nyae* (Cambridge, Mass.: Harvard University Press); Marshall, L. J. (1999), *Nyae Nyae !Kung Beliefs and Rites* (Cambridge, Mass.: Peabody Museum of Archaeology and Ethnology, Harvard University).
57. Marlowe, F. W. (2010), *The Hadza: Hunter-Gatherers of Tanzania* (Berkeley: University of California Press).
58. Pintado Cortina, A. P. (2012), *Los hijos de Riosi y Riablo: Fiestas grandes y resistencia cultural en una comunidad tarahumara de la barranca* (Mexico: Instituto Nacional de Antropología e Historia).
59. Lumholtz, C. (1905), *Unknown Mexico* (London: Macmillan), 558.
60. Mullan, J. (2014), *The Ball in the Novels of Jane Austen*, British Library, London, www.bl.uk.
61. Dietrich, A. (2006), Transient hypofrontality as a mechanism for the psychological effects of exercise, *Psychiatry Research* 145:79–83; Raichlen, D. A., et al. (2012), Wired to run: Exercise-induced endocannabinoid signaling in humans and cursorial mammals with implications for the "runner's high," *Journal of Experimental Biology* 215:1331–36; Liebenberg, L. (2013), *The Origin of Science: On the Evolutionary Roots of Science and Its Implications for Self-Education and Citizen Science* (Cape Town: CyberTracker).

TEN ENDURANCE AND AGING: THE ACTIVE GRANDPARENT AND COSTLY REPAIR HYPOTHESES

1. For a history of people's efforts to avoid death, see Haycock, D. B. (2008), *Mortal Coil: A Short History of Living Longer* (New Haven, Conn.: Yale University Press).
2. Hippocrates, *Hippocrates*, trans. W. H. S. Jones (1953) (London: William Heinemann).
3. Kranish, N., and Fisher, M. (2017), *Trump Revealed: The Definitive Biography of the 45th President* (New York: Scribner).
4. Altman, L. K. (2016), A doctor's assessment of whether Donald Trump's health is "excellent," *New York Times*, Sept. 18, 2016, www.nytimes.com.
5. Ritchie, D. (2016), *The Stubborn Scotsman—Don Ritchie: World Record Holding Ultra Distance Runner* (Nottingham, U.K.: DB).
6. Trump Golf Count, www.trumpgolfcount.com.
7. Blair, S. N., et al. (1989), Physical fitness and all-cause mortality: A prospective study of healthy men and women, *Journal of the American Medical Association* 262:2395–401.
8. Those numbers are "relative risks" that control not just for the effects of smoking and drinking but also for high cholesterol and blood pressure, which are themselves affected by physical activity. So the beneficial effects of exercise are probably even higher than reported. See Blair, S. N., et al.

(1995), Changes in physical fitness and all-cause mortality: A prospective study of healthy and unhealthy men, *Journal of the American Medical Association* 273:1093–98.

9. Willis, B. L., et al. (2012), Midlife fitness and the development of chronic conditions in later life, *Archives of Internal Medicine* 172:1333–40.
10. Muller, M. N., and Wrangham, R. W. (2014), Mortality rates among Kanyawara chimpanzees, *Journal of Human Evolution* 66:107–14; Wood, B. M., et al. (2017), Favorable ecological circumstances promote life expectancy in chimpanzees similar to that of human hunter-gatherers, *Journal of Human Evolution* 105:41–56.
11. Medawar, P. B. (1952), *An Unsolved Problem of Biology* (London: H. K. Lewis).
12. Kaplan, H., et al. (2000), A theory of human life history evolution: Diet, intelligence, and longevity, *Evolutionary Anthropology* 9:156–85.
13. Hawkes, K., et al. (1997), Hadza women's time allocation, offspring provisioning, and the evolution of long postmenopausal life spans, *Current Anthropology* 38:551–77; Meehan, B. (1982), *Shell Bed to Shell Midden* (Canberra: Australian Institute of Aboriginal Studies); Hurtado, A. M., and Hill, K. (1987), Early dry season subsistence strategy of the Cuiva Foragers of Venezuela, *Human Ecology* 15:163–87.
14. Gurven, M., and Kaplan, H. (2007), Hunter-gatherer longevity: Cross-cultural perspectives, *Population and Development Review* 33:321–65.
15. Kim, P. S., Coxworth, J. E., and Hawkes, K. (2012), Increased longevity evolves from grandmothering, *Proceedings of the Royal Society B* 279:4880–84.
16. When hunter-gatherer life spans reached modern levels is difficult to document, but there is evidence that humans were living longer by the time of the Upper Paleolithic, which dates to forty to fifty thousand years ago. See Caspari, R., and Lee, S. H. (2003), Older age becomes common late in human evolution, *Proceedings of the National Academy of Sciences USA* 101:10895–900.
17. Without discounting the importance of grandmothers, we should also note that grandfathers can be valuable. A related hypothesis is the embodied capital hypothesis; according to this idea, an additional selective advantage of longevity is the knowledge and skills that elderly individuals transfer to and use on behalf of younger generations. See Kaplan et al. (2000), Theory of human life history evolution.
18. Hawkes, K., O'Connell, J. F., and Blurton Jones, N. G. (1997), Hadza women's time allocation, offspring provisioning, and the evolution of long postmenopausal life spans. *Current Anthropology* 38:551–77.
19. Marlowe, F. W. (2010), *The Hadza: Hunter-Gatherers of Tanzania* (Berkeley: University of California Press), 160. See also Pontzer, H., et al. (2015), Energy expenditure and activity among Hadza hunter-gatherers, *American Journal of Human Biology* 27:628–37.
20. The sample size of older Hadza women's walking distance is too small to be reliable, but was reported to be 20 percent less in elderly Hadza women compared with younger mothers by Pontzer et al. (2015), Energy expenditure and activity among Hadza hunter-gatherers. Data on Hadza energy expenditure: Raichlen, D. A., et al. (2017), Physical activity patterns and biomarkers of cardiovascular disease risk in hunter-gatherers, *American Journal of Human Biology* 29:e22919. For data on U.S. women, see Tudor-Locke, C., et al. (2013), Normative steps/day values for older adults: NHANES

2005–2006, *Journals of Gerontology Series A: Biological Sciences and Medical Sciences* 68:1426–32; Tudor-Locke, C., Johnson, W. D., and Katzmarzyk, P. T. (2009), Accelerometer-determined steps per day in US adults, *Medicine and Science in Sports and Exercise* 41:1384–91; Tudor-Locke, C., et al. (2012), Peak stepping cadence in free-living adults: 2005–2006 NHANES, *Journal of Physical Activity and Health* 9:1125–29.

21. Raichlen et al. (2017), Physical activity patterns and biomarkers of cardiovascular disease risk in hunter-gatherers.
22. Studenski, S., et al. (2011), Gait speed and survival in older adults, *Journal of the American Medical Association* 305:50–58.
23. Himann, J. E., et al. (1988), Age-related changes in speed of walking, *Medicine and Science in Sports and Exercise* 20:161–66.
24. Pontzer et al. (2015), Energy expenditure and activity among Hadza hunter-gatherers.
25. Walker, R., and Hill, K. (2003), Modeling growth and senescence in physical performance among the Aché of eastern Paraguay, *American Journal of Physical Anthropology* 15:196–208; Blurton-Jones, N., and Marlowe, F. W. (2002), Selection for delayed maturity: Does it take 20 years to learn to hunt and gather?, *Human Nature* 13:199–238.
26. In case you are interested, according to the UN, more than 300,000 people worldwide are older than a hundred. UNDESA (2011), *World Population Prospects: The 2010 Revision* (New York: United Nations), www.unfpa.org.
27. Finch, C. E. (1990), *Longevity, Senescence, and the Genome* (Chicago: University of Chicago Press).
28. Butler, P. G., et al. (2013), Variability of marine climate on the North Icelandic Shelf in a 1357-year proxy archive based on growth increments in the bivalve *Arctica islandica, Palaeogeography, Palaeoclimatology, Palaeoecology* 373:141–51.
29. Hood, D. A., et al. (2011), Mechanisms of exercise-induced mitochondrial biogenesis in skeletal muscle: Implications for health and disease, *Comprehensive Physiology* 1:1119–34; Cobb, L. J., et al. (2016), Naturally occurring mitochondrial-derived peptides are age-dependent regulators of apoptosis, insulin sensitivity, and inflammatory markers, *Aging* 8:796–808.
30. Gianni, P., et al. (2004), Oxidative stress and the mitochondrial theory of aging in human skeletal muscle, *Experimental Gerontology* 39:1391–400; Crane, J. D., et al. (2010), The effect of aging on human skeletal muscle mitochondrial and intramyocellular lipid ultrastructure, *Journals of Gerontology Series A: Biological Sciences and Medical Sciences* 65:119–28; Bratic, A., and Larsson, N. G. (2013), The role of mitochondria in aging, *Journal of Clinical Investigation* 123:951–57.
31. All your cells have the same genome, so you need epigenetic modifications to enable a skin cell to function differently from a neuron or a muscle cell. Some epigenetic modifications appear to be passed on from one generation to the next, making this a nongenetic form of inheritance.
32. Horvath, S. (2013), DNA methylation age of human tissues and cell types, *Genome Biology* 14:R115; Gibbs, W. W. (2014), Biomarkers and ageing: The clock-watcher, *Nature* 508:168–70.
33. Marioni, R. E., et al. (2015), DNA methylation age of blood predicts all-cause mortality in later life, *Genome Biology* 16:25; Perna, L., et al. (2016), Epigenetic age acceleration predicts cancer, cardiovascular, and all-cause mortality in a

German case cohort, *Clinical Epigenetics* 8:64; Christiansen, L., et al. (2016), DNA methylation age is associated with mortality in a longitudinal Danish twin study, *Aging Cell* 15:149–54.

34. He, C., et al. (2012), Exercise-induced BCL2-regulated autophagy is required for muscle glucose homeostasis, *Nature* 481:511–15.
35. This mechanism is especially fascinating because of the molecule mTOR (mammalian target of rapamycin), a protein that senses amino acids and promotes growth. Normally, low mTOR levels are associated with longevity, making it a target of research for extending life, but exercise seems to cause beneficial short-term increases in mTOR. See Efeyan, A., Comb, W. C., and Sabatini, D. M. (2015), Nutrient sensing mechanisms and pathways, *Nature* 517:302–10.
36. Telomere shortening as a cause of senescence is controversial. At birth, each telomere is about ten thousand base pairs long, but 25 percent shorter by age thirty-five and 65 percent shorter when you hit sixty-five. Some studies associate shorter telomeres with higher risk of disease, but others do not. And while exercise helps telomeres become longer (via the action of an enzyme called telomerase), so too does cancer. See Haycock, P. C., et al. (2014), Leucocyte telomere length and risk of cardiovascular disease: Systematic review and meta-analysis, *British Medical Journal* 349:g4227; Mather, K. A., et al. (2011), Is telomere length a biomarker of aging? A review, *Journals of Gerontology Series A: Biological Sciences and Medical Sciences* 66:202–13; Ludlow, A. T., et al. (2008), Relationship between physical activity level, telomere length, and telomerase activity, *Medicine and Science in Sports and Exercise* 40:1764–71.
37. Thomas, M., and Forbes, J. (2010), *The Maillard Reaction: Interface Between Aging, Nutrition, and Metabolism* (Cambridge, U.K.: Royal Society of Chemistry).
38. Kirkwood, T. B. L., and Austad, S. N. (2000), Why do we age?, *Nature* 408:233–38.
39. In fact, some biologists theorize that selection may actually work in reverse as we age because genes have different effects at different stages of life. In a phenomenon known as antagonistic pleiotropy ("pleiotropy" is the technical term for a gene that has multiple effects), a gene that helps us survive and reproduce when we are young might turn out to be harmful when we get older. One infamously extreme example is a mutation that improves immune function earlier in life but causes Huntington's disease (a fatal form of brain degeneration) in middle age. See Williams, G. C. (1957), Pleiotropy, natural selection, and the evolution of senescence, *Evolution* 11:398–411; Eskenazi, B. R., Wilson-Rich, N. S., and Starks, P. T. (2007), A Darwinian approach to Huntington's disease: Subtle health benefits of a neurological disorder, *Medical Hypotheses* 69:1183–89; Carter, A. J., and Nguyen, A. Q. (2011), Antagonistic pleiotropy as a widespread mechanism for the maintenance of polymorphic disease alleles, *BMC Medical Genetics* 12:160.
40. Saltin, B., et al. (1968), Response to exercise after bed rest and after training, *Circulation* 38 (supplement 5): 1–78.
41. McGuire, D. K., et al. (2001), A 30-year follow-up of the Dallas Bedrest and Training Study: I. Effect of age on the cardiovascular response to exercise, *Circulation* 104:1350–57.

42. Ekelund, U., et al. (2019), Dose-response associations between accelerometry measured physical activity and sedentary time and all cause mortality: Systematic review and harmonised meta-analysis, *British Medical Journal* 366:14570.
43. For a summary of these numerous processes (too many to review here), see Foreman, J. (2020), *Exercise Is Medicine: How Physical Activity Boosts Health and Slows Aging* (New York: Oxford University Press).
44. LaForgia, J., Withers, R. T., and Gore, C. J. (2006), Effects of exercise intensity and duration on the excess post-exercise oxygen consumption, *Journal of Sports Science* 24:1247–64.
45. Pedersen, B. K. (2013), Muscle as a secretory organ, *Comprehensive Physiology* 3:1337–62.
46. Clarkson, P. M., and Thompson, H. S. (2000), Antioxidants: What role do they play in physical activity and health?, *American Journal of Clinical Nutrition* 72:637S–646S.
47. Hormesis is a beneficial response from a small dose of something that can be harmful at a high dose. The phenomenon (from the Greek word for rapid motion and eagerness) has led to many controversies, not to mention some quackery. Devotees of hormesis love to quote the maxim "That which does not kill us makes us stronger," but please don't take medical advice from Nietzsche. Small doses of radiation, strychnine, and ricin cause only harm.
48. Examples of these sorts of drugs include metformin, which increases sensitivity of muscle cells to insulin; resveratrol, which triggers enzymes involved in metabolism; AICAR-AMPK, which activates pathways that generate more mitochondria; and irisin, a hormone that promotes bone growth and converts white to brown fat.
49. Pauling, L. C. (1987), *How to Live Longer and Feel Better* (New York: Avon).
50. Bjelakovic, G., et al. (2007), Mortality in randomized trials of antioxidant supplements for primary and secondary prevention: Systematic review and meta-analysis, *Journal of the American Medical Association* 297:842–57.
51. Ristow, M., et al. (2009), Antioxidants prevent health-promoting effects of physical exercise in humans, *Proceedings of the National Academy of Sciences USA* 106:8665–70.
52. Akerstrom, T. C., et al. (2009), Glucose ingestion during endurance training in men attenuates expression of myokine receptor, *Experimental Physiology* 94:1124–31.
53. For a popular review of the current state of both charlatans and serious scientists trying to extend life, see Gifford, W. (2014), *Spring Chicken: Stay Young Forever (or Die Trying)* (New York: Grand Central Publishing).
54. For a review of intermittent fasting, see De Cabo, R., and Mattson, M. P. (2019), Effects of intermittent fasting on health, aging, and disease, *New England Journal of Medicine* 381:2541–51. Note that intermittent fasting, which is short-term calorie restriction, is not the same thing as long-term calorie restriction. Although the stress induced by long-term calorie restriction does lengthen the life span of laboratory mice, there is no good evidence it helps humans or other primates. A highly cited long-term study published in 2009 by Colman and colleagues on macaques in Wisconsin reported benefits from calorie restriction, but the study was flawed because the "control"

monkeys were allowed to eat an unhealthy American-style diet with as much sugar-rich monkey chow as they wanted, leading to high rates of metabolic disease, while the calorie-restricted animals were fed a more normal amount of food. No wonder the calorie-restricted animals were healthier and lived longer. The results were disputed and refuted by a second twenty-three-year-long study conducted at the National Institute on Aging and published in 2012 by Mattison and colleagues. This study compared control monkeys fed a more healthy, normal diet with monkeys fed a low-calorie diet and found no significant extension of life from caloric restriction. See Colman, R. J., et al. (2009), Caloric restriction delays disease onset and mortality in rhesus monkeys, *Science* 325:201–4; Mattison, J. A., et al. (2012), Impact of caloric restriction on health and survival in rhesus monkeys from the NIA study, *Nature* 489:318–21.

55. For a comprehensive survey (including references), see Gurven, M., and Kaplan, H. (2007), Hunter-gatherer longevity: Cross-cultural perspectives, *Population and Development Review* 33:321–65. See also Truswell, A. S., and Hansen, J. D. L. (1976), Medical research among the !Kung, in *Kalahari Hunter-Gatherers,* ed. R. B. Lee and I. DeVore (Cambridge, Mass.: Harvard University Press), 167–94; Gurven, M., et al. (2017), The Tsimane health and life history project: Integrating anthropology and biomedicine, *Evolutionary Anthropology* 26:54–73; Eaton, S. B., Konner, M. J., and Shostak, M. (1988), Stone agers in the fast lane: Chronic degenerative diseases in evolutionary perspective, *American Journal of Medicine* 84:739–49; Eaton, S. B., et al. (1994), Women's reproductive cancers in evolutionary context, *Quarterly Review of Biology* 69:353–67.
56. U.S. Department of Health and Human Services, Centers for Disease Control, *Health, United States, 2016,* table 19, pp. 128–31, www.cdc.gov.
57. Kochanek, K. D., et al. (2017), Mortality in the United States, 2016, NCHS Data Brief, No. 293, Dec. 2017, www.cdc.gov.
58. Lieberman, D. E. (2013), *The Story of the Human Body: Evolution, Health, and Disease* (New York: Pantheon).
59. Vita, A. J., et al. (1998), Aging, health risks, and cumulative disability, *New England Journal of Medicine* 338:1035–41.
60. Mokdad, A., et al. (2004), Actual causes of death in the United States, 2000, *Journal of the American Medical Association* 291:1238–45.
61. For a review, see Chodzko-Zajko, W. J., et al. (2009), American College of Sports Medicine position stand: Exercise and physical activity for older adults, *Medicine and Science in Sports and Exercise* 41:1510–30.
62. Crimmins, E. M., and Beltrán-Sánchez, H. (2011), Mortality and morbidity trends: Is there compression of morbidity?, *Journals of Gerontology* 66:75–86; Cutler, D. M. (2004), *Your Money or Your Life: Strong Medicine for America's Health Care System* (New York: Oxford University Press).
63. Altman, Doctor's assessment of whether Donald Trump's health is "excellent."
64. Herskind, A. M., et al. (1996), The heritability of human longevity: A population-based study of 2872 Danish twin pairs born 1870–1900, *Human Genetics* 97:319–23; Ljungquist, B., et al. (1998), The effect of genetic factors for longevity: A comparison of identical and fraternal twins in the Swedish Twin Registry, *Journals of Gerontology Series A: Biological Sciences and Medical*

Sciences 53:M441–M446; Barzilai, N., et al. (2012), The place of genetics in ageing research, *Nature Reviews Genetics* 13:589–94.

65. Khera, A. V., et al. (2018), Genome-wide polygenic scores for common diseases identify individuals with risk equivalent to monogenic mutations, *Nature Genetics* 50:1219–24.
66. It would be interesting to test if chimpanzees could live as long as humans if they engaged in the kinds of endurance physical activities we know delay senescence in humans: many daily miles and hours of walking along with other moderately vigorous activities like running. Even if such an experiment were ethical, it would be impossible because chimpanzees lack so many of our adaptations for endurance. A chimpanzee could not work day after day as much as a hunter-gatherer, because it would overheat and get fatigued. Perhaps this helps explain why chimpanzees develop high blood pressure and frequently die of heart disease like many (but not all) sedentary humans. See Shave, R. E., et al. (2019), Selection of endurance capabilities and the trade-off between pressure and volume in the evolution of the human heart, *Proceedings of the National Academy of Sciences USA* 116:19905–10.
67. Paffenbarger, R. S., Jr., et al. (1993), The association of changes in physical activity level and other lifestyle characteristics with mortality among men, *New England Journal of Medicine* 328:538–45; Blair et al. (1995), Changes in physical fitness and all-cause mortality; Sui, X., et al. (2007), Estimated functional capacity predicts mortality in older adults, *Journal of the American Geriatrics Society* 55:1940–47.

ELEVEN TO MOVE OR NOT TO MOVE: HOW TO MAKE EXERCISE HAPPEN

1. Physical Activity Guidelines Advisory Committee (2018), *2018 Physical Activity Guidelines Advisory Committee Scientific Report* (Washington, D.C.: U.S. Department of Health and Human Services).
2. To be more precise, as youngsters get older, their enjoyment of physical education becomes more variable because of factors like the quality of their teachers, bullying, and experiences that sometimes reinforce stereotypical hierarchies. See Cardinal, B. J. (2017), Beyond the gym: There is more to physical education than meets the eye, *Journal of Physical Education, Recreation, and Dance* 88:3–5; Cardinal, B. J., et al. (2014), Obesity bias in the gym: An under-recognized social justice, diversity, and inclusivity issue, *Journal of Physical Education, Recreation, and Dance* 85:3–6; Cardinal, B. J., Yan, Z., and Cardinal, M. K. (2013), Negative experiences in physical education and sport: How much do they affect physical activity participation later in life?, *Journal of Physical Education, Recreation, and Dance* 84:49–53; Ladwig, M. A., Vazou, S., and Ekkekakis, P. (2018), "My best memory is when I was done with it": PE memories are associated with adult sedentary behavior, *Translational Journal of the ACSM* 3:119–29.
3. Confusingly, the tennis player Björn Borg is no longer involved with the company that conspicuously puts his name on every product sold.
4. Thaler, R. H., and Sunstein, C. R. (2009), *Nudge: Improving Decisions About Health, Wealth, and Happiness*, 2nd ed. (New York: Penguin).
5. For a review of this phenomenon among Westerners, see McElroy, M. (2002),

Resistance to Exercise: A Social Analysis of Inactivity (Champaign, Ill.: Human Kinetics).

6. The tendency to postpone long-term benefits for short-term alternatives is known as temporal discounting. Economists and psychologists have shown this is a common behavior that leads people to make irrational decisions. See Kahneman, D. (2011), *Thinking, Fast and Slow* (New York: Farrar, Straus and Giroux).
7. Several studies have identified genes associated with avoiding physical activity. These genes can also cause laboratory mice to be less likely to run on wheels in cages. That said, no common genes have been identified (or likely exist) that explain more than a small percentage of the variation in people's inactivity, and estimates of their heritability vary considerably with age and differ between environments. These findings indicate strong, poorly understood interactions between these genes and environments. I doubt anyone in the Paleolithic with these genes was physically inactive, and they are no more deterministic today. See Lightfoot, J. T., et al. (2018), Biological/genetic regulation of physical activity level: Consensus from GenBioPAC, *Medicine and Science in Sports and Exercise* 50:863–73.
8. For a comprehensive review of many hundreds of studies, see part F, chapter 11 of Physical Activity Guidelines Advisory Committee (2018), *2018 Physical Activity Guidelines Advisory Committee Scientific Report.*
9. Elley, C. R., et al. (2003), Effectiveness of counselling patients on physical activity in general practice: Cluster randomised controlled trial, *British Medical Journal* 326:793.
10. Hillsdon, M., Foster, C., and Thorogood, M. (2005), Interventions for promoting physical activity, *Cochrane Database Systematic Reviews* CD003180; Müller-Riemenschneider, F., et al. (2008), Long-term effectiveness of interventions promoting physical activity: A systematic review, *Preventive Medicine* 47:354–68.
11. Physical Activity Guidelines Advisory Committee (2018), *2018 Physical Activity Guidelines Advisory Committee Scientific Report.* The summary of the evidence is in the online supplementary material for part F of chapter 11 at health.gov.
12. Bauman, A. E., et al. (2012), Correlates of physical activity: Why are some people physically active and others are not?, *Lancet* 380:258–71.
13. Many studies find that physical activity levels and socioeconomic status are inversely related. One excellent study is Trost, S. G., et al. (2002), Correlates of adults' participation in physical activity: Review and update, *Medicine and Science in Sports and Exercise* 34:1996–2001. For data showing the difference is almost entirely in terms of leisure-time physical activity, see Stalsberg, R., and Pedersen, A. V. (2018), Are differences in physical activity across socioeconomic groups associated with choice of physical activity variables to report?, *International Journal of Environmental Research and Public Health* 15:922.
14. Marlowe, F. W. (2010), *The Hadza: Hunter-Gatherers of Tanzania* (Berkeley: University of California Press).
15. Christakis, N. (2019), *Blueprint: The Evolutionary Origins of a Good Society* (New York: Little, Brown Spark).
16. For a comprehensive review, see Basso, J. C., and Susuki, W. A. (2017), The

effects of acute exercise on mood, cognition, neurophysiology, and neurochemical pathways: A review, *Brain Plasticity* 2:127–52.

17. Vecchio, L. M., et al. (2018), The neuroprotective effects of exercise: Maintaining a healthy brain throughout aging, *Brain Plasticity* 4:17–52.
18. Knab, A. M., and Lightfoot, J. T. (2010), Does the difference between physically active and couch potato lie in the dopamine system?, *International Journal of Biological Science* 6:133–50; Kravitz, A. V., O'Neal, T. J., and Friend, D. M. (2016), Do dopaminergic impairments underlie physical inactivity in people with obesity?, *Frontiers in Human Neuroscience* 10:514.
19. Nogueira, A., et al. (2018), Exercise addiction in practitioners of endurance sports: A literature review, *Frontiers in Psychology* 9:1484.
20. Young, S. N. (2007), How to increase serotonin in the brain without taking drugs, *Journal of Psychiatry and Neuroscience* 32:394–99; Lerch-Haner, J. K., et al. (2008), Serotonergic transcriptional programming determines maternal behavior and offspring survival, *Nature Neuroscience* 11:1001–3.
21. Babyak, M., et al. (2000), Exercise treatment for major depression: Maintenance of therapeutic benefit at 10 months, *Psychosomatic Medicine* 62:633–38.
22. The best-known endogenous opioid is beta-endorphin, but others include enkephalins and dynorphins.
23. Schwarz, L., and Kindermann, W. (1992), Changes in beta-endorphin levels in response to aerobic and anaerobic exercise, *Sports Medicine* 13:25–36.
24. Dietrich, A., and McDaniel, W. F. (2004), Endocannabinoids and exercise, *British Journal of Sports Medicine* 38:536–41.
25. Hicks, S. D., et al. (2018), The transcriptional signature of a runner's high, *Medicine and Science in Sports and Exercise* 51:970–78.
26. In general, people derive more pleasure from endurance training than intense exercise, but there is plenty of variation. See Ekkekakis, P., Parfitt, G., and Petruzzello, S. J. (2011), The pleasure and displeasure people feel when they exercise at different intensities: Decennial update and progress towards a tripartite rationale for exercise intensity prescription, *Sports Medicine* 41:641–71; Oliveira, B. R., et al. (2013), Continuous and high-intensity interval training: Which promotes higher pleasure?, *PLOS ONE* 8:e79965.
27. Many people hire trainers, but there are almost no standard, externally evaluated qualifications to be a "trainer." Be sure to be a careful consumer and hire someone effective and experienced with a good track record.
28. But usually not enough. It is widely agreed that children should get at least 300 minutes per week, but the worldwide average is 103 minutes a week for primary schools and 100 minutes a week for secondary schools. See North Western Counties Physical Education Association (2014), *World-Wide Survey of School Physical Education: Final Report* (Paris: UNESCO), unesdoc.unesco.org.
29. For seat-belt data, see Glassbrenner, D. (2012), *Estimating the Lives Saved by Safety Belts and Air Bags* (Washington, D.C.: National Highway Traffic Safety Administration), www-nrd.nhtsa.dot.gov.
30. Carlson, S. A., et al. (2018), Percentage of deaths associated with inadequate physical activity in the United States, *Prevention of Chronic Disease* 15:170354.
31. Lee, I. M., et al. (2012), Effect of physical inactivity on major non-communicable diseases worldwide: An analysis of burden of disease and life expectancy, *Lancet* 380:219–29.

32. ReportLinker Insight (2017), Out of shape? Americans turn to exercise to get fit, ReportLinker, May 31, 2017, www.reportlinker.com.
33. Lightfoot, J. T., et al. (2018), Biological/genetic regulation of physical activity level: Consensus from GenBioPAC, *Medicine and Science in Sports and Exercise* 50:863–73.
34. Thaler and Sunstein (2009), *Nudge.*
35. The Child and Adolescent Health Measurement Initiative (CAHMI), *2016 National Survey of Children's Health,* Data Resource Center for Child and Adolescent Health.
36. Katzmarzyk, P. T., et al. (2016), Results from the United States of America's 2016 report card on physical activity for children and youth, *Journal of Physical Activity and Health* 13:S307–S313; National Center for Health Statistics (2006), *National Health and Nutrition Examination Survey* (Hyattsville, Md.: U.S. Department of Health and Human Services, CDC, National Center for Health Statistics); Centers for Disease Control and Prevention (2015), 2015 High School Youth Risk Behavior Surveillance System (Atlanta: U.S. Department of Health and Human Services).
37. World Health Organization, Global Health Observatory (GHO) data, Prevalence of insufficient physical activity, www.who.int.
38. U.S. Department of Health and Human Services (2018), *Physical Activity Guidelines for Children and Adolescents,* health.gov.
39. Hollis, J. L., et al. (2016), A systematic review and meta-analysis of moderate-to-vigorous physical activity levels in elementary school physical education lessons, *Preventive Medicine* 86:34–54.
40. Cardinal (2017), Beyond the gym.
41. Kocian, L. (2011), Uphill push, *Boston Globe,* Oct. 20, 2011, archive.boston.com.
42. Rasberry, C. N., et al. (2011), The association between school-based physical activity, including physical education, and academic performance: A systematic review of the literature, *Preventive Medicine* 52:S10–S20.
43. Mechikoff, R. A., and Estes, S. G. (2006), *A History and Philosophy of Sport and Physical Education: From Ancient Civilizations to the Modern World,* 4th ed. (Boston: McGraw-Hill); Cardinal, B. J., Sorensen, S. D., and Cardinal, M. K. (2012), Historical perspective and current status of the physical education graduation requirement at American 4-year colleges and universities, *Research Quarterly for Exercise and Sport* 83:503–12; Cardinal, B. J. (2017), Quality college and university instructional physical activity programs contribute to *mens sana in corpore sano,* "the good life," and healthy societies, *Quest* 69:531–41.
44. Sparling, P. B., and Snow, T. K. (2002), Physical activity patterns in recent college alumni, *Research Quarterly for Exercise and Sport* 73:200–205.
45. Kim, M., and Cardinal, B. J. (2019), Differences in university students' motivation between a required and an elective physical activity education policy, *Journal of American College Health* 67:207–14; Cardinal, Yan, and Cardinal (2013), Negative experiences in physical education and sport; Ladwig, Vazou, and Ekkekakis (2018), "My best memory is when I was done with it."
46. The claim that students lack enough time or resources to be physically active is even less convincing at residential liberal arts colleges where undergraduates typically don't have to cook, clean, shop, or commute. At Harvard, for example, the students are surrounded by world-class gym facilities and live

on a beautiful campus bisected by the Charles River with paths for walking, biking, and running on both sides.

47. Loprinzi, P. S., and Kane, V. J. (2015), Exercise and cognitive function: A randomized controlled trial examining acute exercise and free-living physical activity and sedentary effects, *Mayo Clinic Proceedings* 90:450–60.

TWELVE HOW MUCH AND WHAT TYPE?

1. World Health Organization (2010), *Global Recommendations on Physical Activity for Health* (Geneva: World Health Organization); Eckel, R. H., et al. (2014), AHA/ACC guideline on lifestyle management to reduce cardiovascular risk: A report of the American College of Cardiology/American Heart Association Task Force on Practice Guidelines, *Circulation* 129:S76–S99; Eckel, R. H., et al. (2014), AHA/ACC guideline on lifestyle management to reduce cardiovascular risk: A report of the American College of Cardiology/American Heart Association Task Force on Practice Guidelines, *Journal of the American College of Cardiology* 63:2960–84; Physical Activity Guidelines Advisory Committee (2018), *2018 Physical Activity Guidelines Advisory Committee Scientific Report* (Washington, D.C.: U.S. Department of Health and Human Services).
2. LaLanneisms, jacklalanne.com.
3. Lee, I. M., and Paffenbarger, R. S., Jr. (1998), Life is sweet: Candy consumption and longevity, *British Medical Journal* 317:1683–84. In the Competing Interests disclosure section of the paper, Paffenbarger and his co-author, I-Min Lee, wrote, "The authors admit to a decided weakness for chocolate and confess to an average consumption of one bar a day each."
4. Paffenbarger, R. S., Jr. (1986), Physical activity, all-cause mortality, and longevity of college alumni, *New England Journal of Medicine* 314:605–13.
5. Paffenbarger, R. S., Jr., et al. (1993), The association of changes in physical-activity level and other lifestyle characteristics with mortality among men, *New England Journal of Medicine* 328:538–45.
6. Pate, R. R. (1995), Physical activity and public health: A recommendation from the Centers for Disease Control and Prevention and the American College of Sports Medicine, *Journal of the American Medical Association* 273:402–7; Physical activity and cardiovascular health: NIH Consensus Development Panel on Physical Activity and Cardiovascular Health (1996), *Journal of the American Medical Association* 276:241–46; U.S. Department of Health and Human Services (1996), *Physical Activity and Health: A Report of the Surgeon General* (Atlanta: U.S. Department of Health and Human Services, Centers for Disease Control and Prevention, National Center for Chronic Disease Prevention and Health Promotion).
7. Physical Activity Guidelines Advisory Committee (2018), *2018 Physical Activity Guidelines Advisory Committee Scientific Report*. Check the report out for yourself at www.hhs.gov.
8. This metric, metabolic equivalents (METs), quantifies dose as the number of calories spent per hour of activity relative to the number of calories spent just sitting for the same amount of time (1 MET, which is about 1 kcal per kilogram per hour). By convention, sedentary activities are defined as less than 1.5 METs, light activities are 1.5–3.0 METs, moderate activities are 3–6 METs, and vigorous activities are greater than 6 METs.

9. Wasfy, M. M., and Baggish, A. L. (2016), Exercise dosage in clinical practice, *Circulation* 133:2297–313. See also Physical Activity Guidelines Advisory Committee (2008), *Physical Activity Guidelines Advisory Committee Report* (Washington, D.C.: U.S. Department of Health and Human Services).
10. Schnohr, P., et al. (2015), Dose of jogging and long-term mortality: The Copenhagen City Heart Study, *Journal of the American College of Cardiology* 65:411–19.
11. Kujala, U. M., et al. (1996), Hospital care in later life among former world-class Finnish athletes, *Journal of the American Medical Association* 276:216–20; Kujala, U. M., Sarna, S., and Kaprio, J. (2003), Use of medications and dietary supplements in later years among male former top-level athletes, *Archives of Internal Medicine* 163:1064–68; Garatachea, N., et al. (2014), Elite athletes live longer than the general population: A meta-analysis, *Mayo Clinic Proceedings* 89:1195–200; Levine, B. D. (2014), Can intensive exercise harm the heart? The benefits of competitive endurance training for cardiovascular structure and function, *Circulation* 130:987–91.
12. Lee, D. C., et al. (2014), Leisure-time running reduces all-cause and cardiovascular mortality risk, *Journal of the American College of Cardiology* 64:472–81.
13. Arem, H., et al. (2015), Leisure time physical activity and mortality: A detailed pooled analysis of the dose-response relationship, *JAMA Internal Medicine* 175:959–67.
14. Wasfy and Baggish (2016), Exercise dosage in clinical practice.
15. Cowles, W. N. (2018), Fatigue as a contributory cause of pneumonia, *Boston Medical and Surgical Journal* 179:555–56.
16. Peters, E. M., and Bateman, E. D. (1983), Ultramarathon running and upper respiratory tract infections: An epidemiological survey, *South African Medical Journal* 64:582–84; Nieman, D. C., et al. (1990), Infectious episodes in runners before and after the Los Angeles Marathon, *Journal of Sports Medicine and Physical Fitness* 30:316–28.
17. Pedersen, B. K., and Ullum, H. (1994), NK cell response to physical activity: Possible mechanisms of action. *Medicine and Science in Sports and Exercise* 26:140–46; Shek, P. N., et al. (1995), Strenuous exercise and immunological changes: A multiple-time-point analysis of leukocyte subsets, CD4/CD8 ratio, immunoglobulin production and NK cell response, *International Journal of Sports Medicine* 16:466–74; Kakanis, M. W., et al. (2010), The open window of susceptibility to infection after acute exercise in healthy young male elite athletes, *Exercise Immunology Review* 16:119–37; Peake, J. M., et al. (2017), Recovery of the immune system after exercise, *Journal of Applied Physiology* 122:1077–87.
18. For a comprehensive review, see Campbell, J. P., and Turner, J. E. (2018), Debunking the myth of exercise-induced immune suppression: Redefining the impact of exercise on immunological health across the lifespan, *Frontiers in Immunology* 9:648.
19. Dhabhar, F. S. (2014), Effects of stress on immune function: The good, the bad, and the beautiful, *Immunology Research* 58:193–210; Bigley, A. B., et al. (2014), Acute exercise preferentially redeploys NK-cells with a highly differentiated phenotype and augments cytotoxicity against lymphoma and multiple myeloma target cells, *Brain, Behavior, and Immunity* 39:160–71.

20. Campbell, J. P., and Turner, J. E. (2018), Debunking the myth of exercise-induced immune suppression.
21. Lowder, T., et al. (2005), Moderate exercise protects mice from death due to influenza virus, *Brain, Behavior and Immunity* 19:377–80.
22. Wilson, M., et al. (2011), Diverse patterns of myocardial fibrosis in lifelong, veteran endurance athletes, *Journal of Applied Physiology* 110:1622–26; La Gerche, A., et al. (2012), Exercise-induced right ventricular dysfunction and structural remodelling in endurance athletes, *European Heart Journal* 33:998–1006; La Gerche, A., and Heidbuchel, H. (2014), Can intensive exercise harm the heart? You can get too much of a good thing, *Circulation* 130:992–1002.
23. Too little exercise is also a risk factor for atrial fibrillation. See Elliott, A. D., et al. (2017), The role of exercise in atrial fibrillation prevention and promotion: Finding optimal ranges for health, *Heart Rhythm* 14:1713–20.
24. Burfoot describes his visit to the doctor and his reaction in an essay in *Runner's World:* Burfoot, A. (2016), I ♥ running, *Runner's World,* Sept. 27, 2016, www.runnersworld.com.
25. Möhlenkamp, S., et al. (2008), Running: The risk of coronary events: Prevalence and prognostic relevance of coronary atherosclerosis in marathon runners, *European Heart Journal* 29:1903–10.
26. Baggish, A. L., and Levine, B. D. (2017), Coronary artery calcification among endurance athletes: "Hearts of Stone," *Circulation* 136:149–51; Merghani, A., et al. (2017), Prevalence of subclinical coronary artery disease in masters endurance athletes with a low atherosclerotic risk profile, *Circulation* 136:126–37; Aengevaeren, V. L., et al. (2017), Relationship between lifelong exercise volume and coronary atherosclerosis in athletes, *Circulation* 136:138–48.
27. DeFina, L. F., et al. (2019), Association of all-cause and cardiovascular mortality with high levels of physical activity and concurrent coronary artery calcification, *JAMA Cardiology* 4:174–81.
28. Rao, P., Hutter, A. M., Jr., and Baggish, A. L. (2018), The limits of cardiac performance: Can too much exercise damage the heart?, *American Journal of Medicine* 131:1279–84.
29. Siscovick, D. S., et al. (1984), The incidence of primary cardiac arrest during vigorous exercise, *New England Journal of Medicine* 311:874–77; Thompson, P. D., et al. (1982), Incidence of death during jogging in Rhode Island from 1975 through 1980, *Journal of the American Medical Association* 247:2535–38.
30. Albert, C. M., et al. (2000), Triggering of sudden death from cardiac causes by vigorous exertion, *New England Journal of Medicine* 343:1355–61; Kim, J. H., et al. (2012), Race Associated Cardiac Arrest Event Registry (RACER) study group: Cardiac arrest during long-distance running races, *New England Journal of Medicine* 366:130–40.
31. Gensel, L. (2005), The medical world of Benjamin Franklin, *Journal of the Royal Society of Medicine* 98:534–38.
32. History of Hoover-ball, Herbert Hoover Presidential Library and Museum, hoover.archives.gov.
33. Black, J. (2013), *Making the American Body: The Remarkable Saga of the Men and Women Whose Feats, Feuds, and Passions Shaped Fitness History* (Lincoln: University of Nebraska Press).

34. A standard back-of-the-envelope estimate of maximum heart rate is 220 minus your age, but this estimate is often inaccurate especially for people who are very fit and for the elderly.
35. You can watch the scene yourself online: archive.org/details/huntersfilmpart2 (be sure to watch part 1 as well: archive.org/details/huntersfilmpart1).
36. For a fun compendium of training methods, check out Wilt, F., ed. (1973), *How They Train*, vol. 2, *Long Distances*, 2nd ed. (Mountain View, Calif.: Tafnews Press).
37. Burgomaster, K. A., et al. (2005), Six sessions of sprint interval training increases muscle oxidative potential and cycle endurance capacity in humans, *Journal of Applied Physiology* 98:1985–90; Burgomaster, K. A., Heigenhauser, G. J., and Gibala, M. J. (2006), Effect of short-term sprint interval training on human skeletal muscle carbohydrate metabolism during exercise and time-trial performance, *Journal of Applied Physiology* 100:2041–47.
38. MacInnis, M. J., and Gibala, M. J. (2017), Physiological adaptations to interval training and the role of exercise intensity, *Journal of Physiology* 595:2915–30.
39. A popular way to mix weight and intense cardio is circuit training. Although circuit training is more aerobic than traditional weight-lifting routines, it rarely gets heart rates above 50 percent of maximum. See Monteiro, A. G., et al. (2008), Acute physiological responses to different circuit training protocols, *Journal of Sports Medicine and Physical Fitness* 48:438–42.
40. Physical Activity Guidelines Advisory Committee (2018), *2018 Physical Activity Guidelines Advisory Committee Scientific Report.*
41. For a good review, see Pate, R. R. (1995), Physical activity and health: Dose-response issues, *Research Quarterly for Exercise and Sport* 66:313–17.

THIRTEEN EXERCISE AND DISEASE

1. Gross, J. (1984), James F. Fixx dies jogging; author on running was 52, *New York Times*, July 22, 1984, www.nytimes.com.
2. Cooper, K. H. (1985), *Running Without Fear: How to Reduce the Risk of Heart Attack and Sudden Death During Aerobic Exercise* (New York: M. Evans).
3. Lieberman, D. E. (2013), *The Story of the Human Body: Evolution, Health, and Disease* (New York: Pantheon).
4. Although widely used, BMI has many problems including not making a distinction between fat and muscle mass. Waist circumference or, even better, the ratio of waist circumference to height is generally a better measure of how much organ visceral fat you have. That said, BMI accurately categorizes people's percentage of body fat about 82 percent of the time. See Dybala, M. P., Brady, M. J., and Hara, M. (2019), Disparity in adiposity among adults with normal body mass index and waist-to-height ratio, *iScience* 21:612–23.
5. Thomas, D. M., et al. (2012), Why do individuals not lose more weight from an exercise intervention at a defined dose? An energy balance analysis, *Obesity Review* 13:835–47; Gomersall, S. R., et al. (2016), Testing the activitystat hypothesis: A randomised controlled trial, *BMC Public Health* 16:900; Liguori, G., et al. (2017), Impact of prescribed exercise on physical activity compensation in young adults, *Journal of Strength and Conditioning Research* 31:503–8.
6. Donnelly, J. E., et al. (2009), Appropriate physical activity intervention strat-

egies for weight loss and prevention of weight regain for adults, *Medicine and Science in Sports and Exercise* 41:459–71.

7. Barry, V. W., et al. (2014), Fitness vs. fatness on all-cause mortality: A meta-analysis, *Progress in Cardiovascular Disease* 56:382–90.
8. Lavie, C. J., De Schutter, A., and Milani, R. V. (2015), Healthy obese versus unhealthy lean: The obesity paradox, *Nature Reviews Endocrinology* 11:55–62; Oktay, A. A., et al. (2017), The interaction of cardiorespiratory fitness with obesity and the obesity paradox in cardiovascular disease, *Progress in Cardiovascular Disease* 60:30–44.
9. Childers, D. K., and Allison, D. B. (2010), The "obesity paradox": A parsimonious explanation for relations among obesity, mortality rate, and aging?, *International Journal of Obesity* 34:1231–38; Flegal, K. M., et al. (2013), Association of all-cause mortality with overweight and obesity using standard body mass index categories: A systematic review and meta-analysis, *Journal of the American Medical Association* 309:71–82.
10. Fogelholm, M. (2010), Physical activity, fitness, and fatness: Relations to mortality, morbidity, and disease risk factors: A systematic review, *Obesity Review* 11:202–21.
11. Hu, F. B., et al. (2004), Adiposity as compared with physical activity in predicting mortality among women, *New England Journal of Medicine* 351:2694–703. Keep in mind these numbers mask some interpretive challenges inherent in large epidemiological studies. One issue is measurement error, which will be higher for self-reported exercise than body weight. That means the estimate of the exercise effect will be more biased and underestimated. In addition, because obesity and exercise are negatively correlated, obesity could be picking up some of the exercise effect. Finally, how one calculates the estimated effects of obesity is complicated by the big difference between being slightly overweight and being highly obese.
12. Slentz, C. A., et al. (2011), Effects of aerobic vs. resistance training on visceral and liver fat stores, liver enzymes, and insulin resistance by HOMA in overweight adults from STRRIDE AT/RT, *American Journal of Physiology: Endocrinology and Metabolism* 301:E1033–E1039; Willis, L. H., et al. (2012), Effects of aerobic and/or resistance training on body mass and fat mass in overweight or obese adults, *Journal of Applied Physiology* 113:1831–37.
13. Türk, Y., et al. (2017), High intensity training in obesity: A meta-analysis, *Obesity Science and Practice* 3:258–71; Carey, D. G. (2009), Quantifying differences in the "fat burning" zone and the aerobic zone: Implications for training, *Journal of Strength and Conditioning Research* 23:2090–95.
14. Gordon, R. (1994), *The Alarming History of Medicine: Amusing Anecdotes from Hippocrates to Heart Transplants* (New York: St. Martin's).
15. Here are the conventions as of 2018: waist circumference more than forty inches in men and thirty-five inches in women; elevated triglycerides 150 milligrams per deciliter of blood (mg/dL) or greater; reduced high-density lipoprotein cholesterol (HDL) less than 40 mg/dL in men or less than 50 mg/dL in women; elevated fasting glucose of 100 mg/dL or greater; blood pressure values of systolic 130 mmHg or higher and/or diastolic 85 mmHg or higher.
16. For a comprehensive review, see Bray, G. A. (2007), *The Metabolic Syndrome and Obesity* (New York: Springer).

17. O'Neill, S., and O'Driscoll, L. (2015), Metabolic syndrome: A closer look at the growing epidemic and its associated pathologies, *Obesity Review* 16:1–12.
18. Type 1 diabetes, also known as juvenile-onset diabetes, arises when the immune system destroys cells in the pancreas that synthesize insulin; gestational diabetes arises toward the end of certain pregnancies from abnormal interactions between the fetus and the mother.
19. Smyth, S., and Heron, A. (2006), Diabetes and obesity: The twin epidemics, *Nature Medicine* 12:75–80; Whiting, D. R., et al. (2011), IDF Diabetes Atlas: Global estimates of the prevalence of diabetes for 2011 and 2030, *Diabetes Research and Clinical Practice* 94:311–21.
20. O'Dea, K. (1984), Marked improvement in carbohydrate and lipid metabolism in diabetic Australian aborigines after temporary reversion to traditional lifestyle, *Diabetes* 33:596–603.
21. Sylow, L., et al. (2017), Exercise-stimulated glucose uptake—regulation and implications for glycaemic control, *Nature Reviews Endocrinology* 13:133–48.
22. Boule, N. G., et al. (2003), Meta-analysis of the effect of structured exercise training on cardiorespiratory fitness in type 2 diabetes mellitus, *Diabetologia* 46:1071–81; Vancea, D. M., et al. (2009), Effect of frequency of physical exercise on glycemic control and body composition in type 2 diabetic patients, *Arquivos Brasileiros de Cardiologia* 92:23–30; Balducci, S., et al. (2010), Effect of an intensive exercise intervention strategy on modifiable cardiovascular risk factors in subjects with type 2 diabetes mellitus: A randomized controlled trial: The Italian Diabetes and Exercise Study (IDES), *Archives of Internal Medicine* 170:1794–803.
23. Sriwijitkamol, A., et al. (2006), Reduced skeletal muscle inhibitor of kappaB beta content is associated with insulin resistance in subjects with type 2 diabetes: Reversal by exercise training, *Diabetes* 55:760–67; Di Loreto, C., et al. (2005), Make your diabetic patients walk: Long-term impact of different amounts of physical activity on type 2 diabetes, *Diabetes Care* 28:1295–302; Umpierre, D., et al. (2011), Physical activity advice only or structured exercise training and association with HbA1c levels in type 2 diabetes: A systematic review and meta-analysis, *Journal of the American Medical Association* 305:1790–99; Umpierre, D., et al. (2013), Volume of supervised exercise training impacts glycaemic control in patients with type 2 diabetes: A systematic review with meta-regression analysis, *Diabetologia* 56:242–51; McInnes, N., et al. (2017), Piloting a remission strategy in type 2 diabetes: Results of a randomized controlled trial, *Journal of Clinical Endocrinology and Metabolism* 102:1596–605.
24. Johansen, M. Y., et al. (2017), Effect of an intensive lifestyle intervention on glycemic control in patients with type 2 diabetes: A randomized clinical trial, *Journal of the American Medical Association* 318:637–46; Reid-Larsen, M., et al. (2019), Type 2 diabetes remission 1year after an intensive lifestyle intervention: A secondary analysis of a randomized clinical trial, *Diabetes Obesity and Metabolism* 21:2257–66.
25. Little, J. P., et al. (2014), Low-volume high-intensity interval training reduces hyperglycemia and increases muscle mitochondrial capacity in patients with type 2 diabetes, *Journal of Applied Physiology* 111:1554–60; Shaban, N., Kenno, K. A., and Milne, K. J. (2014), The effects of a 2 week modified high intensity

interval training program on the homeostatic model of insulin resistance (HOMA-IR) in adults with type 2 diabetes, *Journal of Medicine and Physical Fitness* 54:203–9; Sjöros, T. J., et al. (2018), Increased insulin-stimulated glucose uptake in both leg and arm muscles after sprint interval and moderate intensity training in subjects with type 2 diabetes or prediabetes, *Scandinavian Journal of Medicine and Science in Sports* 28:77–87.

26. Strasser, B., Siebert, U., and Schobersberger, W. (2010), Resistance training in the treatment of the metabolic syndrome: A systematic review and meta-analysis of the effect of resistance training on metabolic clustering in patients with abnormal glucose metabolism, *Sports Medicine* 40:397–415; Yang, Z., et al. (2014), Resistance exercise versus aerobic exercise for type 2 diabetes: A systematic review and meta-analysis, *Sports Medicine* 44:487–99.
27. Church, T. S., et al. (2010), Effects of aerobic and resistance training on hemoglobin A1c levels in patients with type 2 diabetes: A randomized controlled trial, *Journal of the American Medical Association* 304:2253–62.
28. Smith, G. D. (2004), A conversation with Jerry Morris, *Epidemiology* 15:770–73.
29. Morris, J. N., et al. (1953), Coronary heart-disease and physical activity of work, *Lancet* 265:1053–57 and 1111–20.
30. Baggish, A. L., et al. (2008), Training-specific changes in cardiac structure and function: A prospective and longitudinal assessment of competitive athletes, *Journal of Applied Physiology* 104:1121–28.
31. Green, D. J., et al. (2017), Vascular adaptation to exercise in humans: Role of hemodynamic stimuli, *Physiology Reviews* 97:495–528.
32. Thompson, R. C., et al. (2013), Atherosclerosis across 4000 years of human history: The Horus study of four ancient populations, *Lancet* 381:1211–22.
33. Truswell, A. S., et al. (1972), Blood pressures of !Kung bushmen in northern Botswana, *American Heart Journal* 84:5–12; Raichlen, D. A., et al. (2017), Physical activity patterns and biomarkers of cardiovascular disease risk in hunter-gatherers, *American Journal of Human Biology* 29:e22919.
34. Shave, R. E., et al. (2019), Selection of endurance capabilities and the trade-off between pressure and volume in the evolution of the human heart, *Proceedings of the National Academy of Sciences USA* 116:19905–10.
35. Liu, J., et al. (2014), Effects of cardiorespiratory fitness on blood pressure trajectory with aging in a cohort of healthy men, *Journal of the American College of Cardiology* 64:1245–53; Gonzales, J. U. (2016), Do older adults with higher daily ambulatory activity have lower central blood pressure?, *Aging Clinical and Experimental Research* 28:965–71.
36. Kaplan, H., et al. (2017), Coronary atherosclerosis in indigenous South American Tsimane: A cross-sectional cohort study, *Lancet* 389:1730–39.
37. Popkin, B. M. (2015), Nutrition transition and the global diabetes epidemic, *Current Diabetes Reports* 15:64.
38. Jones, D. S., Podolsky, S. H., and Greene, J. A. (2012), The burden of disease and the changing task of medicine, *New England Journal of Medicine* 366:2333–38.
39. Koenig, W., et al. (1999), C-reactive protein, a sensitive marker of inflammation, predicts future risk of coronary heart disease in initially healthy middle-aged men: Results from the MONICA (Monitoring Trends and Deter-

minants in Cardiovascular Disease) Augsburg Cohort Study, 1984 to 1992, *Circulation* 99:237–42; Fryar, C. D., Chen, T., and Li, X. (2012), *Prevalence of Uncontrolled Risk Factors for Cardiovascular Disease: United States, 1999–2010*, National Center for Health Statistics Data Brief, No. 103 (Hyattsville, Md.: U.S. Department of Health and Human Services).

40. The most common measure of general, chronic inflammation is a molecule called CRP (C-reactive protein). A high level of CRP is an independent risk factor that along with high cholesterol and blood pressure increases one's likelihood of developing heart disease. For a review, see Steyers, C. M., 3rd, and Miller, F. J., Jr. (2014), Endothelial dysfunction in chronic inflammatory diseases, *International Journal of Molecular Sciences* 15:11324–49.
41. Lavie, C. J., et al. (2015), Exercise and the cardiovascular system: Clinical science and cardiovascular outcomes, *Circulation Research* 117:207–19.
42. Blair, S. N., et al. (1995), Changes in physical fitness and all-cause mortality: A prospective study of healthy and unhealthy men, *Journal of the American Medical Association* 273:1093–98.
43. Wasfy, M. M., and Baggish, A. L. (2016), Exercise dosage in clinical practice, *Circulation* 133:2297–313.
44. Marceau, M., et al. (1993), Effects of different training intensities on 24-hour blood pressure in hypertensive subjects, *Circulation* 88:2803–11; Tjønna, A. E., et al. (2008), Aerobic interval training versus continuous moderate exercise as a treatment for the metabolic syndrome: A pilot study, *Circulation* 118:346–54; Molmen-Hansen, H. E., et al. (2012), Aerobic interval training reduces blood pressure and improves myocardial function in hypertensive patients, *European Journal of Preventive Cardiology* 19:151–60.
45. Braith, R. W., and Stewart, K. J. (2006), Resistance exercise training: Its role in the prevention of cardiovascular disease, *Circulation* 113:2642–50.
46. Miyachi, M. (2013), Effects of resistance training on arterial stiffness: A meta-analysis, *British Journal of Sports Medicine* 47:393–96; Kraschnewski, J. L., et al. (2016), Is strength training associated with mortality benefits? A 15 year cohort study of US older adults, *Preventive Medicine* 87:121–27; Dankel, S. J., Loenneke, J. P., and Loprinzi, P. D. (2016), Dose-dependent association between muscle-strengthening activities and all-cause mortality: Prospective cohort study among a national sample of adults in the USA, *Archives of Cardiovascular Disease* 109:626–33.
47. Shave et al. (2019), Selection of endurance capabilities and the trade-off between pressure and volume in the evolution of the human heart.
48. Finland is a useful place for this kind of research because it provides free access to health care regardless of income, allows no one to stay in the hospital longer than necessary, and keeps records on everyone. This study thus looked at the effect of senior citizens' previous exercise history on the number and length of their admissions to hospitals for every male athlete who had competed in the Olympic Games or other world championships between 1920 and 1965 against thousands of sedentary controls. The analysis controlled for factors like smoking and whether individuals stopped exercising. The results showed that endurance athletes were about half as likely as sedentary controls to go to the hospital for circulatory diseases, cancers, and respiratory diseases and their incidence of heart attacks was two-thirds lower. In contrast, athletes who competed in power sports had rates of circu-

latory disease, cancers, and respiratory diseases about 30 to 40 percent lower than sedentary Finns, and their rate of heart attacks was actually one-third *higher*. Overall, endurance athletes not only lived the longest but had the least morbidity. See Kujala, U. M., et al. (1996), Hospital care in later life among former world-class Finnish athletes, *Journal of the American Medical Association* 276:216–20. See also Keskimäki, I., and Arro, S. (1991), Accuracy of data on diagnosis, procedures, and accidents in the Finnish hospital discharge register, *International Journal of Health Sciences* 2:15–21.

49. For reviews, see Diamond, J. (1997), *Guns, Germs and Steel: The Fates of Human Societies* (New York: W. W. Norton); and Barnes, E. (2005), *Diseases and Human Evolution* (Albuquerque, N.M.: University of New Mexico Press).
50. Warburton, D. E. R., and Bredin, S. S. D. (2017), Health benefits of physical activity: A systematic review of current systematic reviews, *Current Opinions in Cardiology* 32:541–56; Kostka, T., et al. (2000), The symptomatology of upper respiratory tract infections and exercise in elderly people, *Medicine and Science in Sports and Exercise* 32:46–51; Baik, I., et al. (2000), A prospective study of age and lifestyle factors in relation to community-acquired pneumonia in US men and women, *Archives of Internal Medicine* 160:3082–88.
51. Simpson, R. J., et al. (2012), Exercise and the aging immune system, *Ageing Research Reviews* 11:404–20.
52. Chubak, J., et al. (2006), Moderate-intensity exercise reduces the incidence of colds among postmenopausal women, *American Journal of Medicine* 119:937–42.
53. Nieman, D. C., et al. (1998), Immune response to exercise training and or energy restriction in obese women, *Medicine and Science in Sports and Exercise* 30:679–86.
54. Fondell, E., et al. (2011), Physical activity, stress, and self-reported upper respiratory tract infection, *Medicine and Science in Sports and Exercise* 43:272–79.
55. Baik, I., et al. (2000), A prospective study of age and lifestyle factors in relation to community-acquired pneumonia in US men and women, *Archives of Internal Medicine* 160:3082–88.
56. Grande, A. J., et al. (2015), Exercise versus no exercise for the occurrence, severity, and duration of acute respiratory infections, *Cochrane Database Systematic Reviews* CD010596.
57. Kakanis, M. W., et al. (2010), The open window of susceptibility to infection after acute exercise in healthy young male elite athletes, *Exercise Immunology Review* 16:119–37; Peake, J. M., et al. (2017), Recovery of the immune system after exercise, *Journal of Applied Physiology* 122:1077–87.
58. In this case, the white blood cells measured were neutrophils, which protect primarily against bacterial infections. See Syu, G.-D., et al. (2012), Differential effects of acute and chronic exercise on human neutrophil functions, *Medicine and Science in Sports and Exercise* 44:1021–27.
59. Shinkai, S., et al. (1992), Acute exercise and immune function. Relationship between lymphocyte activity and changes in subset counts. *International Journal of Sports Medicine* 13:452–61; Kakanis, M. W., et al. (2010), The open window of susceptibility to infection after acute exercise in healthy young male elite athletes. *Exercise Immunology Review* 16:119–37.

60. Nieman, D. C. (1994), Exercise, infection, and immunity, *International Journal of Sports Medicine* 15:S131–41.
61. Kruger, K., and Mooren, F. C. (2007), T cell homing and exercise, *Exercise Immunology Review* 13:37–54.
62. Kruger, K., et al. (2008), Exercise-induced redistribution of T lymphocytes is regulated by adrenergic mechanisms, *Brain, Behavior and Immunity* 22:324–38; Bigley, A. B., et al. (2014), Acute exercise preferentially redeploys NK-cells with a highly differentiated phenotype and augments cytotoxicity against lymphoma and multiple myeloma target cells, *Brain, Behavior and Immunity* 39:160–71; Bigley, A. B., et al. (2015), Acute exercise preferentially redeploys NK-cells with a highly differentiated phenotype and augments cytotoxicity against lymphoma and multiple myeloma target cells. Part II: Impact of latent cytomegalovirus infection and catecholamine sensitivity, *Brain, Behavior and Immunity* 49:59–65.
63. Kohut, M. L., et al. (2004), Moderate exercise improves antibody response to influenza immunization in older adults. *Vaccine* 22:2298–306; Smith, T. P., et al. (2004), Influence of age and physical activity on the primary in vivo antibody and T cell–mediated responses in men, *Journal of Applied Physiology* 97:491–98; Schuler, P. B., et al. (2003), Effect of physical activity on the production of specific antibody in response to the 1998–99 influenza virus vaccine in older adults, *Journal of Sports Medicine and Physical Fitness* 43:404; de Araujo, A. L., et al. (2015), Elderly men with moderate and intense training lifestyle present sustained higher antibody responses to influenza vaccine, *Age* 37:105.
64. Montecino-Rodriguez, E., et al. (2013), Causes, consequences, and reversal of immune system aging, *Journal of Clinical Investigation* 123:958–65; Campbell, J. P., and Turner, J. E. (2018), Debunking the myth of exercise-induced immune suppression: Redefining the impact of exercise on immunological health across the lifespan, *Frontiers in Immunology* 9:648.
65. Lowder, T., et al. (2005), Moderate exercise protects mice from death due to influenza virus, *Brain, Behavior and Immunity* 19:377–80.
66. Raso, V., et al. (2007), Effect of resistance training on immunological parameters of healthy elderly women, *Medicine and Science in Sports and Exercise* 39:2152–59.
67. Hannan, E. L., et al. (2001), Mortality and locomotion 6 months after hospitalization for hip fracture: Risk factors and risk-adjusted hospital outcomes, *Journal of the American Medical Association* 285:2736–42.
68. Blurton-Jones, N., and Marlowe, F. W. (2002), Selection for delayed maturity: Does it take 20 years to learn to hunt and gather?, *Human Nature* 13:199–238; Walker, R., and Hill, K. (2003), Modeling growth and senescence in physical performance among the Aché of eastern Paraguay, *American Journal of Physical Anthropology* 15:196–208; Apicella, C. L. (2014), Upper-body strength predicts hunting reputation and reproductive success in Hadza hunter-gatherers, *Evolution and Human Behavior* 35:508–18.
69. Cauley, J. A., et al. (2014), Geographic and ethnic disparities in osteoporotic fractures, *Nature Reviews Endocrinology* 10:338–51.
70. Johnell, O., and Kanis, J. (2005), Epidemiology of osteoporotic fractures, *Osteoporosis International* 16:S3–S7; Johnell, O., and Kanis, J. A. (2006), An estimate of the worldwide prevalence and disability associated with osteo-

porotic fractures, *Osteoporosis International* 17:1726–33; Wright, N. C., et al. (2014), The recent prevalence of osteoporosis and low bone mass in the United States based on bone mineral density at the femoral neck or lumbar spine, *Journal of Bone and Mineral Research* 29:2520–26.

71. Wallace, I. J., et al. (2017), Knee osteoarthritis has doubled in prevalence since the mid-20th century, *Proceedings of the National Academy of Sciences USA* 14:9332–36.
72. Hootman, J. M., et al. (2016), Updated projected prevalence of self-reported doctor-diagnosed arthritis and arthritis-attributable activity limitation among US adults, 2015–2040, *Arthritis and Rheumatology* 68:1582–87.
73. Zurlo, F., et al. (1990), Skeletal muscle metabolism is a major determinant of resting energy expenditure, *Journal of Clinical Investigation* 86:1423–27.
74. A newsworthy example is Ruth Bader Ginsburg, a U.S. Supreme Court justice who in her eighties was still going regularly to the gym to stay strong. See Johnson, B. (2017), *The RBG Workout: How She Stays Strong . . . and You Can Too!* (Boston: Houghton Mifflin Harcourt).
75. Pearson, O. M., and Lieberman, D. E. (2004), The aging of Wolff's "law": Ontogeny and responses to mechanical loading in cortical bone, *Yearbook of Physical Anthropology* 47:63–99.
76. Hernandez, C. J., Beaupre, G. S., and Carter, D. R. (2003), A theoretical analysis of the relative influences of peak BMD, age-related bone loss, and menopause on the development of osteoporosis, *Osteoporosis International* 14:843–47.
77. Males are less susceptible to this problem because enzymes in bone cells convert testosterone into estrogen, thus preventing as much bone loss. Drops in testosterone levels, however, also affect men's bones.
78. Kannus, P., et al. (2005), Non-pharmacological means to prevent fractures among older adults, *Annals of Medicine* 37:303–10; Rubin, C. T., Rubin, J., and Judex, S. (2013), Exercise and the prevention of osteoporosis, in *Primer on the Metabolic Bone Diseases and Disorders of Mineral Metabolism*, ed. C. J. Rosen (Hoboken, N.J.: Wiley), 396–402.
79. Chakravarty, E. F., et al. (2008), Long distance running and knee osteoarthritis: A prospective study, *American Journal of Preventive Medicine* 35:133–38.
80. Berenbaum, F., et al. (2018), Modern-day environmental factors in the pathogenesis of osteoarthritis, *Nature Reviews Rheumatology* 14:674–81.
81. Shelburne, K. B., Torry, M. R., and Pandy, M. G. (2006), Contributions of muscles, ligaments, and the ground-reaction force to tibiofemoral joint loading during normal gait, *Journal of Orthopedic Research* 24:1983–90.
82. Karlsson, K. M., and Rosengren, B. E. (2012), Physical activity as a strategy to reduce the risk of osteoporosis and fragility fractures, *International Journal of Endocrinology and Metabolism* 10:527–36.
83. Felson, D. T., et al. (1988), Obesity and knee osteoarthritis: The Framingham Study, *Annals of Internal Medicine* 109:18–24; Wluka, A. E., Lombard, C. B., and Cicuttini, F. M. (2013), Tackling obesity in knee osteoarthritis, *Nature Reviews Rheumatology* 9:225–35; Leong, D. J., and Sun, H. B. (2014), Mechanical loading: Potential preventive and therapeutic strategy for osteoarthritis, *Journal of the American Academy of Orthopedic Surgery* 22:465–66.
84. Kiviranta, I., et al. (1988), Moderate running exercise augments glycosaminoglycans and thickness of articular cartilage in the knee joint of young

beagle dogs, *Journal of Orthopedic Research* 6:188–95; Säämänen, A.-M., et al. (1988), Running exercise as a modulatory of proteoglycan matrix in the articular cartilage of young rabbits, *International Journal of Sports Medicine* 9:127–33; Wallace, I. J., et al. (2019), Physical inactivity and knee osteoarthritis in guinea pigs, *Osteoarthritis and Cartilage* 27:1721–28; Urquhart, D. M., et al. (2008), Factors that may mediate the relationship between physical activity and the risk for developing knee osteoarthritis, *Arthritis Research and Therapy* 10:203; Semanik, P., Chang, R. W., and Dunlop, D. D. (2012), Aerobic activity in prevention and symptom control of osteoarthritis, *Physical Medicine and Rehabilitation* 4:S37–S44.

85. Gao, Y., et al. (2018), Muscle atrophy induced by mechanical unloading: Mechanisms and potential countermeasures, *Frontiers in Physiology* 9:235.
86. See Aktipis, A. (2020), *The Cheating Cell: How Evolution Helps Us Understand and Treat Cancer* (Princeton, N.J.: Princeton University Press).
87. Stefansson, V. (1960), *Cancer: Disease of Civilization?* (New York: Hill and Wang); Eaton, S. B., et al. (1994), Women's reproductive cancers in evolutionary context, *Quarterly Review of Biology* 69:353–67; Friborg, J. T., and Melby, M. (2008), Cancer patterns in Inuit populations, *Lancet Oncology* 9:892–900; Kelly, J., et al. (2008), Cancer among the circumpolar Inuit, 1989–2003: II. Patterns and trends, *International Journal of Circumpolar Health* 67:408–20; David, A. R., and Zimmerman, M. R. (2010), Cancer: An old disease, a new disease, or something in between?, *Nature Reviews Cancer* 10:728–33.
88. Rigoni-Stern, D. A. (1842), Fatti statistici relativi alle malattie cancerose, *Giornale per Servire ai Progressi della Patologia e della Terapeutica* 2:507–17.
89. Greaves, M. (2001), *Cancer: The Evolutionary Legacy* (Oxford: Oxford University Press).
90. Cancer Incidence Data, Office for National Statistics and Welsh Cancer Incidence and Surveillance Unit (WCISU), www.statistics.gov.uk and www.wcisu.wales.nhs.uk.
91. Ferlay, J., et al. (2018), Estimating the global cancer incidence and mortality in 2018: GLOBOCAN sources and methods, *International Journal of Cancer* 144:1941–53.
92. Arem, H., et al. (2015), Leisure time physical activity and mortality: A detailed pooled analysis of the dose-response relationship, *JAMA Internal Medicine* 175:959–67.
93. Kyu, H. H., et al. (2016), Physical activity and risk of breast cancer, colon cancer, diabetes, ischemic heart disease, and ischemic stroke events: Systematic review and dose-response meta-analysis for the Global Burden of Disease Study, *British Medical Journal* 354:3857; Li, T., et al. (2016), The dose-response effect of physical activity on cancer mortality: Findings from 71 prospective cohort studies, *British Journal of Sports Medicine* 50:339–45; Friedenreich, C. M., et al. (2016), Physical activity and cancer outcomes: A precision medicine approach, *Clinical Cancer Research* 22:4766–75; Moore, S. C., et al. (2016), Association of leisure-time physical activity with risk of 26 types of cancer in 1.44 million adults, *JAMA Internal Medicine* 176:816–25.
94. Friedenreich, C. M., and Orenstein, M. R. (2002), Physical activity and cancer prevention: Etiologic evidence and biological mechanisms, *Journal of Nutrition* 132:3456S–3464S. For a comprehensive review, see section F of Physical Activity Guidelines Advisory Committee (2018), *2018 Physical Activ-*

ity Guidelines Advisory Committee Scientific Report (Washington, D.C.: U.S. Department of Health and Human Services).

95. Jasienska, G., et al. (2006), Habitual physical activity and estradiol levels in women of reproductive age, *European Journal of Cancer Prevention* 15:439–45.
96. Eaton et al. (1994), Women's reproductive cancers in evolutionary context; Morimoto, L. M., et al. (2002), Obesity, body size, and risk of postmenopausal breast cancer: The Women's Health Initiative (United States), *Cancer Causes and Control* 13:741–51.
97. Il'yasova, D., et al. (2005), Circulating levels of inflammatory markers and cancer risk in the health aging and body composition cohort, *Cancer Epidemiology Biomarkers and Prevention* 14:2413–18; McTiernan, A. (2008), Mechanisms linking physical activity with cancer, *Nature Reviews Cancer* 8:205–11.
98. San-Millán, I., and Brooks, G. A. (2017), Reexamining cancer metabolism: Lactate production for carcinogenesis could be the purpose and explanation of the Warburg Effect, *Carcinogenesis* 38:119–33; Moore et al. (2016), Association of leisure-time physical activity with risk of 26 types of cancer in 1.44 million adults.
99. Coussens, L. M., and Werb, Z. (2002), Inflammation and cancer, *Nature* 420:860–67.
100. Jakobisiak, M., Lasek, W., and Golab, J. (2003), Natural mechanisms protecting against cancer, *Immunology Letters* 90:103–22.
101. Bigley, A. B., and Simpson, R. J. (2015), NK cells and exercise: Implications for cancer immunotherapy and survivorship, *Discoveries in Medicine* 19:433–45.
102. For reviews, see Brown, J. C., et al. (2012), Cancer, physical activity, and exercise, *Comprehensive Physiology* 2:2775–809; Pedersen, B. K., and Saltin, B. (2015), Exercise as medicine—evidence for prescribing exercise as therapy in 26 different chronic diseases, *Scandinavian Journal of Medicine and Science in Sports* 25:S1–S72; Stamatakis, E., et al. (2018), Does strength-promoting exercise confer unique health benefits? A pooled analysis of data on 11 population cohorts with all-cause, cancer, and cardiovascular mortality endpoints, *American Journal of Epidemiology* 187:1102–12.
103. Smith, M., Atkin, A., and Cutler, C. (2017), An age old problem? Estimating the impact of dementia on past human populations, *Journal of Aging and Health* 29:68–98.
104. Brookmeyer, R., et al. (2007), Forecasting the global burden of Alzheimer's disease, *Alzheimer's and Dementia* 3:186–91.
105. Selkoe, D. J. (1997), Alzheimer's disease: From genes to pathogenesis, *American Journal of Psychiatry* 5:1198; Niedermeyer, E. (2006), Alzheimer disease: Caused by primary deficiency of the cerebral blood flow, *Clinical EEG Neuroscience* 5:175–77.
106. Baker-Nigh, A., et al. (2015), Neuronal amyloid-β accumulation within cholinergic basal forebrain in ageing and Alzheimer's disease, *Brain* 138:1722–37.
107. Shi, Y., et al. (2017), ApoE4 markedly exacerbates tau-mediated neurodegeneration in a mouse model of tauopathy, *Nature* 549:523–27. Another interesting hypothesis possibly related to astrocyte function is that reduced sleep increases vulnerability to Alzheimer's by compromising maintenance functions such as plaque removal. See Neese, R. M., Finch, C. E., and Nunn, C. L. (2017), Does selection for short sleep duration explain human vulnerability to Alzheimer's disease?, *Evolution in Medicine and Public Health* 2017:39–46.

108. Trumble, B. C., et al. (2017), Apolipoprotein E4 is associated with improved cognitive function in Amazonian forager-horticulturalists with a high parasite burden, *FASEB Journal* 31:1508–15. ApoE4, by the way, is a gene involved in cholesterol synthesis in the liver, but it appears to affect cholesterol-rich plaque formation in the brain as well. For a review, see Carter, D. B. (2005), The interaction of amyloid-ß with ApoE, *Subcellular Biochemistry* 38:255–72.
109. For more information, see Rook, G. A. W. (2019), *The Hygiene Hypothesis and Darwinian Medicine* (Basel: Birkhäuser).
110. Stojanoski, B., et al. (2018), Targeted training: Converging evidence against the transferable benefits of online brain training on cognitive function, *Neuropsychologia* 117:541; National Academies of Sciences, Engineering, and Medicine (2017), *Preventing Cognitive Decline and Dementia: A Way Forward* (Washington, D.C.: National Academies Press).
111. Hamer, M., and Chida, Y. (2009), Physical activity and risk of neurodegenerative disease: A systematic review of prospective evidence, *Psychological Medicine* 39:3–11.
112. Buchman, A. S., et al. (2012), Total daily physical activity and the risk of AD and cognitive decline in older adults, *Neurology* 78:1323–29.
113. Forbes, D., et al. (2013), Exercise programs for people with dementia, *Cochrane Database Systematic Reviews* 12:CD006489.
114. For an excellent review, see Raichlen, D. A., and Polk, J. D. (2013), Linking brains and brawn: Exercise and the evolution of human neurobiology, *Proceedings of the Royal Society B: Biological Science* 280:20122250.
115. Choi, S. H., et al. (2018), Combined adult neurogenesis and BDNF mimic exercise effects on cognition in an Alzheimer's mouse model, *Science* 361:eaan8821.
116. Weinstein, G., et al. (2014), Serum brain-derived neurotrophic factor and the risk for dementia: The Framingham Heart Study, *JAMA Neurology* 71:55–61.
117. Giuffrida, M. L., Copani, A., and Rizzarelli, E. (2018), A promising connection between BDNF and Alzheimer's disease, *Aging* 10:1791–92.
118. Paillard, T., Rolland, Y., and de Souto Barreto, P. (2015), Protective effects of physical exercise in Alzheimer's disease and Parkinson's disease: A narrative review, *Journal of Clinical Neurology* 11:212–19.
119. Adlard, P. A., et al. (2005), Voluntary exercise decreases amyloid load in a transgenic model of Alzheimer's disease, *Journal of Neuroscience* 25:4217–21; Um, H. S., et al. (2008), Exercise training acts as a therapeutic strategy for reduction of the pathogenic phenotypes for Alzheimer's disease in an NSE/APPsw-transgenic model, *International Journal of Molecular Medicine* 22:529–39; Belarbi, K., et al. (2011), Beneficial effects of exercise in a transgenic mouse model of Alzheimer's disease-like Tau pathology, *Neurobiology of Disease* 43:486–94; Leem, Y. H., et al. (2011), Chronic exercise ameliorates the neuroinflammation in mice carrying NSE/htau23, *Biochemical and Biophysical Research Communications* 406:359–65.
120. Panza, G. A., et al. (2018), Can exercise improve cognitive symptoms of Alzheimer's Disease?, *Journal of the American Geriatrics Society* 66:487–95; Paillard, Rolland, and de Souto Barreto (2015), Protective effects of physical exercise in Alzheimer's disease and Parkinson's disease.
121. Buchman et al. (2012), Total daily physical activity and the risk of AD and cognitive decline in older adults.

122. Juvenal, a frequently angry man who was venting his spleen at fellow Roman citizens, actually wrote, *"Orandum est ut sit mens sana in corpore sano"* (You should pray for a healthy mind in a healthy body), in *Satires* 10.356–64. If you read the whole passage, you'll see he meant that instead of praying for a long life, Romans should pray for virtues like courage and perseverance. Over time, the line has been divorced from its context and its meaning changed to suggest that a healthy mind comes from a healthy body. Juvenal might have valued physical activity, but I suspect he would not have appreciated how his line has been hijacked.
123. Farris, S. G., et al. (2018), Anxiety and Depression Association of America Conference 2018, Abstract S1-094, 345R, and 315R; see also Melville, N. A. (2018), Few psychiatrists recommend exercise for anxiety disorders, *Medscape,* April 10, 2018, www.medscape.com.
124. Nesse, R. M. (2019), *Good Reasons for Bad Feelings: Insights from the Frontiers of Evolutionary Psychiatry* (New York: Dutton).
125. Kessler, R. C., et al. (2007), Lifetime prevalence and age-of-onset distributions of mental disorders in the World Health Organization's World Mental Health Survey Initiative, *World Psychiatry* 6:168–76; Ruscio, A. M., et al. (2017), Cross-sectional comparison of the epidemiology of DSM-5 generalized anxiety disorder across the globe, *JAMA Psychiatry* 74:465–75; Colla, J., et al. (2006), Depression and modernization: A cross-cultural study of women, *Social Psychiatry and Psychiatric Epidemiology* 41:271–79; Vega, W. A., et al. (2004), 12-month prevalence of DSM-III-R psychiatric disorders among Mexican Americans: Nativity, social assimilation, and age determinants, *Journal of Nervous and Mental Disease* 192:532; Lee, S., et al. (2007), Lifetime prevalence and inter-cohort variation in DSM-IV disorders in metropolitan China, *Psychological Medicine* 37:61–71.
126. Twenge, J. M., et al. (2010), Birth cohort increases in psychopathology among young Americans, 1938–2007: A cross-temporal meta-analysis of the MMPI, *Clinical Psychology Review* 30:145–54.
127. Twenge, J. M., et al. (2019), Age, period, and cohort trends in mood disorder indicators and suicide-related outcomes in a nationally representative dataset, 2005–2017, *Journal of Abnormal Psychology* 128:185–99.
128. Chekroud, S. R., et al. (2018), Association between physical exercise and mental health in 1.2 million individuals in the USA between 2011 and 2015: A cross-sectional study, *Lancet Psychiatry* 5:739–47.
129. There are hundreds of studies, but here are a few recent meta-analyses: Morres, I. D., et al. (2019), Aerobic exercise for adult patients with major depressive disorder in mental health services: A systematic review and meta-analysis, *Depression and Anxiety* 36:39–53; Stubbs, B., et al. (2017), An examination of the anxiolytic effects of exercise for people with anxiety and stress-related disorders: A meta-analysis, *Psychiatry Research* 249:102–8; Schuch, F. B., et al. (2016), Exercise as a treatment for depression: A meta-analysis adjusting for publication bias, *Journal of Psychiatric Research* 77:42–51; Josefsson, T., Lindwall, M., and Archer, T. (2014), Physical exercise intervention in depressive disorders: Meta-analysis and systematic review, *Scandinavian Journal of Medicine and Science in Sports* 24:259–72; Wegner, M., et al. (2014), Effects of exercise on anxiety and depression disorders: Review of meta-analyses and neurobiological mechanisms, *CNS and Neurological Disorders—Drug Targets*

13:1002–14; Asmundson, G. J., et al. (2013), Let's get physical: A contemporary review of the anxiolytic effects of exercise for anxiety and its disorders, *Depression and Anxiety* 30:362–73; Mammen, G., and Falkner, G. (2013), Physical activity and the prevention of depression: A systematic review of prospective studies, *American Journal of Preventive Medicine* 45:649–57; Stathopoulou, G., et al. (2006), Exercise interventions for mental health: A quantitative and qualitative review, *Clinical Psychology: Science and Practice* 13:179–93.

130. For a comparison of exercise with other treatments, see Cooney, G. M., et al. (2013), Exercise for depression, *Cochrane Database Systematic Reviews* CD004366.
131. Lin, T. W., and Kuo, Y. M. (2013), Exercise benefits brain function: The monoamine connection, *Brain Science* 3:39–53.
132. Maddock, R. J., et al. (2016), Acute modulation of cortical glutamate and GABA content by physical activity, *Journal of Neuroscience* 36:2449–57.
133. Meyer, J. D., et al. (2019), Serum endocannabinoid and mood changes after exercise in major depressive disorder, *Medicine and Science in Sports and Exercise* 51:1909–17.
134. Thomas, A. G., et al. (2012), The effects of aerobic activity on brain structure, *Frontiers in Psychology* 3:86; Schulkin, J. (2016), Evolutionary basis of human running and its impact on neural function, *Frontiers in Systems Neuroscience* 10:59.
135. Duclos, M., and Tabarin, A. (2016), Exercise and the hypothalamo-pituitary-adrenal axis, *Frontiers in Hormone Research* 47:12–26.
136. Machado, M., et al. (2006), Remission, dropouts, and adverse drug reaction rates in major depressive disorder: A meta-analysis of head-to-head trials, *Current Medical Research Opinion* 22:1825–37.
137. Hillman, C. H., Erickson, K. I., and Kramer, A. F. (2008), Be smart, exercise your heart: Exercise effects on brain and cognition, *Nature Reviews Neuroscience* 9:58–65; Raichlen, D. A., and Alexander, G. E. (2017), Adaptive capacity: An evolutionary neuroscience model linking exercise, cognition, and brain health, *Trends in Neuroscience* 40:408–21.
138. Knaepen, K., et al. (2010), Neuroplasticity—exercise-induced response of peripheral brain-derived neurotrophic factor: A systematic review of experimental studies in human subjects, *Sports Medicine* 40:765–801; Griffin, É. W., et al. (2011), Aerobic exercise improves hippocampal function and increases BDNF in the serum of young adult males, *Physiology and Behavior* 104:934–41; Schmolesky, M. T., Webb, D. L., and Hansen, R. A. (2013), The effects of aerobic exercise intensity and duration on levels of brain-derived neurotrophic factor in healthy men, *Journal of Sports Science and Medicine* 12:502–11; Saucedo-Marquez, C. M., et al. (2015), High-intensity interval training evokes larger serum BDNF levels compared with intense continuous exercise, *Journal of Applied Physiology* 119:1363–73.
139. Ekkekakis, P. (2009), Let them roam free? Physiological and psychological evidence for the potential of self-selected exercise intensity in public health, *Sports Medicine* 39:857–88; Jung, M. E., Bourne, J. E., and Little, J. P. (2014), Where does HIT fit? An examination of the affective response to high-intensity intervals in comparison to continuous moderate- and continuous vigorous-intensity exercise in the exercise intensity-affect continuum, *PLOS*

ONE 9:e114541; Meyer, J. D., et al. (2016), Influence of exercise intensity for improving depressed mood in depression: A dose-response study, *Behavioral Therapy* 47:527–37.

140. Stathopoulou et al. (2006), Exercise interventions for mental health; Pedersen and Saltin (2015), Exercise as medicine—evidence for prescribing exercise as therapy in 26 different chronic diseases.

Index

Page numbers in *italics* refer to illustrations or notes.

A NOTE ABOUT THE AUTHOR

Daniel E. Lieberman is a paleoanthropologist at Harvard University, where he is the Edwin M. Lerner II Professor of Biological Sciences in the Department of Human Evolutionary Biology. He was educated at Harvard University and Cambridge University, and is best known for his research on the evolution of the human body, especially regarding running and other physical activities. He lives in Cambridge, Massachusetts, where he also enjoys running.

A NOTE ON THE TYPE

This book was set in Scala, a typeface designed by the Dutch designer Martin Majoor (b. 1960) in 1988 and released by the FontFont foundry in 1990. While designed as a fully modern family of fonts containing both a serif and a sans serif alphabet, Scala retains many refinements normally associated with traditional fonts.

Composed by North Market Street Graphics,
Lancaster, Pennsylvania

Printed and bound by Berryville Graphics,
Berryville, Virginia

Designed by Cassandra J. Pappas